Coastal Engineering and Fluid–Structure Interactions

Coastal Engineering and Fluid–Structure Interactions

Guest Editors

Miaohua Mao
Junliang Gao

MDPI

Basel • Beijing • Wuhan • Barcelona • Belgrade • Novi Sad • Cluj • Manchester

Guest Editors

Miaohua Mao
Yantai Institute of Coastal Zone Research
Chinese Academy of Sciences
Yantai
China

Junliang Gao
School of Naval Architecture and Ocean Engineering
Jiangsu University of Science and Technology
Zhenjiang
China

Editorial Office
MDPI AG
Grosspeteranlage 5
4052 Basel, Switzerland

This is a reprint of the Special Issue, published open access by the journal *Water* (ISSN 2073-4441), freely accessible at: https://www.mdpi.com/journal/water/special_issues/U9P32Z0A55.

For citation purposes, cite each article independently as indicated on the article page online and as indicated below:

Lastname, A.A.; Lastname, B.B. Article Title. *Journal Name* **Year**, *Volume Number*, Page Range.

ISBN 978-3-7258-7627-3 (Hbk)
ISBN 978-3-7258-7628-0 (PDF)
https://doi.org/10.3390/books978-3-7258-7628-0

Contents

About the Editors

Miaohua Mao

Miaohua Mao got a BS from Zhejiang University and achieved a PhD degree at University of Maryland in 2018. He works at Yantai Institute of Coastal Zone Research (YIC), Chinese Academy of Sciences (CAS) as a Professor and Ph.D. advisor. His research interests include, but are not limited to, nearshore hydrodynamics, numerical modelling, and observation of wave-current interaction. Dr. Mao has published more than 20 papers as first/corresponding author in the prestigious top journals of physical oceanography and numerical modelling, including *Journal of Physical Oceanography* and *Journal of Geophysical Research: Oceans and Ocean Modelling*, and presented over 20 times at the international academic conferences (as a Keynote Speaker, Session Chair, and Outstanding Poster Presenter, etc.). Dr. Mao is the reviewer for the NSFC, CAS International Cooperation, Shandong Province, Guangdong Province, Beijing City proposals and Ph.D. theses from universities; He is also a reviewer of peer-reviewed journals such as *Progress in Oceanography, Ocean Modelling, Water Resources Research, ECSS, Marine Pollution Bulletin, Ocean and Coastal Management, JMSE*, etc., and were awarded as the outstanding contribution in reviewing ECSS of year 2018, and 2023 Outstanding Reviewer Award of JMSE. Guest Editor of the SCI journal *Water*. He is the Principal Investigator (PI) of several projects, including the Natural Science Foundation of China (NSFC), the Chinese Academy of Sciences (CAS), etc.

Junliang Gao

Junliang Gao is a professor and the head of the Department of Ocean Engineering at the School of Naval Architecture and Ocean Engineering, Jiangsu University of Science and Technology. He was selected into "The Qing-Lan Project" of Jiangsu Province. In May 2015, he graduated from the State Key Laboratory of Coastal and Offshore Engineering of the Dalian University of Technology with a doctor's degree in Engineering. He is a member of the China Shipbuilding Engineering Society, the China Navigation Society, the American Society of Mechanical Engineers, and other academic organisations. Since the graduate student stage, he has been engaged in research on the basic theory of ocean engineering, including harbour resonance, island and reef hydrodynamics, tsunami waves, the development of wave propagation evolution models, fluid–solid interactions based on computational fluid dynamics (CFD), etc. He has published more than 60 papers (including more than 40 SCI papers and more than 10 EI papers) in Chinese and English academic journals, such as *Coastal Engineering, Renewable Energy, Energy Conversion and Management, Ocean Engineering, and Engineering Mechanics*, and has applied for more than 20 patents (7 have been authorised). He served as the communication review expert of the NSFC and as a reviewer of more than 20 journals at home and abroad, such as *Coastal Engineering, Ocean Modeling, Ocean Engineering*, and the *Journal of Computational Science*.

water

MDPI

Editorial

Coastal Engineering and Fluid–Structure Interactions

Miaohua Mao [1,2,*] and Junliang Gao [3]

1 Yantai Institute of Coastal Zone Research, Chinese Academy of Sciences, Yantai 264003, China
2 Shandong Key Laboratory of Coastal Zone Environmental Processes and Ecological Security, Yantai 264003, China
3 School of Naval Architecture and Ocean Engineering, Jiangsu University of Science and Technology, Zhenjiang 212100, China
* Correspondence: mhmao@yic.ac.cn

1. Introduction

Fluid–structure interaction (FSI) in coastal engineering is a core topic in the interdisciplinary fields of harbor, coastal, offshore, and structural engineering. It involves the dynamic effects of fluid loads, including waves, ocean currents, and winds on structures (e.g., harbors, breakwaters, cross-sea bridges, offshore wind turbine foundations, and subsea pipelines) and their responses and counteractions. Recently, significant progress has been made in theoretical and modeling techniques, experimental and monitoring technologies, and engineering applications.

Advancements in theory and model developments include the following: (1) traditional linear potential flow theories (e.g., the Morrison equation, diffraction and radiation theories), which remain widely used but have evolved into higher-order boundary element methods and fully nonlinear models (e.g., higher-order spectral methods), enabling more accurate simulations of wave breaking, overtopping, and highly nonlinear waves [1]. (2) Computational Fluid Dynamics (CFD) methods (e.g., VOF, SPH, LES) have gradually matured, enabling the processes of complex turbulence, vortex-induced vibrations, and coupled interactions between waves, structures, and sea beds (e.g., fluid–structure-soil coupling and gas–liquid–solid three-phase coupling) [2]. (3) High-fidelity CFD-FSI coupling employs the partitioned coupling method (e.g., MPCCI, Fluent + ABAQUS) or the immersed boundary method to achieve interactions between transient flow and large structural deformation [3]. The computational efficiency is significantly enhanced by the acceleration of GPU using the OpenFOAM integrated solvers with high-resolution [4]. Combining Large Eddy Simulation (LES) with potential flow theory achieve cross-scale simulations from wave propagation in open waters to the interaction with local structures in coastal engineering [5]. (4) Under stochastic wave loads, Proper Orthogonal Decomposition and deep learning (e.g., LSTM, CNN) methods are used in surrogate models to accelerate long-term dynamic response analysis [6]. Uncertainty quantification and reliability analysis are introduced for material and geometric uncertainties, structural failure probability, and fatigue life through Monte Carlo methods and polynomial chaos expansion [7].

Progress in experimental and monitoring technologies include (1) advanced experimental facilities: Large wave tanks (e.g., multi-directional wave generators), wind–wave-current interaction basins, and centrifuge model tests can simulate complex environmental loads and foundation conditions [8]; particle image velocimetry and laser Doppler velocimetry are employed for precise measurement of flow fields and vortex structures around structures [9]; and (2) on-site monitoring and digital twins: Long-term health monitoring of actual engineering structures (e.g., coastal and offshore wind turbines and bridges)

https://doi.org/10.3390/w18020206

are conducted based on fiber-optic sensing, the Internet of Things, and satellite remote sensing. Digital twin technology integrates monitoring data and numerical models to achieve real-time structural state diagnosis and predictive maintenance [10].

Applications of engineering fields primarily include three scenarios: (1) Offshore renewable energy structures: Coupled wave–current–wind dynamic analysis of fixed and floating coastal and offshore wind turbine foundations, particularly the stability of tension leg platforms and semi-submersible platforms under severe sea conditions. Studies include flow-structure interactions and energy capture efficiency optimization of tidal and wave energy devices [11]. (2) Coastal protection and ecological engineering: This includes research on the wave dissipation mechanisms, hydraulic performance of permeable and porous media structures (e.g., ecological reefs and vegetated wetlands), with a focus on disaster prevention and ecological restoration, and assessing the overtopping and breaching risks of breakwaters and seawalls in response to sea level rise and typhoon storm surges [12]. (3) Sea-crossing transportation infrastructure: This scenario focuses on the wave impact and local scour protection for deepwater bridge pile foundations, and hydrodynamic stability during the sinking and splicing of submarine tunnel pipe sections [13].

Studies on the coastal engineering and fluid–structure interactions face multiple aspects of challenges for future research. Under extreme environments and climate adaption (e.g., rouge waves, tsunamis, and tropical cyclones), research on the survivability and resilience design of structures is important [14]. Research needs to be further developed for analysis methods for multi-hazard coupling effects (e.g., complex disaster scenarios of seismic-wave interaction and sea–ice–wave coupling loads). Intelligence and sustainable development are key to coastal engineering and FSI. Therefore, one future direction could optimize FSI model parameters with AI and develop adaptive structures with autonomous control (e.g., intelligent damping) [15]. Another direction is to develop durability assessment methods for green materials (e.g., composites and eco-concrete) in complex fluid environments involving corrosion and biofouling.

2. Findings Reported in This Special Issue

Coastal engineers have designed and constructed protected structures to cope with coastal dynamics, including wave behavior, storm surge, sediment transport, erosion, and sea level changes. Therefore, understanding coastal hydrodynamic environments and fluid–structure interaction is an important issue in coastal engineering. The research topics in the field of coastal engineering include broad scopes such as (1) coastal dynamic environments of winds, waves, currents, sea ice; (2) sediment transport in the changing morphology of coastal, estuarine, and offshore regions; (3) the technical and functional design of coastal and harbor structures; (4) fluid–structure interactions, including conventional hard and nature-based soft structures; (5) innovations in research methods and techniques, including mathematical and numerical modeling, laboratory and field observations, and experiments. This Special Issue invites papers including, but not limited to, the abovementioned topics. This Special Issue comprises 13 articles, including 12 research articles and 1 review article.

Wang et al. (contribution 1) apply the Volume of Fluid (VOF) method in OpenFOAM to study the impact of a submerged shell dike on dissipating broken waves before a breakwater. When the broken wave heights decrease and the dike's radii increase, the peak pressure decreases. The relationship is established between the broken wave pressures and the dimensionless parameters as a function of broken wave and breakwater heights and the dike's radii. They propose the equations for estimating broken wave pressures on various points along the breakwater.

https://doi.org/10.3390/w18020206

Zhao et al. (contribution 2) use OpenFOAM v2206 to discuss model simulations from four turbulence models and three mesh types. It reveals that the stabilized $k - \omega$ shear stress transfer turbulence model better simulates the complex wave evolution process on the cube and effectively captures the wave free surface, wave run-up, and reflection coefficient, and it is selected in the hydrodynamic model with new ecological hollow cubes.

Chen et al. (contribution 3) study the incipient motion and sediment scouring near submarine cables with five grain sizes. They find that the relative flow velocity, scour rate, and sediment erosion increase with the increasing Froude number. The modified formula is reliable in assessing the submarine cable exposure risks, providing insights into developing effective protection strategies, and enhancing cable stability in complex marine conditions. Understanding sediment dynamics and submarine cable stability is important to developing effective protection strategies in the dynamic marine conditions.

Peng and Yin (contribution 4) analyze the distribution and anomalies of total suspended sediment with the satellite-derived residual surface currents. Coastal water masses show obvious seasonal variations. It varies spatially for the offshore transport pathway, which extends to the shelf edge restricted by the Taiwan Warm Current, and shifts from northeastward to eastward for the persistent transport pathway. The Hangzhou Bay's pathway is related to the tidal mixing, and the Yangtze River estuary's path follows Yangtze River Diluted Water. These crucial observational insights are beneficial to the material cycling model in the East China Sea shelf.

Qian et al. (contribution 5) use the commercial finite element method to model the submarine cable's pullback process with the horizontal directional drilling technology. The cable's tension increases with the incident angle and crossing length. During its pullback operation, the cable could be locked by the extremely large velocity. Further investigations indicate that the permissible values could be important for similar engineering projects.

Hou et al. (contribution 6) use satellite observations to study the effects of consecutive typhoons on chlorophyll-a concentrations in the South China Sea. In 2006, after Typhoon Durian, strong vertical mixing and upwelling rapidly enhanced chlorophyll-a concentrations. These enhancements were significant in the path for Typhoon Utor. The changing marine environment brought by the first typhoon modulates the effects of successive typhoons on marine ecology, and this provides insights into understanding a typhoon's effects on marine productivity.

Ghaderi et al. (contribution 7) use CFD simulations in a rectangular channel to examine the effects of four distinct converging configurations of guide-banks on the unsteady flow's propagation. The flow experiences a depth increase as it encounters the converging geometries. This forms a hydraulic jump for the resulting waves' progress upstream. When floods interact with the topography, the contracting channel results in pronounced initial water elevation rises and deeper reflected waves. Less time is needed to influence upstream for the waves generated by the trapezoidal configuration.

The increased water elevation becomes wider with longer configurations in the converging zone. This work helps understand hydraulic processes during dam failure scenarios.

Fan et al. (contribution 8) use the physical system to investigate the mechanical properties of soft soils and geotextile pipe and bags. The soil's compression curve shows strong nonlinearity and high apparent compressibility from silt. The deviatoric stress in the soft soil increases when the consolidation pressure rises, indicating a typical behavior of strain-hardening. Based on the modeling results, the structural horizontal displacement is substantially impacted by the reclamation construction process. It centers at the toe of the slope for the significant stress, while the maximum tensile stress occurs in the central part of the pipe-bag structure. The drainage board's installation effectively speeds up the pore pressure dissipation, achieving near-complete consolidation in one year. This work

https://doi.org/10.3390/w18020206

lays a theoretical foundation and practical guidance for the assessment of the stability and safety of the geotextile pipe and bag structures on soft soil foundations.

Zhang et al. (contribution 9) apply a two-dimensional model to analyze the effects of the submerged dike orientation and height on flow and flux. For the alongshore flow, the obvious velocity variations occur at 1/5 of the dike length. Flow velocity changes with the increase in its distance to the submerged dike. The dividing value of 0.7 is the ratio between the submerged dike height and water depth. The cross-dike flux increases with the orientation angle. These findings support the establishment of a theoretical basis supporting future integrated management of coastal zones.

Fang et al. (contribution 10) analyze rainfall characteristics from 1986 to 2017 at three typical irrigation stations in East China. High utilization is observed during extreme rainfall events, and larger variability during moderate rainfalls. The Support Vector Regression model is used to predict daily effective precipitation based on rainfall, antecedent precipitation index, drought days, and extreme rainfall indicators. Results indicate that the model captures the nonlinear relationship satisfactorily between effective precipitation and rainfall characteristics. They conclude that the machine learning method can be used as an alternative tool for the existing estimation models for water-saving crop cultivation and irrigation scheduling.

Liu et al. (contribution 11) collect the long-term Sentinel-2 Multispectral Instrument and Sentinel-1 Synthetic Aperture Radar data, constructing a high-resolution time series of Apparent Inundation Frequency obtained through Unmanned Aerial Vehicle surveys. Results show that the average inundation probability in the Luanhe Estuary shows a fluctuated, but overall, upward trend from 2016 to 2025. The quantification demonstrates that microtopography is the major controlling factor influencing fine-scale tidal inundation variations. This work suggests a reliable framework for the assessment of coastal vulnerability. The high-resolution quantitative data offers scientific support for the strategies of geomorphological management and disaster mitigation.

Ma and Zhang (contribution 12) apply the Reynolds-averaged Navier–Stokes (RANS) model to investigate the correlation between the disturbance parameters and wave features. The $k - \varepsilon$ model and immersed boundary method (IBM) are used to resolve the flow turbulence and fluid–structure interaction (FSI), and the free surface is tracked using the VOF method. Results indicate that the proposed model captures various wave patterns satisfactorily. The wave evolution strongly depends on the disturbance duration and its width. Shorter durations trigger earlier soliton fission, and longer widths accelerate phase celerity. This work concludes that the disturbance parameters are critical in governing soliton formation and energy propagation patterns that are vital in disaster forecasting.

Jiang et al. (contribution 13) review sandy beach resilience (e.g., resistance, recovery, and adaptation). They examine the Deep Learning (DL) method in monitoring and forecasting external forcing, which includes typhoon tracks and storm surge peaks. The DL is applied to forecast beach processes, including medium- and long-term shoreline evolution. Governance options are expanded by DL methods and multi-scenario generation. From the perspective of management and decision support, policy adoption and risk communication are enhanced by interpretable features with uncertainty quantification. The DL method surpasses traditional models by reducing the observation–model–decision cycle, expanding analysis for various scenarios, and improving the governance transparency. However, it remains as a challenge for the cross-domain generalization, robustness in extreme situations, and the data governance. This work reviews DL as a potential technology to improve sandy beach resilience while offering a theoretical basis for future studies.

https://doi.org/10.3390/w18020206

3. Conclusions and Future Directions

The articles collected in the Special Issue "Coastal Engineering and Fluid Structure Interactions" include keywords of coastal and nearshore engineering, fluid–structure interactions (FSIs), extreme events (e.g., typhoon and seismic loads), numerical model (e.g., CFD, VOF, and OpenFOAM) and observations, Deep Learning (DL) method, Immersed Boundary Method (IBM), nonlinear process, climate change, human activities, submerged cables, and coastal protection. These keywords are aligned with the theme of this Special Issue but have some extensions related to coastal engineering and FSI. Studies of coastal engineering and FSI have evolved from deterministic analysis to stochastic, nonlinear, and multiscale coupling, with increasingly close integration of numerical simulations, experimental techniques, and field monitoring. Future trends will focus on extreme climate adaptability, intelligent operation and maintenance, and interdisciplinary integrated innovation, providing critical support for the safety, economy, and sustainable development of coastal and marine engineering.

Author Contributions: Conceptualization, M.M. and J.G.; writing—original draft preparation, M.M.; writing—review and editing, M.M. and J.G.; project administration, M.M. All authors have read and agreed to the published version of the manuscript.

Funding: This research received no external funding.

Acknowledgments: I would like to express my deep appreciation to all the editors, authors, and reviewers who contributed to this Special Issue.

Conflicts of Interest: The authors declare no conflicts of interest.

List of Contributions:

1. Wang, N.; Wang, G.; Zhang, H.; Li, X. Investigation of Broken Wave Dissipation Effects of Submerged Shell Dike in Front of Breakwater. *Water* **2025**, *17*, 609.
2. Zhao, H.; Ye, J.; Wang, K.; Zhou, Y.; Zeng, Z.; Li, Q.; Zhao, X. Selection of a Turbulence Model for Wave Evolution on a New Ecological Hollow Cube. *Water* **2025**, *17*, 1149.
3. Chen, F.; Yang, W.; Liu, F.; Zhu, L.; Sun, Z. Experimental Study of Sediment Incipient Velocity and Scouring in Submarine Cable Burial Areas. *Water* **2025**, *17*, 1310.
4. Peng, Y.; Yin, W. Spatial–Temporal Distribution of Offshore Transport Pathways of Coastal Water Masses in the East China Sea Based on GOCI-TSS. *Water* **2025**, *17*, 1370.
5. Qian, G.; Kang, W.; Cong, Y.; Liu, Z. A Numerical Study on the Pullback Process of a Submarine Cable Based on Trenchless Directional Drilling Technology. *Water* **2025**, *17*, 1517.
6. Hou, X.; Ruan, Z.; Li, B.; Wang, Y. Effects of Successive Typhoon Durian and Typhoon Utor on Chlorophyll-*a* Response in South China Sea. *Water* **2025**, *17*, 1567.
7. Ghaderi, A.; Shahini, H.; Mohammadnezhad, H.; Hamidifar, H.; Pu, J.H. Hydraulic Response of Dam-Break Flood Waves to Converging Channel Geometries: A Numerical Investigation. *Water* **2025**, *17*, 2593.
8. Fan, P.; Ren, H.; Zhang, X.; Li, W.; Guo, W. Investigation into the Working Behavior of Geotextile Pipe-Bag Systems on Soft Soil Foundations in the Ningde Port Industrial Zone, China. *Water* **2025**, *17*, 3063.
9. Zhang, X.; Zhang, Y.; Deng, Y.; Li, X.; Guan, B. Two-Dimensional Numerical Analysis of Submerged Dike Hydrodynamics. *Water* **2025**, *17*, 3455.
10. Fang, H.; Weng, Z.; Hu, M.; Feng, X.; Liu, Q. Rainfall Characteristics and Effective Precipitation Analysis of Rice Growth Period in Typical Areas in East China. *Water* **2025**, *17*, 3542.
11. Liu, Y.; Ni, P.; Ma, W.; Zhang, Q.; Hu, Q.; Ling, Z. Microtopography Governs Tidal Inundation Frequency in the Luanhe Estuarine Salt Marsh: A Decadal Assessment Integrating Sentinel Data and UAV Photogrammetry. *Water* **2025**, *17*, 3559.

https://doi.org/10.3390/w18020206

12. Ma, H.-P.; Zhang, H.-X. Nonlinear Water Waves Induced by Vertical Disturbances Through a Navier–Stokes Solver with the Implementation of the Immersed Boundary Method. *Water* **2025**, *17*, 3573.
13. Jiang, Y.; Zhou, Y.; Zhang, J. Deep Learning-Driven Sandy Beach Resilience Assessment: Integrating External Forcing Forecasting, Process Simulation, and Risk-Informed Decision Support. *Water* **2025**, *17*, 3383.

References

1. Chen, X.; Zhu, R.; Ma, C.; Fan, J. Computations of Linear and Nonlinear Ship Waves by Higher-Order Boundary Element Method. *Ocean Eng.* **2016**, *114*, 142–153. [CrossRef]
2. Ye, T.; Pan, D.; Huang, C.; Liu, M. Smoothed Particle Hydrodynamics (SPH) for Complex Fluid Flows: Recent Developments in Methodology and Applications. *Phys. Fluids* **2019**, *31*, 011301. [CrossRef]
3. Aziz, M.A.; Abdullah, M.Z.; Khor, C.Y.; Fairuz, Z.M.; Iqbal, A.M.; Mazlan, M.; Rasat, M.S.S. Thermal Fluid-Structure Interaction in the Effects of Pin-Through-Hole Diameter during Wave Soldering. *Adv. Mech. Eng.* **2014**, *6*, 275735. [CrossRef]
4. Piscaglia, F.; Ghioldi, F. GPU Acceleration of CFD Simulations in OpenFOAM. *Aerospace* **2023**, *10*, 792. [CrossRef]
5. Lohmann, I.P.; Fredsøe, J.; Sumer, B.M.; Christensen, E.D. Large Eddy Simulation of the Ventilated Wave Boundary Layer. *J. Geophys. Res. Ocean.* **2006**, *111*, C06036. [CrossRef]
6. Xu, Y.; Sha, Y.; Wang, C.; Cao, W.; Wei, Y. Comparative Studies of Predictive Models for Unsteady Flow Fields Based on Deep Learning and Proper Orthogonal Decomposition. *Ocean Eng.* **2023**, *272*, 113935. [CrossRef]
7. Sudret, B. Meta-Models for Structural Reliability and Uncertainty Quantification. *arXiv* **2012**, arXiv:1203.2062. [CrossRef]
8. Ji, X.; Zou, L.; Yang, Z.; Wang, D.; Bingham, H.B. Numerical Research on the Interaction of Multi-Directional Random Waves with an Offshore Wind Turbine Foundation. *Ocean Eng.* **2022**, *250*, 111029. [CrossRef]
9. Pedersen, N.; Larsen, P.S.; Jacobsen, C.B. Flow in a Centrifugal Pump Impeller at Design and Off-design Conditions—Part I: Particle Image Velocimetry (PIV) and Laser Doppler Velocimetry (LDV). Measurements. *J. Fluids Eng.* **2003**, *125*, 61–72. [CrossRef]
10. Liu., Y.; Zhang, J.M.; Min, Y.T.; Yu, Y.; Lin, C.; Hu, Z.Z. A Digital Twin-Based Framework for Simulation and Monitoring Analysis of Floating Wind Turbine Structures. *Ocean Eng.* **2023**, *283*, 115009. [CrossRef]
11. Hu, Y.; Yang, J.; Baniotopoulos, C.; Wang, X.; Deng, X. Dynamic Analysis of Offshore Steel Wind Turbine Towers Subjected to Wind, Wave and Current Loading during Construction. *Ocean Eng.* **2020**, *216*, 108084. [CrossRef]
12. Suh, S.W.; Lee, M.H. Analysis of Typhoon-Induced Wave Overtopping Vulnerability due to Sea Level Rise Using a Coastal–Seawall–Terrestrial Seamless Grid System. *J. Mar. Sci. Eng.* **2023**, *11*, 2114. [CrossRef]
13. Li, J.; Lian, J.; Guo, Y.; Wang, H.; Yang, X. Numerical Study on Scour Protection Effect of Monopile Foundation Based on Disturbance Structure. *Ocean Eng.* **2022**, *248*, 110856. [CrossRef]
14. Baragamage, D.S.; Wu, W. A Three-Dimensional Fully-Coupled Fluid-Structure Model for Tsunami Loading on Coastal Bridges. *Water* **2024**, *16*, 189. [CrossRef]
15. Wu, Y.; Cheng, Z.; McConkey, R.; Lien, F.S.; Yee, E. Modelling of Flow-Induced Vibration of Bluff Bodies: A Comprehensive Survey and Future Prospects. *Energies* **2022**, *15*, 8719. [CrossRef]

https://doi.org/10.3390/w18020206

water

MDPI

Article

Nonlinear Water Waves Induced by Vertical Disturbances Through a Navier–Stokes Solver with the Implementation of the Immersed Boundary Method

Hai-Ping Ma * and Hong-Xia Zhang

School of Civil and Architectural Engineering, Panzhihua University, Panzhihua 617000, China; swzhxdd@163.com
* Correspondence: mahaiping@pzhu.edu.cn

Abstract

Nonlinear water waves (NWWs) can be generated by the vertical bottom disturbance, which represents the conceptual processes of the rise of seabed rupture under seismic loads. To explore the correlation between the disturbance parameters and the wave features, a Reynolds-averaged Navier–Stokes (RANS) model is applied, with the flow turbulence and fluid–structure interaction (FSI) being resolved by the k–ε model and the immersed boundary method (IBM), respectively. The free surface is tracked using the volume of fluid (VOF) method. After validating against the theoretical solutions and experimental results, the effects of disturbance duration and bulk on the wave features at the source region (the generation stage) and offshore direction (the propagation stage) are systematically discussed. The fixed maximal vertical displacement is considered, with four moving durations and five disturbance widths being simulated, resulting in four disturbance velocities and five disturbance bulks. The results indicate that the proposed RANS model can accurately create various wave patterns (including the linear, solitary, and tsunami-like waves) generated by bottom disturbances. Special attentions are paid to the tsunami-like wave. The wave evolution exhibits strong dependence on disturbance duration and width, with shorter durations triggering earlier soliton fission and longer widths accelerating phase celerity. These findings highlight the critical role of disturbance parameters in governing soliton formation and energy propagation patterns, which are vital in disaster forecasting.

Keywords: vertical bottom disturbance; nonlinear water wave; Reynolds-averaged Navier–Stokes equations; immersed boundary method; arriving time; tsunami-like wave

1. Introduction

Nonlinear water waves (NWWs) have garnered significant attention due to their complex generation mechanisms and associated coastal hazards. They can be generated by tide or current flowing over submarine topography (e.g., Semenov and Wu [1]). Another fundamental generation mechanism is sudden bottom disturbance, which has been extensively investigated through laboratory experiments and numerical simulations (e.g., Derakhti et al. [2]; Gao et al. [3]). Among these, vertical bottom disturbances represent a canonical scenario, conceptually modeling seabed earthquakes (Synolakis & Bernard [4]; Reeve et al. [5]). Research on this mechanism has revealed that it can generate diverse and often extreme wave forms, including solitary waves, tsunami-like waves, and undular bores (e.g., Fang et al. [6]; Jing et al. [7]; Gao et al. [8,9]). A thorough understanding of the generation and evolution of such NWWs is, therefore, of considerable engineering significance for coastal hazard mitigation.

Theoretical investigations of NWWs commonly assume an inviscid and incompressible fluid. Under these assumptions, several governing equations have been derived, including the forced Korteweg–de Vries (fKdV) equation (Hammack & Segur [10]; Tinti & Bortolucci [11]; Infeld et al. [12]), the Green–Naghdi (G–N) equations (Nadiga et al. [13]; Duan et al. [14]), and the Boussinesq or generalized Boussinesq (B/g-B) equations (Madsen et al. [15]; Løvholt et al. [16]). Analytical solutions indicate that NWWs generated by bottom disturbances are influenced by both the velocity and bulk of the disturbance, findings that are corroborated numerically by Shen and Chan [17] and Jin et al. [18]. Nevertheless, the inviscid assumption inherently introduces discrepancies in wave height, phase celerity, wave form, breaking pattern, etc., which are attributable to the flow separation and energy dissipation (Whittaker et al. [19]; Jin et al. [18]; Ebrahimi & Boroomand [20]; Mi et al. [21]). These discrepancies between potential solutions and real waves (either generated from experiments or simulations by high-fidelity numerical solvers) have been widely noted (Zhang & Chwang [22]; Whittaker et al. [19]; Jin et al. [23]). In a related context, Meyla and Stepanyants [24] theoretically examined the scattering of gravity–capillary waves over a bottom step. They discovered an extra evanescent mode unique to gravity–capillary wave scattering, which is essential for satisfying the additional boundary condition at the surface introduced by capillary effects. Collectively, while existing theoretical studies provide a foundational framework for understanding NWWs, notable gaps remain between idealized models and real wave dynamics.

The real waves, inherently encompassing key fluid properties, such as viscosity, turbulence, and vorticity, are distinguished from idealized potential flow descriptions. When fluid viscosity is considered, artificial Rayleigh damping has been widely adopted in theoretical analyses (Ockendon et al. [25]; Kim et al. [26]) to partially account for viscous effects. While this approach can qualitatively represent viscous damping, determining an appropriate value for the Rayleigh damping coefficient remains challenging, as it is influenced by multiple factors such as wall roughness, surface dissipation, and fluid–structure interaction (Kim et al. [26]; Ibrahim [27]; Mao et al. [28]; Jin et al. [29–31]). Nevertheless, with careful calibration, the discrepancy between theoretical predictions and experimental (or simulated) waves, particularly those generated by fluid–structure interaction (FSI), can be rendered negligible. This has motivated a series of subsequent studies incorporating artificial Rayleigh damping. For example, Wang et al. [32] introduced an artificial viscosity term into the dynamic free surface boundary condition to analyze viscous effects on the hydrodynamic performance of a fixed oscillating water column wave energy converter. In a more advanced approach, Zhang et al. [33] developed a fully nonlinear sloshing model enhanced by a machine learning strategy, in which a neural network was trained to adaptively calibrate the artificial damping coefficient to evaluate the damping ratio.

Despite these advances, the capability of these artificial damping-based methods to capture truly viscous effects remains limited. To the best of our knowledge, Wu et al. [33] derived an exact solution based on the linearized Navier–Stokes equations. However, significant deviations among the theoretical results, experimental data, and high-fidelity numerical simulations (fully nonlinear potential solver and Navier–Stokes model) persist (Zhang et al. [33]; Jin and Lin [34]; Jin et al. [35,36]), underscoring the ongoing difficulty in accurately representing the effects of liquid viscosity in theoretical models.

An alternative approach for accurately exploring wave responses with realistic viscosity is through laboratory experiments. Pioneering experimental work was conducted by Hammack and Segur [10], who generated tsunami-like waves in a flume. Jamin et al. [37] performed experiments in a circular wave basin, creating nonlinear waves via vertical bottom disturbances. Slunyaev et al. [38] provided accurate reconstructions of dynamic pressure fields from surface displacement data across all tested models using experimental methods. More recently, Reeve et al. [4] validated the applicability of linear theory through 32 sets of experiments. It is important to note, however, that most experiments are conducted at relatively small scales and can be cost-prohibitive, and limited parametric flexibility also restricts the comprehensiveness of the experiments.

High-fidelity numerical modeling offers another powerful alternative. When thoroughly validated, numerical simulations are particularly suitable for investigating the fundamental nature of NWWs, as they circumvent the scale effects inherent in physical experiments. Early simulations primarily relied on depth-averaged Navier–Stokes equations, commonly referred to as shallow-water equations (SWEs) (Lee et al. [39]; Bai and Cheung [40]; Whittaker et al. [19]; Gao et al. [41]; Jin et al. [42]). For instance, Dragani [43] applied an SWE model to study long ocean wave generation in the coastal waters of Buenos Aires province, Argentina. Auclair et al. [44] employed a non-hydrostatic and non-Boussinesq model to investigate gravity wave generation, discussing the influence of bottom topography on wave formation. Nevertheless, it should be noted that these inviscid, depth-averaged models struggle to accurately capture strongly nonlinear wave dynamics, particularly in the near-field generation region (Wassim et al. [45]; Jin et al. [18]).

Because of the fast development of computational technologies, directly solving the Navier–Stokes equations with appropriate numerical methods to resolve FSI offers a rigorous approach to revealing the intrinsic behavior of realistic waves (e.g., Gao et al. [46]; Mi et al. [47]). Shen and Chan [17] developed a combined immersed boundary method (IBM) and a Navier–Stokes model to study NWWs induced by vertical motions, revealing a clear relationship between the arrival time and disturbance duration. Wang and Chan [48] proposed a RANS model to investigate wave generation mechanisms and derived explicit predictive equations for wavelength and amplitude. Xie and Du [49] implemented a tsunami generation procedure using the interFoam solver to simulate leading-depression tsunami waves and analyze their run-up behaviors. More recently, Jin et al. [18] identified the disturbance velocity as a dominant parameter governing stable wave profiles. While these studies have broadly outlined NWW generation via bottom movements, the reported wave characteristics remain scattered across parameter spaces, and a systematic portrayal of wave generation and propagation processes is still lacking.

To address the research gap, the present study investigates the characteristics of NWWs generated by vertically moving bottom disturbances using a RANS model, in which flow turbulence and FSI are resolved by the k–ε model and IBM, respectively. Wave dynamics during both generation and propagation phases are analyzed in detail. The influences of movement duration and disturbance width are examined, revealing that both parameters critically govern the arrival time of NWWs. The remainder of this paper is organized as follows: Section 2 introduces the governing equations and numerical methodology; Section 3 presents benchmark tests and model validation; Section 4 discusses the features of NWWs induced by vertical disturbances; and Section 5 summarizes the key findings.

2. Governing Equations and Numerical Methodology

The OpenFOAM version proposed by Mi et al. [21] is utilized to simulate NWWs induced by upward-moving bottom disturbances with varying movement duration and disturbance bulk. The turbulence and the interaction between the fluid and structure are implemented by the k–ε model and IBM, respectively. The governing equations are solved using a two-step projection method for momentum conservation, and free surface dynamics are captured using the volume of fluid (VOF) technique.

2.1. Governing Equations

The fluid media are tap water and air, with densities of 998 kg/m^3 and 1.0 kg/m^3, respectively. Considering that the fluid and structure motions are at a low speed, it is acceptable that both fluids are assumed to be incompressible. Then, the basic governing equations of the flow motions are formulated as the continuity equation and the Reynolds-averaged Navier–Stokes (RANS) equation. The k–ε model is adopted to solve the turbulence. The total equations are summarized as follows:

$$\frac{\partial u_i}{\partial x_i} = 0 \tag{1}$$

$$\frac{\partial u_i}{\partial t} + \frac{\partial u_i u_j}{\partial x_j} = -\frac{1}{\rho}\frac{\partial p}{\partial x_i} + f_i + \frac{1}{\rho}\frac{\partial}{\partial x_j}\left(\mu\frac{\partial u_i}{\partial x_j} - \rho u'_i u'_j\right) + (F_{\mathrm{IBM}})_i \tag{2}$$

where I and j = 1 and 2 for the two-dimensional (2D) flow. u_i and p are the mean velocity and fluid pressure, respectively; ρ is the fluid density; f_i is the gravitational acceleration; and μ is the kinematic viscosity. F_{IBM} is the immersed boundary force, which represents the hydrodynamic forces exerted on the flow by the vertical bottom disturbance. $\rho u'_i u'_j$ stands for the Reynolds stress, which can be resolved through the two-equation models as follows:

$$\frac{\partial(\rho k)}{\partial t} + \frac{\partial(\rho k u_j)}{\partial x_j} = \frac{\partial}{\partial x_j}\left[\left(\mu + \frac{\mu_t}{\sigma_k}\right)\frac{\partial k}{\partial x_j}\right] + P_k - \rho\varepsilon \tag{3}$$

$$\frac{\partial}{\partial t}(\rho\varepsilon) + \frac{\partial}{\partial x_j}(\rho\varepsilon u_i) = \frac{\partial}{\partial x_j}\left[\left(\mu + \frac{\mu_t}{\sigma_\varepsilon}\right)\frac{\partial \varepsilon}{\partial x_j}\right] + C_{1\varepsilon}\frac{\varepsilon}{k}P_k - G_{2\varepsilon}\rho\frac{\varepsilon^2}{k} - R \tag{4}$$

$$R = \frac{C_\mu \rho \eta^3 (1 - \eta/\eta_0)}{1 + \beta\eta^3}\frac{\varepsilon^2}{k}, \eta = \frac{Sk}{\varepsilon}, \mu_t = \rho C_u \frac{k^2}{\varepsilon} \tag{5}$$

where k and ε are the turbulent kinetic energy and the dissipation rate, respectively. μ_t is the turbulent viscosity, and $P_k = \mu_t S^2$ represents the productions due to the mean velocity shear, where S is the average strain rate tensor. σ_k is the Prandtl number corresponding to k, with a value of 0.7179, and σ_ε is the Prandtl number corresponding to ε, with a value of 0.7179. C_u is a constant of 0.0845. The empirical constants $C_{1\varepsilon}$ and $C_{2\varepsilon}$, η_0 and β, are respectively set as 1.42, 1.68, 4.38, and 1.92.

2.2. Numerical Methodology

This two-step projection method can effectively uncouple the pressure from the momentum equation. This strategy transforms the coupled problem into sequential, computationally tractable steps, significantly enhancing numerical stability and efficiency. The details of this method are divided into the following two steps:

$$\text{Step 1}: \frac{\tilde{u}_i^{n+1} - u_i^n}{\Delta t} = -u_j^n\frac{\partial u_i^n}{\partial x_j} + \frac{1}{\rho^n}\frac{\partial}{\partial x_j}\left(\mu\frac{\partial u_i^n}{\partial x_j} - \rho\left\langle (u_i^n)'\left(u_j^n\right)'\right\rangle\right) \tag{6}$$

$$\text{Step 2}: \frac{u_i^{n+1} - \tilde{u}_i^{n+1}}{\Delta t} = -\frac{1}{\rho^n}\frac{\partial p^{n+1}}{\partial x_i} + f_i^{n+1} + (F_{\mathrm{IBM}})_i^{n+1} \tag{7}$$

Step 1 is to first compute an intermediate velocity (denoted by $\tilde{u}_i^{n+1}$), which may not satisfy the continuity equation of Equation (1), and then Step 2 is to subsequently correct it to the final flow field via a pressure projection. Taking the divergence of Equation (7) and substituting Equation (1) yields the pressure Poisson equation (PPE) as follows:

$$\frac{\partial}{\partial x_i}\left(\frac{1}{\rho^n}\frac{\partial p^{n+1}}{\partial x_i}\right) = \frac{1}{\Delta t}\frac{\partial \widetilde{u}_i^{n+1}}{\partial x_i} + \frac{\partial f_i^{n+1}}{\partial x_i} + \frac{\partial (F_{\mathrm{IBM}})_i^{n+1}}{\partial x_i} \tag{8}$$

It is noted that the hydrodynamic force $(F_{\mathrm{IBM}})_i^{n+1}$ is only non-zero at the fluid–solid interface, which can be estimated from Equation (7) as follows:

$$(F_{\mathrm{IBM}})_i^{n+1} = \frac{u_i^{n+1} - \widetilde{u}_i^{n+1}}{\Delta t} + \frac{1}{\rho^n}\frac{\partial p^{n+1}}{\partial x_i} - f_i^{n+1} \tag{9}$$

At the (n + 1)-th time step, the velocity u_i^{n+1} is unknown. To address this, a new velocity $\hat{u}_i^{n+1}$ is introduced to replace u_i^{n+1} in Equation (9). This velocity is interpolated between the predicted fluid velocity in the interior fluid cells and the zero-velocity condition enforced on the fluid–solid interface. The forcing term is implemented as an implicit source within the momentum equation, coupled with the pressure–velocity correction procedure to maintain stability at practical time steps.

3. Benchmark Tests and Model Validation

This section presents the validation of the proposed model against established theoretical solutions and experimental and numerical data from the literature. Three benchmark cases of gravity waves are considered: linear, solitary, and tsunami-like waves.

3.1. Generation of Linear Waves

The generation of a linear wave (LW) is based on the principle of volume displacement, which assumes the crest volume above the still water depth (defined as h) equals the water confined by the wavemaker (Galvin [50]). Following this principle, a numerical underwater wavemaker is implemented to simulate an LW generated by a rectangular upthrust. The general formula can be expressed as follows in Equation (10):

$$\int_0^t \int_0^l y(x,t)dxdt = 2\int_0^t \eta(x,t)dt \tag{10}$$

where l denotes the upthrust width, $y(x,t)$ represents the upthrust displacement, and $\eta(t)$ expresses the free surface displacement. The schematic diagram of the underwater wavemaker is shown in Figure 1. Factor 2 is used on the account that two systematic wave trains are generated on both sides. To avoid wave reflection from the left and right boundaries, the transmissive boundary condition is applied, which is as follows:

$$\frac{\partial \varphi}{\partial t} + U_n\frac{\partial \varphi}{\partial n} + \frac{c}{L}\varphi = 0 \tag{11}$$

where φ denotes various wave characteristics at the boundary, including the mean velocity, the mean free surface displacement, the turbulent kinetic energy, the turbulent dissipation rate, etc. U_n is the normal component of velocity at the boundary. c and L are the characteristic wave speed (equaling to $(gh)^{0.5}$, where g is the gravitational acceleration) and the characteristic length scale (set as the wavelength in the present study), respectively.

To excite a linear wave whose free surface displacement satisfies $\eta(x,t) = H/2\sin(2\pi/\mathrm{T} \times t)$, where H and T denote the wave height and wave period, respectively, a prescribed bottom upthrust motion $y(x,t)$ is applied. To minimize startup transients, the displacement follows a smoothed trajectory given by $y(x,t)$ = S × [1 − cos(2π/T × t)] with a corresponding velocity of $2\pi S/\mathrm{T} \times \sin(2\pi/\mathrm{T} \times t)$, where S represents half of the stroke. A linear wave with H = 0.0068 m and T = 1.0 s is expected in a domain of h = 0.1 m. Based on the linear dispersion relationship, kh = 0.68, where k stands for the wave number, meaning an intermediate water depth. The computational domain of 50.0 m × 0.115 m is discretized into 2500 × 230 uniform grids with $\Delta x = 0.02$ m and $\Delta z = 0.0005$ m. Radiation boundary conditions are imposed at both sides of the domain to avoid wave reflection, and the time step is automatically adjusted to ensure numerical stability. The upthrust is a rectangular shape that is initially arranged at the middle of the domain in x axis, with $l = 0.2$ m, and S is estimated to be 0.01 m.

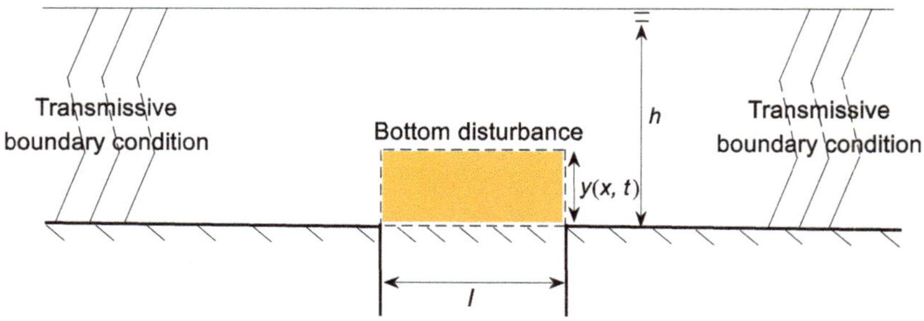

Figure 1. Schematic diagram of an underwater wavemaker.

Figure 2 shows the comparisons of the free surface profiles at different instants t/T = 1, 3, 5, 7, and 9 between the present numerical results and analytical wave profiles. From the figure, two successive wave trains have been generated on both sides; the leading waves at both onshore (negative x) and offshore (positive x) directions are smaller than the waves near the wavemaker zone. Excellent agreement has been guaranteed between the present numerical result and the analytical solution, implying the accuracy of the present model in modeling the LW with small amplitude, which further indicates the negligible numerical dissipation.

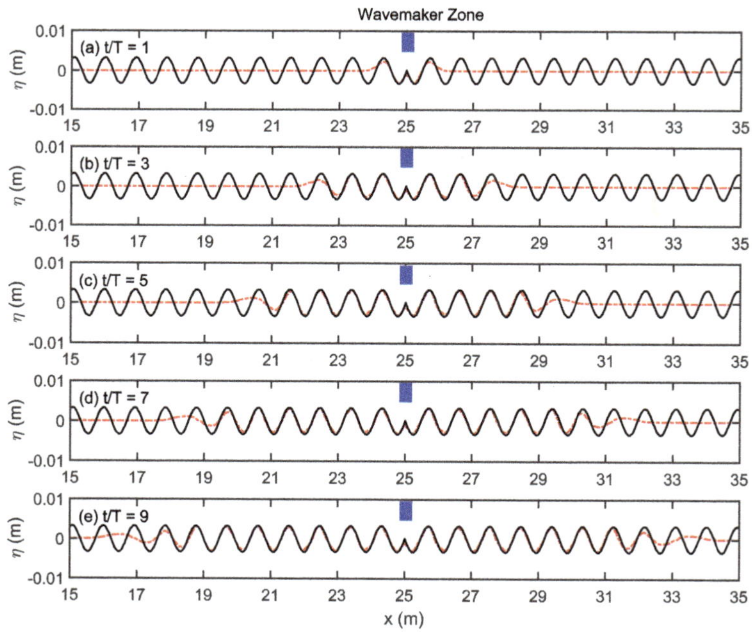

Figure 2. Comparisons of the free surface profiles between the present numerical results (dash–dot line) and analytical wave profiles (solid line) at different instants: (**a**) $t/\mathrm{T} = 1$; (**b**) $t/\mathrm{T} = 3$; (**c**) $t/\mathrm{T} = 5$; (**d**) $t/\mathrm{T} = 7$; (**e**) $t/\mathrm{T} = 9$.

3.2. Generation of Tsunami-like Waves

The configuration of a rectangular bottom disturbance in water of depth of h is illustrated in Figure 3a, where b_{up} and $\zeta(t)$ are the disturbance width and displacement, respectively, and t_{up} stands for the rise time of the disturbance. This setup replicates the seminal experiments of Hammack and Segur [10] on tsunami-like wave generation by vertical bottom motion. Shen and Chan [17] reconsidered this problem through an in-house code Navier–Stokes model, with the bottom disturbance being implemented through IBM. The schematic diagram of the experimental setup is shown in Figure 3a, where three wave gauges are installed at 0.61 m, 1.61 m, and 9.61 m away from the left boundary wall. h and b_{up} are set as 0.05 m and 0.61 m, respectively. The disturbance motion lasts 1.0 s, with the corresponding movement velocity and disturbance displacement shown in Figure 3b,c.

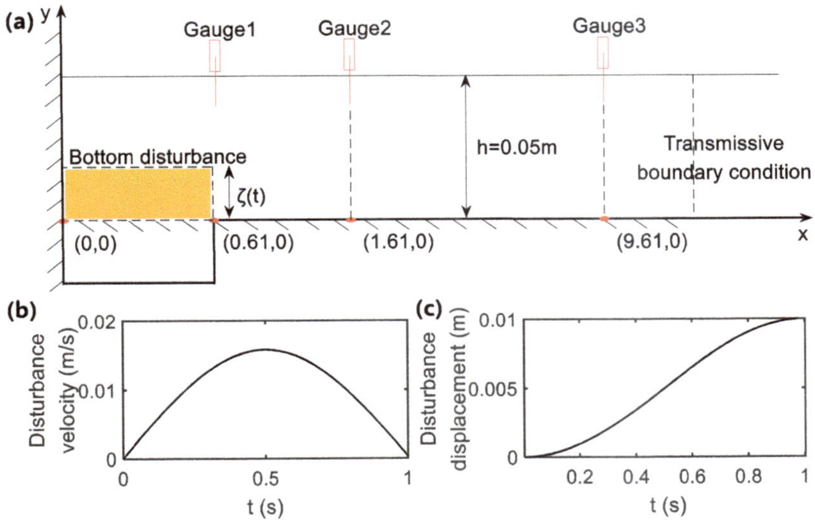

Figure 3. Generation of the tsunami-like wave: (**a**) schematic diagram of the fluid domain and bottom disturbance; (**b**) movement velocity; (**c**) disturbance displacement.

The computation domain 10.0 m $\times$ 0.06 m is discretized into 2000 $\times$ 150 uniform grids with $\Delta x = 0.005$ m and $\Delta z = 0.0004$ m. The time step is adaptively adjusted to ensure numerical stability. The free surface displacements at three wave gauges are presented in Figure 4, alongside the experimental data of Hammack and Segur [10] and the numerical results of Shen and Chan [17] for comparison. The discrepancies between the present model and that of Shen and Chan [17] are minimal, particularly in the near-field generation region, although the present model is slightly ahead of that of Shen and Chan [17] at the far region, $x = 9.61$ m, and overall agreement remains good. Both numerical results slightly underestimate the experimental data of Hammack and Segur [10] in the generating region, but the phase shift is unapparent, which implies good numerical conservation. In the far region, $x = 9.61$ m, the experimental data lag both numerical results, and this may be mainly due to accumulative errors in the numerical simulations.

Figure 5 illustrates the free surface profiles and corresponding pressure distributions at successive instants: $t = 2.0$ s, 4.0 s, 6.0 s, 8.0 s, 10.0 s, and 12.0 s. After the initial 1.0 s, gravity is the only restoring force without other disturbances. By $t = 2.0$ s, there exists a bulk of water proceeding offshore (positive x). Subsequently, the leading bulk undergoes fission under gravitational effects, gradually splitting into successive soliton trains in the leading crest, behaving like a dispersive wave, which is also called the tsunami-like wave

(Jing et al. [51]). The water in the generating zone restores calm promptly after the wave leaves the generating region, as reflected by the hydrostatic pressure distribution observed in the later stages.

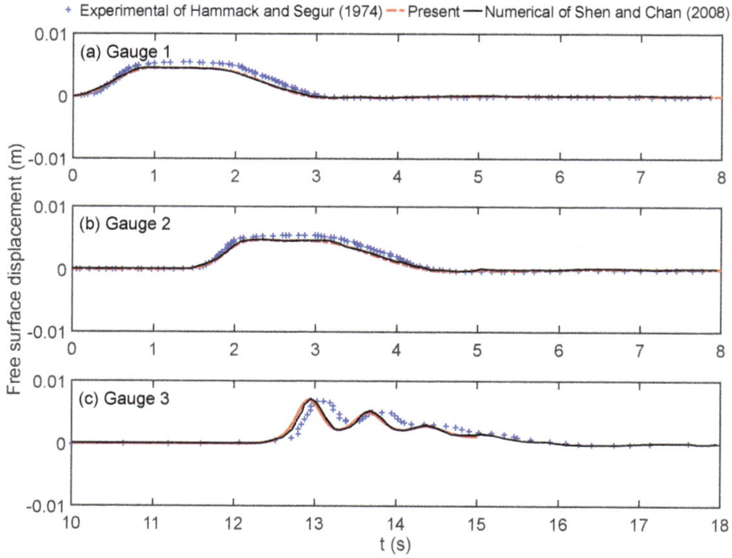

Figure 4. Comparisons of time series free surface displacements among the present numerical result (dash–dot line), numerical result of Shen and Chan [17] (solid line), and experimental data of Hammack and Segur [10] (plus) at three gauges: (**a**) Gauge 1; (**b**) Gauge 2; (**c**) Gauge 3.

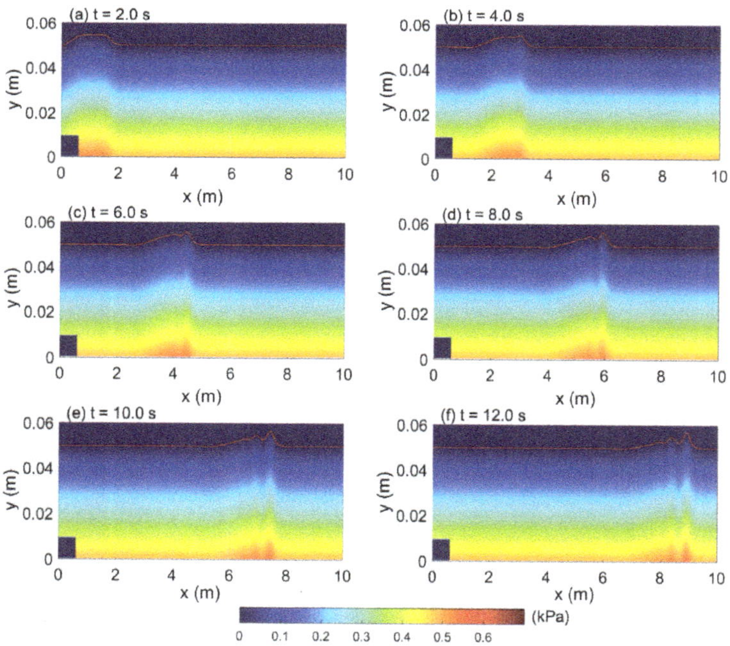

Figure 5. Free surface snapshots and pressure distributions at different instants: (**a**) t = 2.0 s; (**b**) t = 4.0 s; (**c**) t = 6.0 s; (**d**) t = 8.0 s; (**e**) t = 10.0 s; (**f**) t = 12.0 s.

4. Effects of Bottom Disturbance on Wave Features

Tsunami-like waves represent a common manifestation of seabed deformation during marine earthquakes. Seismic loading can trigger parts of the seabed to move vertically, which also represents the conceptual process of seabed upthrust, serving as a key generation mechanism for such waves. Studies have identified that both the movement velocity and disturbance bulk (or width) dominate the generated wave (Derakhti et al. [2]; Shen and Chan [17]). Building on the validation in Section 3.2, this part extends the investigation to NWWs generated by an upward-moving bottom disturbance.

4.1. Disturbance Settings

The field investigation and numerical simulation have revealed that the arrival time of the tsunami and tsunami-like wave is strongly dependent on the total duration of the seabed movement (Derakhti et al. [2]; Shen and Chan [17]), which is correlated to the movement velocity. Additionally, the disturbance bulk is another dominant parameter (Hammack and Segur [10]; Wu [52]). The larger bulk means more water is prone to move under the influence of seabed disturbance. To systematically investigate these effects, the following two sub-sections examine the influences of the two parameters on the generated waves.

In the subsequent simulations, a fixed disturbance displacement of 0.01 m with four disturbance durations is adopted, with whom velocity and displacement are displayed in Figure 6. The two solid lines in the figure denote the benchmark case discussed in Section 3.2. In addition, five disturbance widths are simulated and presented in Section 4.3. In total, twenty numerical cases are modeled to comprehensively elucidate the role of disturbance velocity and bulk in governing wave characteristics.

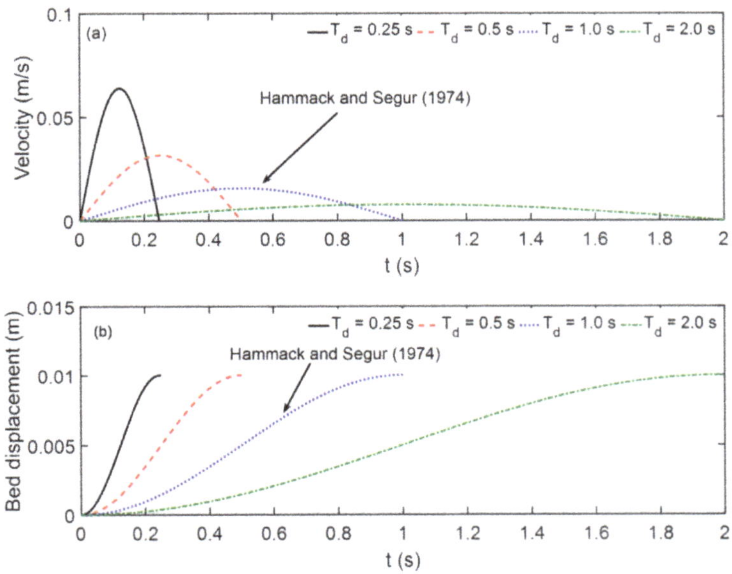

Figure 6. Time series movements of a bottom disturbance: (**a**) velocity; (**b**) displacement [10].

4.2. Influence of Disturbance Velocity

To evaluate the influence of disturbance velocity on wave generation, four distinct velocities are considered, as illustrated in Figure 6, among which the case of $T_d = 1.0$ s is a reference case, which has yet to be verified. All cases share the same maximum vertical displacement, while the movement duration varies from 0.25 s to 2.0 s, resulting in maximum disturbance velocities (U_{max}) ranging approximately from 0.0079 m/s to 0.064 m/s.

The comparisons of time series free surface displacements at Gauge 1, Gauge 2, and Gauge 3 under four disturbance velocities are displayed in Figure 7. Near the source region (Gauge 1), the surface elevation exhibits a non-isosceles trapezoidal shape in all cases, except for the longest duration (T_d = 2.0 s), which retains a solitary-like profile. The peak surface elevations are consistently slightly less than half of the maximum bottom displacement, consistent with earlier studies (Hammack and Segur [10]; Shen and ChaN [17]; Wu [52]). As the wave propagates to Gauge 2 (x = 1.61 m), the leading wave begins to fission into multiple solitons in the shorter-duration cases (T_d = 0.25 s and 0.5 s), accompanied by an amplitude increase relative to Gauge 1. The case of T_d = 1.0 s maintains a trapezoidal form, while that of T_d = 2.0 s continues to show a solitary-like profile with little change in wave height from Gauge 1.

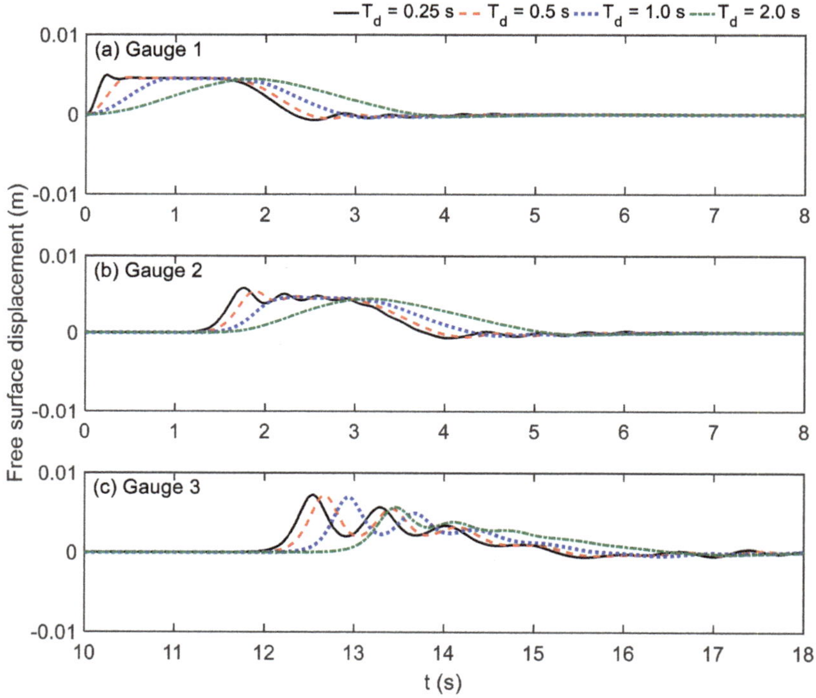

Figure 7. Comparisons of time series free surface displacements among various T_d = 0.25 s, 0.5 s, 1.0 s, and 2.0 s at three gauges: (**a**) Gauge 1; (**b**) Gauge 2; (**c**) Gauge 3.

At the far-field Gauge 3, clear soliton trains are observed in all cases, confirming the eventual formation of a tsunami-like wave. The amplitude of the leading soliton is higher than that recorded at the previous gauges. Both the arrival time and amplitude of the leading soliton vary systematically with T_d. Shorter disturbance durations result in earlier arrival and larger phase celerity, indicating that faster disturbance motion generates faster nonlinear waves and thus an earlier tsunami onset. The wave amplitude decreases with increasing T_d. Moreover, the time intervals between the leading solitons of consecutive cases are approximately 0.13 s, 0.26 s, and 0.50 s, roughly half of the differences in disturbance duration, which highlights the critical role of movement duration in tsunami wave forecasting.

The wave energy evolutions across various gauges and under different disturbance durations are accordingly discussed, as displayed in the wavelet spectrogram results of Figure 8. The results reveal distinct spatiotemporal patterns of energy concentration in the frequency–time domain at each measurement location. As T_d increases from 0.25 s to 2.0 s, a clear delay in wave energy migration is observed at Gauge 3, as marked by dashed lines in both the temporal and frequency dimensions. Specifically, for T_d = 0.25 s, the dominant energy at Gauge 3 is concentrated around a relatively high frequency and early time, whereas for T_d = 2.0 s, it shifts toward lower frequencies and later times. In contrast, Gauge 1 and Gauge 2 display less pronounced energy variation with T_d, implying that wave energy transformation becomes most pronounced in the far field (Gauge 3) under extended disturbance durations. This spatial and parametric disparity in energy distribution underscores the strong dependence of wave energy propagation on both the measurement location and the disturbance duration. Furthermore, the results indicate that wave celerity in the generation region varies significantly with T_d, a finding essential for understanding nonlinear wave dynamics.

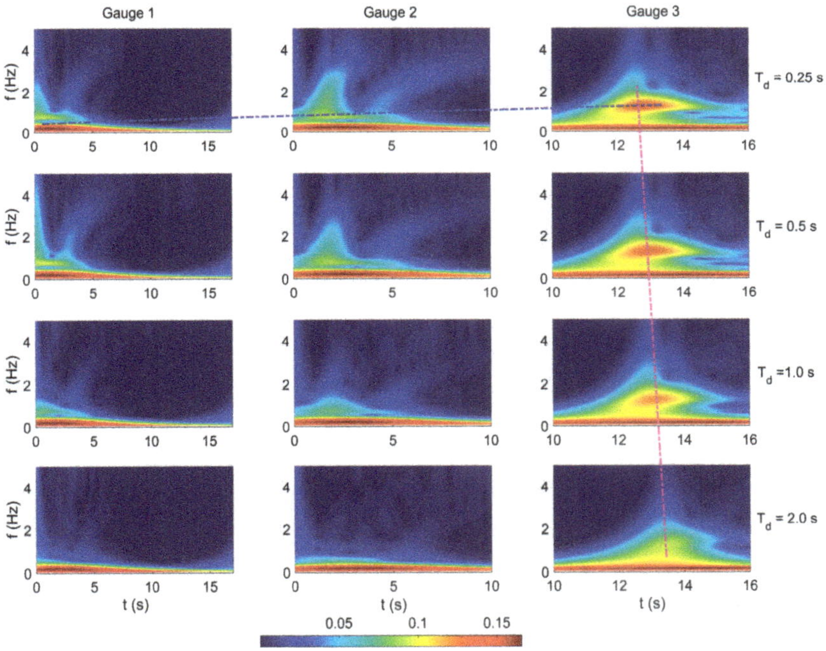

Figure 8. Time series wavelet spectrogram results for the corresponding free surface displacements across different wave gauges (left, middle, and right columns for Gauge 1, Gauge 2, and Gauge 3, respectively) and under varied disturbance durations (first, second, third, and fourth rows for T_d = 0.25 s, 0.5 s, 1.0 s, and 2.0 s, respectively). The blue and pink dashed lines represent the trends of spatial-temporal energy transfer.

To further elucidate the wave characteristics, Figure 9 compares the free surface profiles at several instants (t = 0.2, 0.5, 1.0, 4.0, 8.0, and 12.0 s) across the four disturbance durations, with the case of T_d = 1.0 s included as a reference (previously validated in Figure 4). As shown in the figures, the shortest-duration case (T_d = 0.25 s) consistently exhibits the largest wave height and phase celerity of the leading crest. In all cases, the initial elevation undergoes fission, breaking up into several solitons at the wave front. Both the number of solitons and the instant at which fission initiates increase as the disturbance duration T_d decreases.

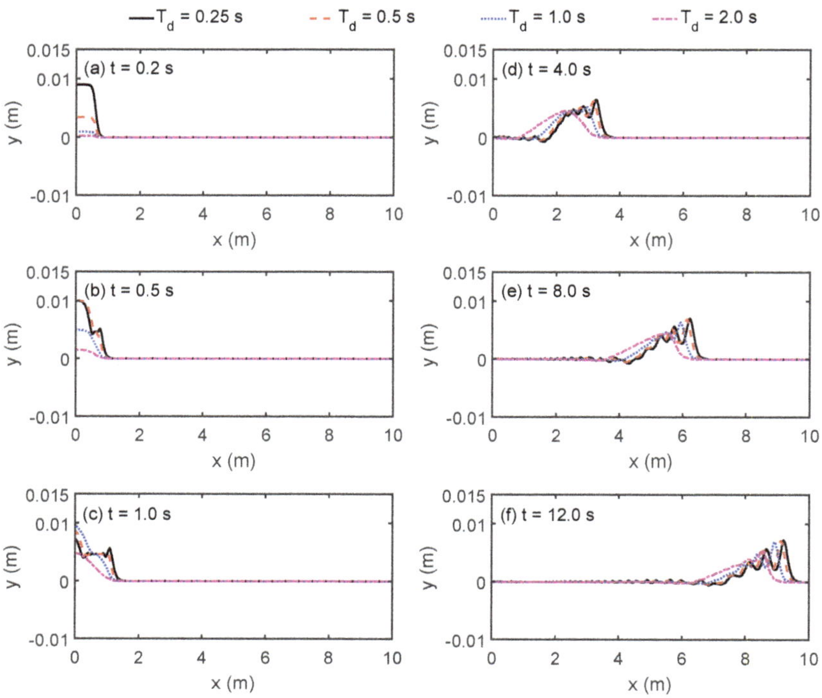

Figure 9. Comparisons of the free surface profiles among cases with various T_d = 0.25 s, 0.5 s, 1.0 s (the reference case), and 2.0 s at different instants: (**a**) t = 0.2 s; (**b**) t = 0.5 s; (**c**) t = 1.0 s; (**d**) t = 4.0 s; (**e**) t = 8.0 s; (**f**) t = 12.0 s.

4.3. Influence of Disturbance Bulk

Following the analysis of a fixed bottom width L_{up} = 0.61 m in Section 3.2, this section examines nonlinear waves generated by disturbances with varying widths. Five different widths are considered, namely, 0.2 m, 0.4 m, 0.61 m, 0.8 m, and 1.0 m, implying a series of disturbance bulks. To emphasize nonlinear effects, the shortest disturbance duration T_d = 0.25 s is selected, as it produces the most pronounced wave height and highest phase celerity (see Figure 6). Accordingly, the case with T_d = 0.25 s and L_{up} = 0.61 m serves as the reference, which has not yet been discussed.

Figure 10 gives the comparisons of time series of free surface displacements at four gauges, including an additional gauge at x = 6.61 m to better capture intermediate-field behavior. Near the generation region, the maximal wave heights for the two cases with L_{up} = 0.8 m and L_{up} = 1.0 m can approach 0.01 m, whereas the remaining three cases remain below half of the maximum disturbance displacement, consistent with trends in Figure 7. As the waves pass over Gauge 2 (see Figure 10b), wave heights of the two widest disturbances with L_{up} = 0.8 m and L_{up} = 1.0 m decrease to a value comparable to the other cases. All profiles behave as the non-isosceles trapezoid, except for the case with L_{up} = 0.2 m, the minimal upthrust width, exhibiting an irregular solitary-like feature. In the far fields, x = 6.61 m and x = 9.61 m, well-defined soliton trains emerge for $L_{up} \geq 0.4$ m. Among these, the case with L_{up} = 0.4 m shows a slightly reduced leading wave height. In contrast, the narrowest case (L_{up} = 0.2 m) maintains a solitary wave profile throughout propagation, particularly evident in the far field x = 9.61 m. Furthermore, the phase celerity of the leading wave grows systematically with the disturbance width L_{up}.

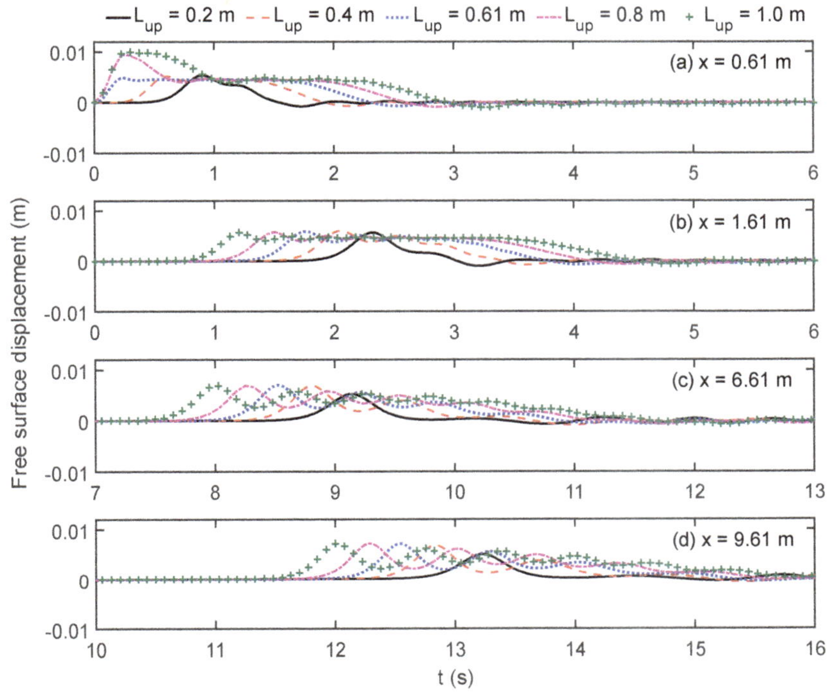

Figure 10. Comparisons of time series of the free surface displacements among various L_{up} = 0.2 m, 0.4 m, 0.61 m (reference case), 0.8 m, and 1.0 m at four gauges: (**a**) x = 0.61 m; (**b**) x = 1.61 m; (**c**) x = 6.61 m; (**d**) x = 9.61 m.

To elucidate the influence of disturbance width on wave energy evolutions across distinct spatial locations, the wavelet spectrogram results of three cases (L_{up} = 0.2 m, 0.61 m, 1.0 m) are selected for further discussions, as displayed in Figure 11. The energy distribution in the frequency–time domain presents remarkable spatial–temporal heterogeneity. As L_{up} increases from 0.2 m to 1.0 m, the wave energy migration (marked by dashed lines) at downstream gauges (especially Gauge 3) exhibits a systematic shift in both frequency and time. Specifically, for L_{up} = 1.0 m, the dominant energy at Gauge 3 is concentrated at relatively lower frequencies and later times compared to cases with narrower disturbance widths. In contrast, upstream gauges (Gauge 1 and Gauge 2) show less significant energy variation with L_{up}, indicating that the modulation of disturbance width exerts a more pronounced effect on wave energy evolution. This spatial divergence in energy dynamics across gauges and L_{up} values underscores the critical role of disturbance bulk in governing wave energy propagation, which is essential for advancing the understanding of the wave-bottom interaction mechanism.

A comparison of wave characteristics among the five cases is demonstrated in Figure 12, with the case of L_{up} = 0.61 m serving as the reference (previously shown as the solid line in Figure 9, T_d = 0.25 s and L_{up} = 0.61 m). The case of L_{up} = 1.0 m displays the largest wave height and the highest phase celerity. The comparisons also imply that the phase celerity increases with the increasing disturbance width L_{up}, as well as the timing of wave fission. In contrast, the narrowest disturbance case (L_{up} = 0.2 m) gradually develops into an irregular solitary wave, exhibiting transient spatiotemporal features of a strain of oscillating tails rather than a flat tail as the standard solitary wave, which is consistent with tsunami-like waves documented in field observations of Kawai et al. [53].

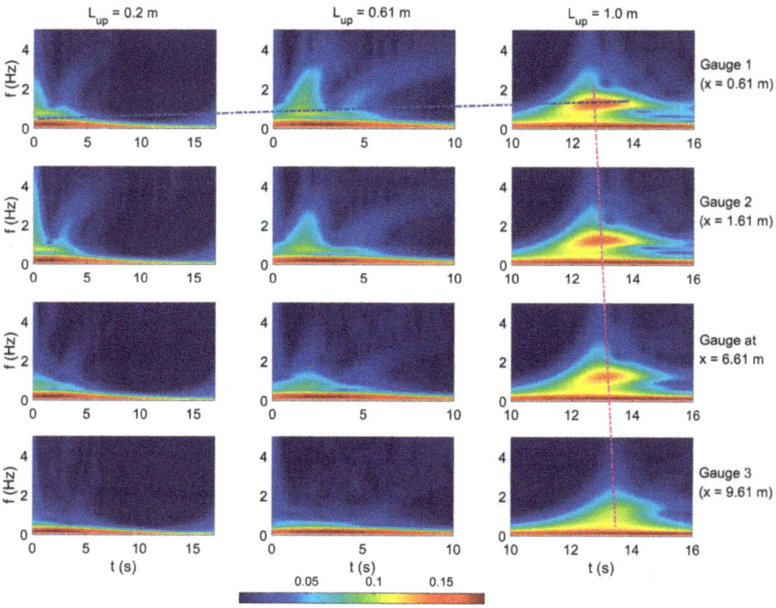

Figure 11. Time series wavelet spectrogram results for the corresponding free surface displacements across different wave gauges (first, second, third, and fourth rows, for Gauge 1, Gauge 2, gauge at x = 6.61 m, and Gauge 3, respectively) and under varied disturbance widths (left, middle, and right columns for L_{up} = 0.2 m, 0.61 m, and 1.0 m, respectively). The blue and pink dashed lines represent the trends of spatial-temporal energy transfer.

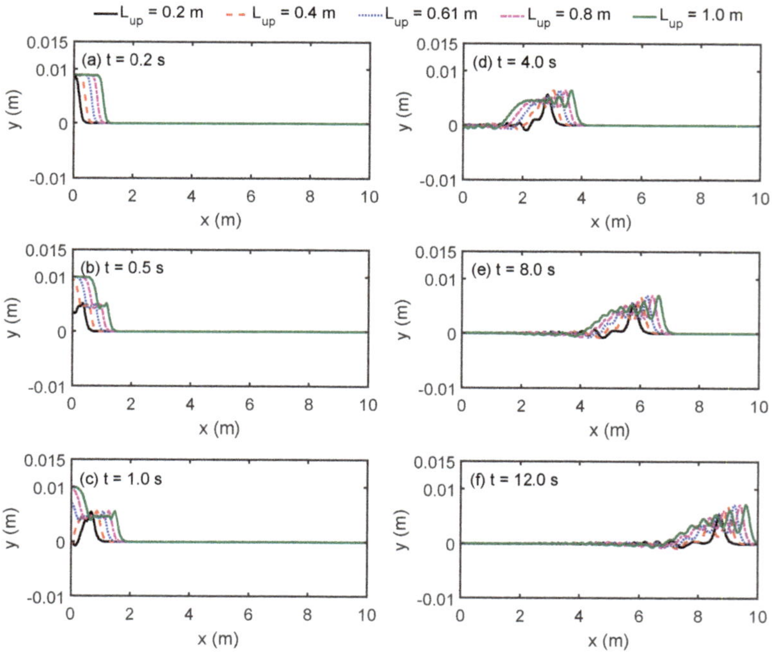

Figure 12. Comparisons of the free surface profiles among cases with various L_{up} = 0.2 m, 0.4 m, 0.61 m (the reference case), 0.8 m, and 1.0 m at different instants: (**a**) t = 0.2 s; (**b**) t = 0.5 s; (**c**) t = 1.0 s; (**d**) t = 4.0 s; (**e**) t = 8.0 s; (**f**) t = 12.0 s.

5. Conclusions

This study numerically investigates the generation and propagation of nonlinear water waves (NWWs) by vertical bottom disturbances, employing a Reynolds-averaged Navier–Stokes (RANS) solver. The model incorporates the k–ε model and the immersed boundary method (IBM) to represent the moving seabed disturbance, with the free surface captured by the volume of fluid (VOF) method. The model demonstrates robust accuracy in capturing the key features of resultant wave fields. The key findings are summarized as follows:

1. The model demonstrates high accuracy across a defined parameter space, simulating bottom disturbances with non-dimensional velocity amplitudes ($U_d/\sqrt{gh}$) ranging from 0.113 to 0.091 and non-dimensional widths ($L_{up}/0.61$) ranging from 0.328 to 1.639, validating its capability in capturing essential wave features across a broad range of disturbance velocities and widths.
2. The generated wave field exhibits strong dependence on both the disturbance duration T_d and width L_{up}. For instance, the decrease in T_d and the growth of L_{up} can independently lead to an increase in phase celerity and wave height of the leading soliton. All simulated cases evolve into dispersive wave trains whose leading crest undergoes fission into successive solitons, a hallmark of a tsunami-like wave.
3. Shorter disturbance durations result in earlier fission of the leading crest into soliton trains and higher phase celerities. This inverse relationship between disturbance duration and wave celerity provides crucial insight for wave forecasting applications.
4. Larger disturbance widths generate nonlinear waves in a near-linear increase in the phase celerity of the leading wave. The amplitude of the leading soliton decreases with increasing T_d but increases with expanding L_{up}, revealing competing mechanisms governing wave amplitude evolution.
5. Wave energy evolution demonstrates distinct spatiotemporal patterns, with the main wave energy nonlinearly migrating from higher frequencies to lower frequencies in the offshore direction (the prorogating direction) for longer disturbances. This spectral evolution underscores the critical role of both disturbance duration and width in governing wave energy propagation characteristics.

These findings substantially advance the understanding of tsunami-like wave generation mechanisms and provide valuable insights for predicting wave behaviors in scenarios involving seabed disturbances, with relevance to coastal hazard assessment and early warning systems.

Author Contributions: Conceptualization and methodology, H.-P.M.; validation, visualization, and writing—original draft, H.-P.M. and H.-X.Z. All authors have read and agreed to the published version of the manuscript.

Funding: This research was funded by the Doctoral research start-up fee of Panzhihua University, grant number 035200243; the Open Fund project of the State Key Laboratory of Comprehensive Utilization of Vanadium and Titanium Resources, grant number 035001648; the Laboratory for Comprehensive Development and Utilization of Industrial Solid Waste, grant number 035300420; and the School Funding of Panzhihua University, grant number 035300968.

Data Availability Statement: Data are available upon request due to restrictions on personal privacy.

Conflicts of Interest: The authors declare no conflicts of interest.

References

1. Semenov, Y.A.; Wu, G.X. Free-surface gravity flow due to a submerged body in uniform current. *J. Fluid Mech.* **2020**, *883*, A60. [CrossRef]
2. Derakhti, M.; Dalrymple, R.A.; Okal, E.A.; Synolakis, C.E. Temporal and topographic source effects on tsunami generation. *J. Geophys. Res. Ocean.* **2019**, *124*, 5270–5288. [CrossRef]
3. Gao, J.L.; Ma, X.Z.; Dong, G.H.; Chen, H.Z.; Liu, Q.; Zang, J. Investigation on the effects of Bragg reflection on harbor oscillations. *Coast. Eng.* **2021**, *170*, 103977. [CrossRef]
4. Synolakis, C.E.; Bernard, E.N. Tsunami science before and beyond boxing day 2004. *Philos. Trans. R. Soc. A Math. Phys. Eng. Sci.* **2006**, *364*, 2231–2265. [CrossRef]
5. Reeve, D.E.; Horrillo-Caraballo, J.; Karunarathna, H.; Wang, X. Experimental study of wave trains generated by vertical bed movements. *Appl. Ocean Res.* **2024**, *147*, 103971. [CrossRef]
6. Fang, K.; Liu, Z.; Sun, J.; Xie, Z.; Zheng, Z. Development and validation of a two-layer Boussinesq model for simulating free surface waves generated by bottom motion. *Appl. Ocean Res.* **2020**, *94*, 101977. [CrossRef]
7. Jing, H.; Gao, Y.; Liu, C.; Hou, J. Far-field characteristics of linear water waves generated by a submerged landslide over a flat seabed. *J. Mar. Sci. Eng.* **2020**, *8*, 196. [CrossRef]
8. Gao, J.; Ji, C.; Gaidai, O.; Liu, Y.; Ma, X. Numerical investigation of transient harbor oscillations induced by N-waves. *Coast. Eng.* **2017**, *125*, 119–131. [CrossRef]
9. Gao, J.; Ma, X.; Chen, H.; Zang, J.; Dong, G. On hydrodynamic characteristics of transient harbor resonance excited by double solitary waves. *Ocean Eng.* **2021**, *219*, 108345. [CrossRef]
10. Hammack, J.L.; Segur, H. The Korteweg-de Vries equation and water waves. Part 2. Comparison with experiments. *J. Fluid Mech.* **1974**, *65*, 289–314. [CrossRef]
11. Tinti, S.; Bortolucci, E. Energy of water waves induced by submarine landslides. *Pure Appl. Geophys.* **2000**, *157*, 281–318. [CrossRef]
12. Infeld, E.; Karczewska, A.; Rowlands, G.; Rozmej, P. Exact cnoidal solutions of the extended KdV equation. *Acta Phys. Pol. A* **2018**, *133*, 1191–1199. [CrossRef]
13. Nadiga, B.T.; Margolin, L.G.; Smolarkiewicz, P.K. Different approximations of shallow fluid flow over an obstacle. *Phys. Fluids* **1996**, *8*, 2066–2077. [CrossRef]
14. Duan, W.Y.; Wang, Z.; Zhao, B.B.; Ertekin, R.C.; Kim, J.W. Steady solution of the velocity field of steep solitary waves. *Appl. Ocean Res.* **2018**, *73*, 70–79. [CrossRef]
15. Madsen, P.A.; Bingham, H.B.; Schäffer, H.A. Boussinesq-type formulations for fully nonlinear and extremely dispersive water waves: Derivation and analysis. *Proc. R. Soc. A* **2003**, *459*, 1075–1104. [CrossRef]
16. Løvholt, F.; Pedersen, G.; Harbitz, C.B.; Glimsdal, S.; Kim, J. On the characteristics of landslide tsunamis. *Philos. Trans. R. Soc. A Math. Phys. Eng. Sci.* **2015**, *373*, 20140376. [CrossRef]
17. Shen, L.W.; Chan, E.S. Numerical simulation of fluid-structure interaction using a combined volume of fluid and immersed boundary method. *Ocean Eng.* **2008**, *35*, 939–952. [CrossRef]
18. Jin, X.; Qin, Y.Y.; Tao, Y.; Lin, P. Numerical modeling of transient nonlinear water waves generated by horizontally moving bottom disturbances. *Phys. Fluids* **2024**, *36*, 113125. [CrossRef]
19. Whittaker, C.N.; Nokes, R.I.; Lo, H.Y.; Liu, P.F.; Davidson, M.J. Physical and numerical modelling of tsunami generation by a moving obstacle at the bottom boundary. *Environ. Fluid Mech.* **2017**, *17*, 929–958. [CrossRef]
20. Ebrahimi, A.; Boroomand, B. Simulation of nonlinear free surface waves using a fixed grid method. *J. Appl. Fluid Mech.* **2023**, *16*, 2054–2068. [CrossRef]
21. Mi, S.; Wang, M.; Avital, E.J.; Williams, J.J.; Chatjigeorgiou, I.K. An implicit Eulerian–Lagrangian model for flow-net interaction using immersed boundary method in OpenFOAM. *Ocean Eng.* **2022**, *264*, 112843. [CrossRef]
22. Zhang, D.; Chwang, A.T. Numerical study of nonlinear shallow water waves produced by a submerged moving disturbance in viscous flow. *Phys. Fluids* **1996**, *8*, 147–155. [CrossRef]
23. Jin, X.; Dai, C.; Huang, H.L.; Qin, Y.Y.; Tao, Y.; Zhang, F.G. A comparative study of RNN, CNN, LSTM and GRU for random sea wave forecasting. *Ocean Eng.* **2025**, *338*, 122013. [CrossRef]
24. Meylan, M.H.; Stepanyants, Y.A. Scattering of gravity-capillary waves on a bottom step. *Phys. Fluids* **2024**, *36*, 011701. [CrossRef]
25. Ockendon, H.; Ockendon, J.R.; Waterhouse, D.D. Multi-mode resonance in fluids. *J. Fluid Mech.* **1996**, *315*, 317–344. [CrossRef]
26. Kim, M.W.; Koo, W.; Hong, S.Y. Numerical analysis of various artificial damping schemes in a three-dimensional numerical wave tank. *Ocean Eng.* **2013**, *75*, 165–173. [CrossRef]
27. Ibrahim, R.A. Recent advances in physics of fluid parametric sloshing and related problems. *J. Fluids Eng.* **2015**, *137*, 090801. [CrossRef]
28. Mao, H.; He, Y.; Wu, G.; Lin, J.; Ji, R. Study of liquid viscosity effects on hydrodynamic forces on an oscillating circular cylinder underwater using OpenFOAM. *Symmetry* **2021**, *13*, 1806. [CrossRef]

29. Jin, X.; Liu, M.M.; Zou, Y.J.; Luo, M.; Yang, F.; Wang, L. Numerical simulation of Faraday waves in a rectangular tank and damping mechanism of internal baffles. *J. Fluids Struct.* **2022**, *109*, 103503. [CrossRef]
30. Jin, X.; Xue, M.A.; Lin, P.Z. Experimental and numerical study of nonlinear modal characteristics of Faraday waves. *Ocean Eng.* **2021**, *221*, 108554. [CrossRef]
31. Jin, X.; Huang, H.L.; Tao, Y.; Zhang, F.G.; Luo, M.; Fan, C.Y.; Tang, T. A systematic review on mechanism and regulation strategy of marine hydrodynamic noise: Advances, challenges, and perspectives. *Ocean Eng.* **2025**, *330*, 121202. [CrossRef]
32. Wang, R.; Ning, D.; Zhang, C.; Zou, Q.; Liu, Z. Nonlinear and viscous effects on the hydrodynamic performance of a fixed OWC wave energy converter. *Coast. Eng.* **2018**, *131*, 42–50. [CrossRef]
33. Zhang, C.; Tan, J.; Ning, D. Machine learning strategy for viscous calibration of fully-nonlinear liquid sloshing simulation in FLNG tanks. *Appl. Ocean Res.* **2021**, *114*, 102737. [CrossRef]
34. Wu, G.X.; Taylor, R.E.; Greaves, D.M. The effect of viscosity on the transient free surface waves in a two-dimensional tank. *J. Eng. Math.* **2001**, *40*, 77–90. [CrossRef]
35. Jin, X.; Lin, P.Z. Viscous effects on liquid sloshing under external excitations. *Ocean Eng.* **2019**, *171*, 695–707. [CrossRef]
36. Jin, X.; Tang, J.B.; Tang, X.C.; Mi, S.; Wu, J.X.; Liu, M.M.; Huang, Z.L. Effect of viscosity on sloshing in a rectangular tank with intermediate liquid depth. *Exp. Therm. Fluid Sci.* **2020**, *118*, 110148. [CrossRef]
37. Jamin, T.; Gordillo, L.; Ruiz-Chavarría, G.; Berhanu, M.; Falcon, E. Experiments on generation of surface waves by an underwater moving bottom. *Proc. R. Soc. A* **2015**, *471*, 20150069. [CrossRef]
38. Slunyaev, A.V.; Kokorina, A.V.; Klein, M. Nonlinear dynamic pressure beneath waves in water of intermediate depth: Theory and experiment. *Eur. J. Mech. B Fluids* **2022**, *94*, 155–170. [CrossRef]
39. Lee, J.J.; Skjelbreia, J.E.; Raichlen, F. Measurements of velocities in solitary waves. *J. Waterw. Port Coast. Ocean. Div.* **1982**, *108*, 200–218. [CrossRef]
40. Bai, Y.; Cheung, K.F. Depth-integrated free-surface flow with a two-layer non-hydrostatic formulation. *Int. J. Numer. Methods Fluids* **2012**, *69*, 411–429. [CrossRef]
41. Gao, J.; Ma, X.; Zang, J.; Dong, G.; Ma, X.; Zhu, Y.; Zhou, L. Numerical investigation of harbor oscillations induced by focused transient wave groups. *Coast. Eng.* **2020**, *158*, 103670. [CrossRef]
42. Jin, X.; Huang, H.L.; Xu, X.K.; Qin, Y.Y.; Luo, M.; Wen, Y. Assessment of offshore wind and wave energy resources for combined exploitation in the East China Sea. *Energy* **2025**, *323*, 135730. [CrossRef]
43. Dragani, W.C. Numerical experiments on the generation of long ocean waves in coastal waters of the Buenos Aires province, Argentina. *Cont. Shelf Res.* **2007**, *27*, 699–712. [CrossRef]
44. Auclair, F.; Bordois, L.; Dossmann, Y.; Duhaut, T.; Paci, A.; Ulses, C.; Nguyen, C. A non-hydrostatic non-Boussinesq algorithm for free-surface ocean modelling. *Ocean Model.* **2018**, *132*, 12–29. [CrossRef]
45. Wassim, E.; Zheng, B.; Shang, Y. A parallel two-grid method based on finite element approximations for the 2D/3D Navier-Stokes equations with damping. *Eng. Comput.* **2024**, *40*, 541–554. [CrossRef]
46. Gao, J.; Wu, Y.; Song, Z.; He, M. Influences of low velocity uniform current on characteristics of gap resonance occurring between two adjacent fixed bodies. *Mar. Struct.* **2026**, *106*, 103961. [CrossRef]
47. Mi, C.; Gao, J.; Song, Z.; Liu, Y. Hydrodynamic wave forces on two side-by-side barges subjected to nonlinear focused wave groups. *Ocean Eng.* **2025**, *317*, 120056. [CrossRef]
48. Wang, H.E.; Chan, I.C. Numerical investigation of wave generation characteristics of bottom-tilting flume wavemaker. *J. Mar. Sci. Eng.* **2020**, *8*, 769. [CrossRef]
49. Xie, P.; Du, Y. Tsunami wave generation in Navier-Stokes solver and the effect of leading trough on wave run-up. *Coast. Eng.* **2023**, *182*, 104293. [CrossRef]
50. Galvin, C.J. *Wave Height Prediction for Wave Generators in Shallow Water*; Technical Memorandum No. 4; U.S. Army Coastal Engineering Research Center: Duck, NC, USA, 1964.
51. Jing, H.; Chen, G.; Liu, C.; Wang, W.; Zuo, J. Dispersive effects of water waves generated by submerged landslide. *Nat. Hazards* **2020**, *103*, 1917–1941. [CrossRef]
52. Wu, C.S. A modified volume-of-fluid/hybrid Cartesian immersed boundary method for simulating free-surface undulation over moving topographies. *Comput. Fluids* **2019**, *179*, 91–111. [CrossRef]
53. Kawai, H.; Satoh, M.; Kawaguchi, K.; Seki, K. Characteristics of the 2011 Tohoku tsunami waveform acquired around Japan by NOWPHAS equipment. *Coastal. Eng. J.* **2013**, *55*, 1350008. [CrossRef]

water

MDPI

Article

Investigation of Broken Wave Dissipation Effects of Submerged Shell Dike in Front of Breakwater

Na Wang, Gang Wang, Hui Zhang and Xing Li *

Fisheries Engineering Institute, Chinese Academy of Fishery Sciences, Beijing 100141, China;
18511730108@163.com (N.W.); wangg@cafs.ac.cn (G.W.); 13127050013@163.com (H.Z.)
* Correspondence: 15011527056@163.com

Abstract: In this study, the effects of a submerged shell dike in front of a breakwater on dissipating broken waves were studied. The dissipation effects of different broken wave heights and the submerged shell dike were investigated through numerical simulations. High-precision wave gauges and pressure sensors were used to collect data. Numerical simulations were performed using OpenFOAM software, based on the Volume of Fluid (VOF) method, to simulate broken waves. The time-histories of broken wave heights simulated by the numerical model were validated by physical experiment results, and the proportion of errors was less than 5.6%. The results show that the broken wave exerted on different positions of the breakwater shows a different time-history of pressures, and the peak pressure decreases with the decreasing broken wave height (from 0.342 to 0.227 m in the model) and increasing radii of the submerged shell dike (from 0.03 m to 0.18 m in the model). Through dimensional analysis, the relationship between the broken wave pressures and the dimensionless parameters related to broken wave height, breakwater height, and the radii of the submerged shell dike were established. Following the attenuation of the broken wave by the submerged shell dike, the equations for estimating broken wave pressures on various points along the breakwater were proposed. These equations are functions of the broken wave height, the radius of the submerged shell dike, and the height of the breakwater.

Keywords: breakwater; shell dike; broken wave; OpenFOAM; wave dissipation

1. Introduction

Amidst global climate change and rising extreme weather events, coastal protection engineering has gained prominence. The interaction of hurricane-generated extreme waves with broken waves has led to significant damage to offshore structures [1].

Submerged breakwaters are hydraulic structures built below the designed water level, serving as effective measures for coastal protection. Jiang et al. [2] examined wave trapping by submerged breakwaters near a vertical one. Two modes are found: reflection and trapping. The distance between them affects local hydrodynamics and stresses. More porous submerged breakwaters reduce wave force on the vertical breakwater and liquefaction around it. Hassanpour et al. [3] experimentally compared low-crested and submerged breakwaters with emerged ones using existing formulae and reference configurations, finding few reliable estimates of the transmission coefficient, higher errors in emerged breakwaters, and highlighting differences in transmitted wave energy. Physical model tests on a submerged horizontal plate for solitary wave control showed that indirect reflection dominates, dissipation rises with friction and nonlinearity, and specific parameters peak.

Modified factors reveal the best classification and prediction methods [4]. Recent studies have increasingly focused on the impact of submerged breakwaters on wave propagation and vortex generation. For instance, Li et al. [5] developed a meshless computational framework based on the Localized Method of Fundamental Solutions (LMFS) to simulate the nonlinear wave propagation around trapezoidal submerged breakwaters in a three-dimensional numerical wave tank, demonstrating its accuracy through comparison with analytical solutions and experimental data. Xu et al. [1] proposed a new type of semicircular submerged breakwater designed to reduce wave loads on coastal bridges, showing superior energy dissipation efficiency compared to traditional designs through high-precision numerical simulations. Additionally, Xuan et al. [6] studied the hydrodynamic performance of hollow triangle breakwaters, highlighting their effectiveness in altering wave transmission characteristics and kinetic energy dissipation patterns under various liquid properties.

Recent studies highlight the crucial role of mathematical models in designing breakwaters. These models are essential for understanding the complexity and broken nature of nearshore waves exacerbated by extreme weather. Zhao et al. [7] proposed a Constraint Interpolation Profile (CIP) viscous flow model, coupled with the Finite Element Method for structural deformation analysis, to investigate the wave interaction with a fluid-filled membrane submerged breakwater. Setyandito et al. [8] analyzed flow velocity in wave energy conversion systems within perforated breakwaters, showing that steeper slopes enhance energy generation. Yin et al. [9] employed OpenFOAM for simulations of pile breakwaters under solitary waves, providing practical design guidance. Guler et al. [10] assessed tsunami attacks on rockfill dam breakwaters using OpenFOAM's IHFOAM solver, demonstrating the model's practical engineering applications. Marisol and vanGent [11] simulated wave overtopping over rock-armored breakwaters with OpenFOAM, validating physical model trends and design guidelines. These studies collectively underscore the power of mathematical models in understanding complex fluid dynamics and optimizing coastal defense structures.

As concepts such as marine ecology, environmental awareness, and sustainable coastal development have gained increasing attention, there has been a shift towards emphasizing the harmonious integration of engineering structures with natural ecosystems, leading to new trends in the development of breakwater structures and the emergence of various ecological submerged breakwaters. Chowdhury et al. [12] found that oyster reefs can effectively reduce erosion at the front edge of tidal flats, with factors such as the duration of submergence and water depth significantly impacting wave attenuation. Specifically, oyster reefs submerged for less than 50% of the time can achieve up to 68% reduction in wave height. Zhuang et al. [13] used smartphone photogrammetry to create high-resolution DEMs of oyster reefs, revealing morphological differences between live and degraded reefs and emphasizing the role of oysters in local hydrodynamics and reef growth. Michalzik et al. [14] investigated grass covers for dikes, revealing minor losses in resistance after one storm surge season, with herb- and legume-dominated covers developing more slowly than grass-dominated ones, this having negligible effects on dike safety. Shinn et al. [15] monitored oyster reefs made of shell, concrete, and stone over seven years. The reefs created a habitat but did not fully achieve coastal protection goals like material stability or wave attenuation. These studies underscore the importance of ecological wave attenuation for sustainable coastal protection.

Due to the good wave-breaking effect of the shell reef, in this paper, the effect of the wave dissipation of submerged shell dikes in front of a typical breakwater is explored. A numerical model was established and verified based on physical experiment results. Broken wave pressures on the typical breakwater for different wave conditions and submerged shell dikes were recorded, and the equations of the pressures on the breakwaters were proposed for different points.

2. Materials and Methods

2.1. Broken Wave Flume

The experimental setup of the broken wave flume consists of a polygonal reservoir and a test flume. As illustrated in Figure 1, the test flume was equipped with a sluice gate, a slope, a platform, a vent pipe, and a drainage area. The reservoir, an irregular polygon, had a maximum length of 11.5 m, a maximum width of 11.0 m, and a height of 0.8 m, with a maximum working water depth of 0.7 m and a total capacity of 105 m^3. The water level in the reservoir was dynamically maintained by controlling both the pump and the sluice gate.

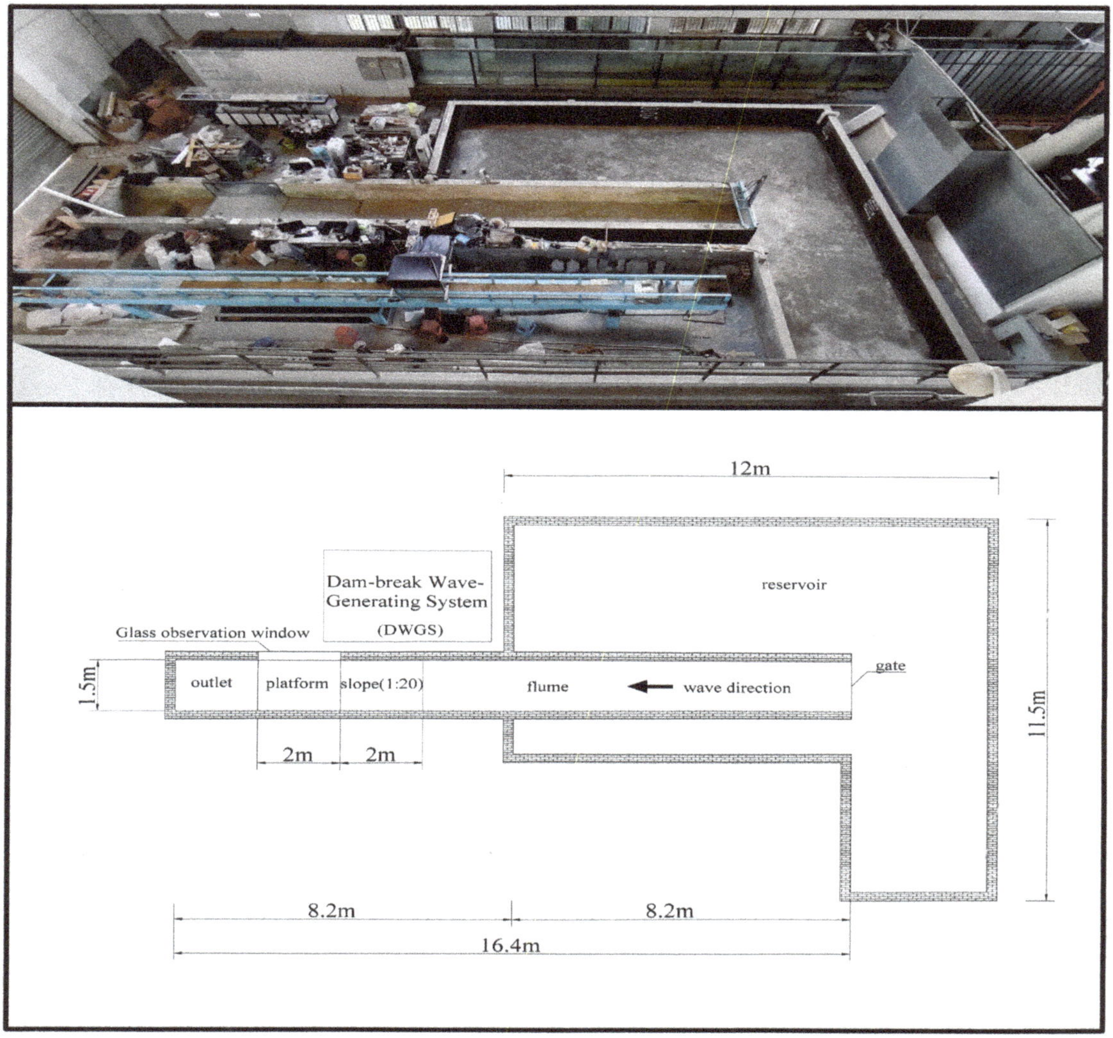

Figure 1. Photograph and schematic of broken wave flume.

Based on the experimental requirements, the reservoir water levels were set to 0.5 m, 0.6 m, and 0.7 m. The test flume, measuring 16.5 m in length, 1.5 m in width, and 0.8 m in height, was connected to the reservoir via the sluice gate. The sidewalls of the flume were constructed of blue bricks, and the bottom was polished concrete. To prevent leakage, the sides and bottom of the flume were coated with a waterproof asphalt layer. Additionally, a stainless-steel slope (2 m in length and 1.5 m in width) with a gradient of 1:20 was installed between the flume and the shoreline to simulate the breaking process of the broken waves upon reaching the shore. A transparent glass window was installed behind the slope to facilitate the observation of the interaction between the broken waves and the model structures.

When propagating near the shore, the broken waves could arrive as a crest; thus, the broken waves impacted the coastal structures above the initial water level. This experiment was conducted under wet-bed conditions to simulate the impact of broken waves. To maintain wet-bed conditions in front of the breakwater, a 1.5 m wide and 0.2 m thick steel plate was installed at the end of the flume to raise the bed. Given that the flume's width is 1.5 m and the broken waves heights are relatively small, the flume has a high width-to-depth ratio. Under these conditions, the sidewall effects can be considered negligible, and the flow outside the near-wall region can be regarded as two-dimensional.

2.2. Experimental Apparatus

The physical model experiment simulated broken waves in the test flume using the Dam-break Wave-Generating System (DWGS), which consists of a computer control system, gate control box, hydraulic system, and sluice gate. This system was developed by Fuzhou University in Fuzhou, China. The sluice gate is a flat gate that can move vertically with high flexibility. To generate the dam-break wave, the computer control system sends precise instructions to the gate control box, dictating the gate's lifting time and opening height. Specifically, the gate lifting time is 0.28 s, and the opening height is 0.35 m. For an RWL of 0.4 m, the gate lifting time is less than the theoretical dam-break time, indicating that the gate's lifting time and height do not significantly affect the wave's propagation.

When the waves enter shallow water, longer wavelengths occur after breaking up. To generate and maintain a stable broken wave profile within the flume (with a length of more than 10 m), the gate is lifted to a specified height and remains open for 5 s. This operation ensures that the dam-break wave formation and the subsequent broken waves' simulation meet the precise requirements and standards of the experiment.

For accurate wave height measurements, the LG-2 high-precision wave gauges were employed. This sophisticated instrument is capable of precisely measuring wave heights within a range of 0 to 70 cm and boasts an impressive data recording frequency of 2000 Hz, ensuring comprehensive and detailed capture of even the most dynamic fluid surface variations. This high-resolution recording ensures precise and real-time data, meeting the stringent accuracy demands of broken wave research. Prior to the experiment, the wave gauges were fully calibrated by placing them in environments with different water depths, establishing a linear relationship (KC value) between the electrical signal output and the water depth.

2.3. Numerical Method

The motion of fluid can be described by the Navier–Stokes equations. In this study, the Reynolds-Averaged Navier–Stokes (RANS) equations, along with the SST k-ω turbulence model, are employed to describe the fluid flow. In this model, the fluid is an incompressible viscous flow, and its governing equations for mass and momentum conservation are given by Equations (1) and (2), respectively.

$$\frac{\partial \overline{u}_i}{\partial x_i} = 0 \tag{1}$$

$$\frac{\partial}{\partial t}(\rho\omega) + \frac{\partial}{\partial x_j}(\rho\omega\overline{u}_j) = \frac{\partial}{\partial x_j}\left(\Gamma_\omega \frac{\partial \omega}{\partial x_j}\right) + G_\omega - Y_\omega + D_\omega + S_\omega \tag{2}$$

In these equations, $\overline{u}_i$ and $\overline{u}_j$ represent the mean velocities in the i and j directions, respectively; u_i' and u_j' denote the fluctuating velocities in the i and j directions, respectively; x_i and x_j are the spatial coordinates along the i and j directions; $\overline{p}$ represents the mean pressure; μ is the dynamic viscosity coefficient; ρ is the fluid density; t represents time.

The SST k-ω turbulence diffusion equations are given as below:

$$\frac{\partial}{\partial t}(\rho k) + \frac{\partial}{\partial x_j}(\rho k\overline{u}_j) = \frac{\partial}{\partial x_j}\left(\Gamma_k \frac{\partial k}{\partial x_j}\right) + G_k - Y_k + S_k \tag{3}$$

In the equations, k represents the turbulent kinetic energy; u_i and u_j denote the velocities in the i and j directions, respectively, where $u_i = \overline{u}_i + u_i'$ and $u_j = \overline{u}_j + u_j'$; the terms Γ_ω and Γ_k correspond to the effective diffusion terms for k and ω; G_k and G_ω are the production terms for k and ω; Y_k and Y_ω denote the dissipation terms for k and ω; D_ω represents the transverse diffusion term; and S_k and S_ω are the user-defined source terms.

2.4. Numerical Model Setup

In this study, a numerical model (as depicted in Figure 2) of the experimental facility was established using the computational fluid dynamics software OpenFOAM v2036 (released in June 2023). The scale of the numerical model is 1:1 relative to the physical model, and 1:25 relative to the actual size in reality. Five different reservoir water levels (RWLs) were set at 0.2 m, 0.25 m, 0.3 m, 0.35 m, and 0.4 m to generate broken waves of varying intensities. The breakwater model (as depicted in Figure 3) has a height of 0.46 m, a width of 1.5 m, and a slope of 1:1.5. The submerged shell dike models were configured as R3×6, R6×3, R9×2, and R18×1, with each shell mound having a base length of 0.36 m and a width of 1.5 m, oriented along the direction of the waves. By examining the wave attenuation effects of the shell mounds at the five different RWLs, five operational scenarios (Case 1 to Case 5) were established, as summarized in Table 1.

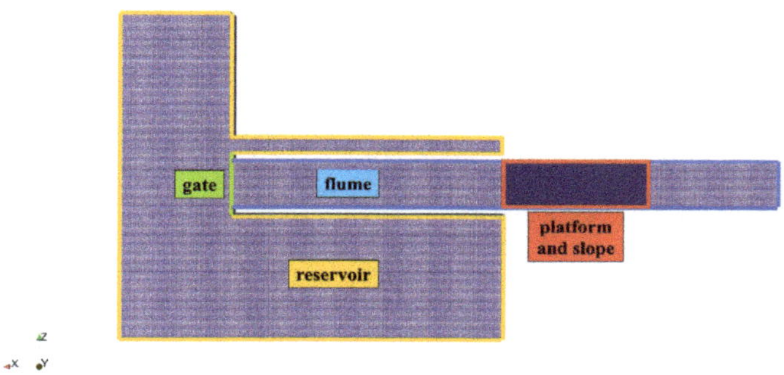

Figure 2. Numerical flume model (background grid spacing of 5 cm, with a locally densified grid of 2.5 cm in the flume area).

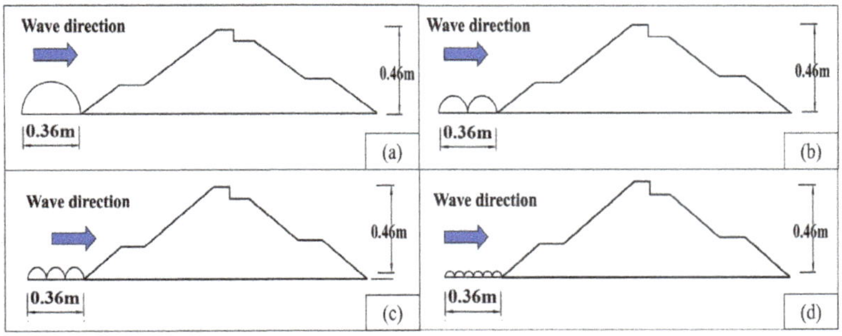

Figure 3. Numerical model setup for breakwater and submerged shell dike: (**a**) $R = 0.18$ m; (**b**) $R = 0.09$ m; (**c**) $R = 0.06$ m; (**d**) $R = 0.03$ m.

Table 1. Reservoir water level for different broken wave generating conditions.

Conditions	RWL (m)	Description
Case1	0.20	Weak wave
Case2	0.25	Minor weak wave
Case3	0.30	Moderate wave
Case4	0.35	Minor strong wave
Case5	0.40	Strong wave

The numerical model consists of a uniform grid with a cell size of 0.05 m. The bottom, sides, and sluice gate boundaries of the model are set as no-slip boundary conditions, while the top boundary is designated as an atmospheric boundary. Within the model, the roughness coefficient K_s of the flume bottom is set to 0.0025, while the other surfaces are assumed to be smooth. This study employs the InterFoam solver, which is an incompressible multiphase solver based on the Volume of Fluid method for capturing interfaces and simulating free-surface flows. The solver utilizes the finite volume method, where the computational domain is represented by multiple control volumes or elements located at the centroids of the computational nodes. The transport equations are then discretized and integrated over these control volumes in both spatial and temporal domains.

In the numerical model, water level monitoring points that correspond to the locations of the wave gauges used in the physical model experiments were established (as shown in Figure 4). This arrangement aims to validate the reliability of the broken waves generated by the numerical model. To obtain pressure data from the monitoring points, pressure probes at specific locations on the coastal protection model were installed (as illustrated in Figure 5). This setup allows for the measurement of pressure values at the corresponding points, enabling an assessment of the pressure conditions experienced by the coastal structure.

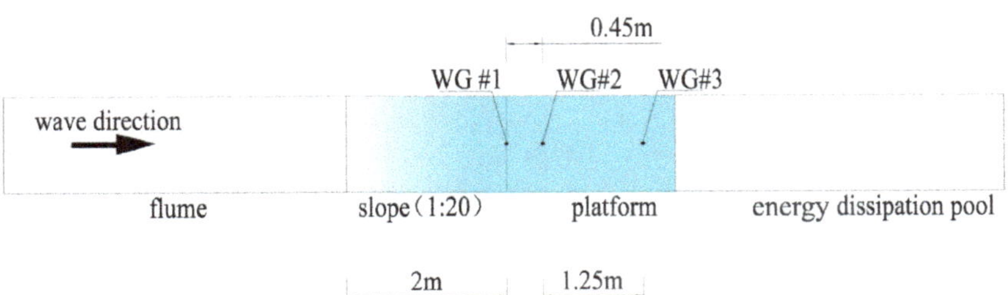

Figure 4. Wave gauge locations in the physical model flume.

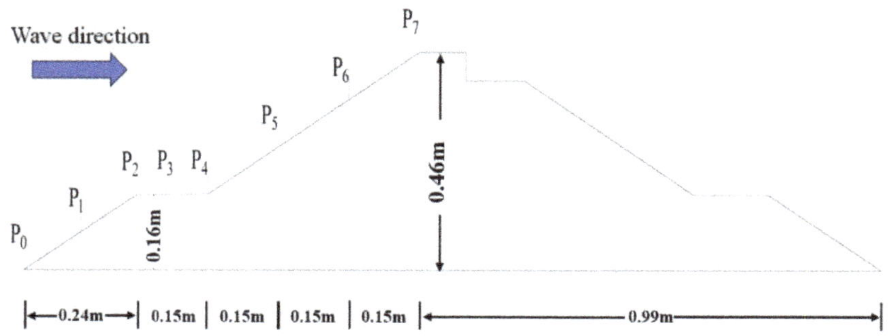

Figure 5. Pressure measurement points on breakwater.

3. Results

3.1. Comparison of Broken Wave Between Numerical and Physical Flume

There are a total of six numerical simulation test conditions used for validation, with reservoir water levels (RWLs) set to 0.5, 0.6, and 0.7 m, and initial water levels (IWLs) set to 0.05 and 0.1 m. The test results are shown in Table 2. The results indicate that as RWL increases, the broken wave height (h_s) also increases, and similarly, as IWL increases, the broken wave height also rises accordingly.

Table 2. Numerical model validation test results.

RWL (m)	IWL (m)	Broken Wave Height (m)
0.5	0.05	0.227
0.6	0.05	0.270
0.7	0.05	0.299
0.5	0.1	0.286
0.6	0.1	0.326
0.7	0.1	0.342

The comparison between the numerical simulation results and physical model experiments demonstrates a high degree of consistency across different wave conditions. Figure 6 illustrates the variation in wave height over time for different reservoir water levels (RWLs) and initial water levels (IWLs). Each subfigure presents time series data of wave height, where the numerical simulation results are validated against experimental measurements, providing a comprehensive assessment of the numerical model's accuracy.

The numerical model successfully captures the key dynamic characteristics of the broken wave, including the rapid rise in wave height following dam failure and the gradual stabilization as the wave propagates. Overall, the maximum discrepancy between the numerical simulation and experimental results does not exceed 5.6%, indicating that the model's error is within an acceptable range. However, minor discrepancies in peak wave height and the steady-state phase are observed, which may be attributed to simplifications in the numerical model or uncertainties in the physical experiments.

For instance, in the case where RWL = 0.7 m and IWL = 0.05 m, both the numerical and experimental results show similar trends in the initial wave height rise, although the experimental data slightly overestimate the peak height. As the IWL increases, these differences diminish, with better alignment between the numerical and experimental waveforms, as observed in the case of RWL = 0.7 m and IWL = 0.1 m.

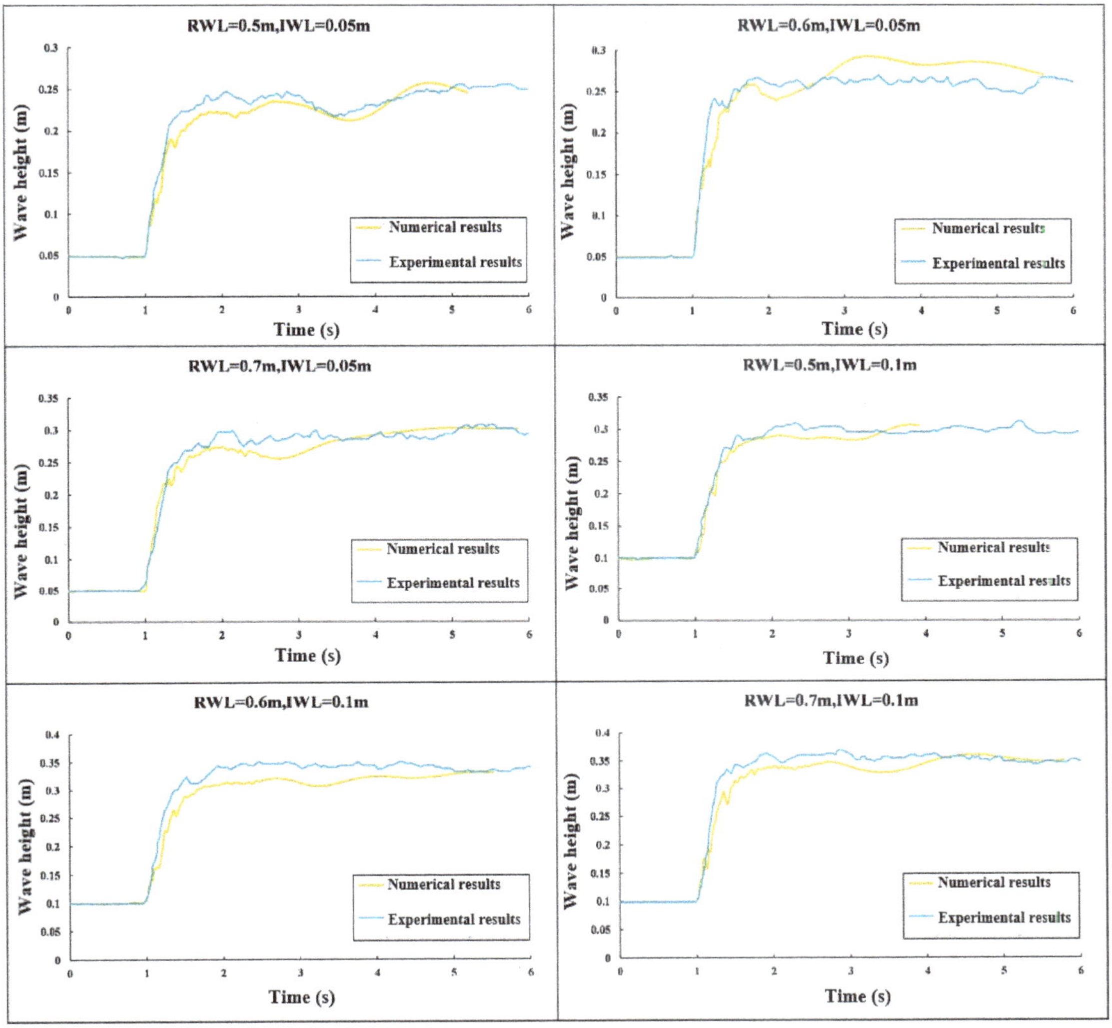

Figure 6. Comparison of numerical simulation results with physical model test results.

These comparisons confirm that the numerical flume is capable of accurately reproducing the broken wave phenomena observed in physical modeling, demonstrating that it is a reliable tool for investigating such events under various hydraulic conditions.

Given the successful validation of the numerical model of the broken wave flume, the different cases of the broken wave heights (h_s) required for this study (rounded to two decimal places) are listed as follows in Table 3:

Table 3. The results of numerical simulation experiments on broken wave heights.

Case	RWL (m)	h_s (m)
1	0.20	0.16
2	0.25	0.18
3	0.30	0.21
4	0.35	0.23
5	0.40	0.25

3.2. Time-Histories of Pressures for Different Points on Breakwater

Investigations of time-histories of pressures for different points on the breakwater show that all cases follow similar patterns. Case 3 (h_s = 0.21) can be taken as an example, for the period between 6 and 8 s, and for which the pressure time-history curves are shown in Figure 7. Here, significant pressure fluctuations at various measurement points indicate that when the broken waves passed over the submerged shell dike and reached the slope, a substantial impact force was generated. P_0 and P_1 are located at the toe of the slope, closest to the wave's point of incidence. These positions received the direct impact of the waves, resulting in the highest pressures. For the R3 submerged shell dike condition, the peak pressure at P_0 and P_1 approached or exceeded 3000 kPa, indicating that the wave energy dissipated minimally upon reaching these points. As the radius of the submerged breakwater increased, the wave energy was more effectively dissipated during transmission, resulting in a gradual reduction in pressure at these points under larger breakwater radii. For the R18 submerged shell dike condition, the overall pressure values were the lowest, and the peak pressure at points such as P_0 and P_1 significantly decreased, demonstrating that this shell dike radius was the most effective in dissipating broken wave energy. At positions P_6 and P_7, the pressure remained nearly stable, suggesting that the higher locations were less affected by broken wave impacts.

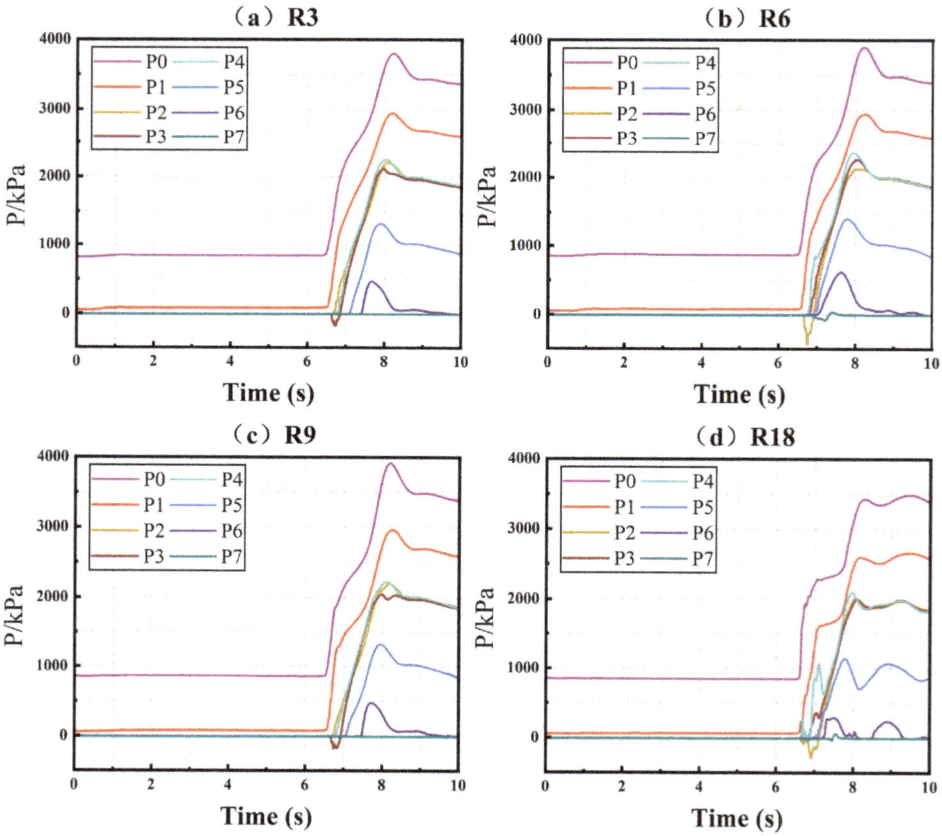

Figure 7. For the conditions of case 3, the pressure time-history curves for each point are as follows: (**a**) represents the R3 submerged shell dike, (**b**) represents the R6 submerged shell dike, (**c**) represents the R9 submerged shell dike, and (**d**) represents the R18 submerged shell dike.

3.3. Pressure for Different Broken Wave Heights

Investigations of pressure for different broken wave heights show that all cases follow similar patterns. The R6 submerged shell dike condition can be taken as an example, with the analysis for five different broken wave heights shown in Figure 8. P_1 is located near the toe of the slope, where the wave impact force is relatively high. As the wave heights increase, the pressure peak rises significantly. When h_s is 0.25 m, the pressure peak approaches 4000 kPa, whereas the pressure is approximately 1500 kPa at an h_s of 0.16 m. This indicates that as the wave heights increase, this leads to a significant increase in pressure at P_1.

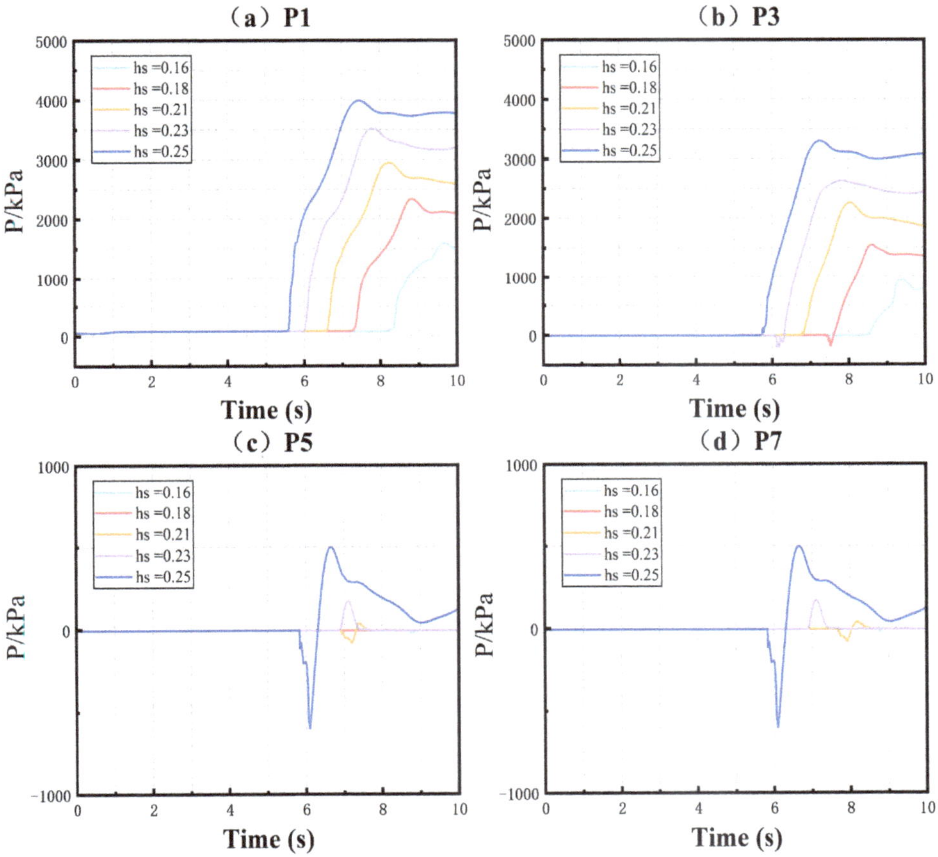

Figure 8. The pressure at various points for different impact conditions in the R6 scenario: (**a**) represents point P_1, (**b**) represents point P_3, (**c**) represents point P_5, and (**d**) represents point P_7.

P_3 is located in the middle of the slope, farther from the toe, where the wave energy has partially dissipated during transmission. As the wave heights increase, the pressure at P_3 also rises, but the pressure peak is slightly lower compared to P_1, indicating that the wave energy weakens as it reaches P_3.

P_5 is located higher up on the slope, where the impact of the waves is minimal. As shown in Figure 8, P_5 is affected by the waves in cases of higher h_s (e.g., h_s = 0.25 m), but the overall pressure remains low, with peak values below 1000 kPa. In some instances, negative pressure occurs, which could be attributed to the formation of negative pressure zones during wave reflection or dissipation at higher elevations.

P_7 is located at the top of the slope and is minimally affected by direct wave impacts. Due to its position far from the direct impact of the waves, the wave energy has largely dissipated by the time it reaches this point.

It should be noted that the formation of negative pressure in waves is a complex phenomenon resulting from the interaction of wave reflection and energy dissipation, particularly when waves encounter obstacles or varying topography. When waves meet elevated terrains or structural barriers, a portion of their energy reflects back, creating zones of reduced pressure as the reflected waves interfere with incoming ones. Additionally, the energy dissipation at higher altitudes, caused by factors such as air resistance and thermal effects, further contributes to the decrease in local pressure. This interplay between reflected waves and dissipative forces leads to the formation of negative pressure zones, which can significantly influence various meteorological and environmental conditions.

3.4. Pressure for Different Radii of Submerged Shell Dike

Investigations of pressure for different radii of submerged shell dike show that all cases follow similar patterns. Taking wave heights of 0.21 m as an example, the different radii of the submerged shell dikes (R3, R6, R9, R18) affect the pressure at four measurement points (P1, P3, P5, P7), as shown in Figure 9.

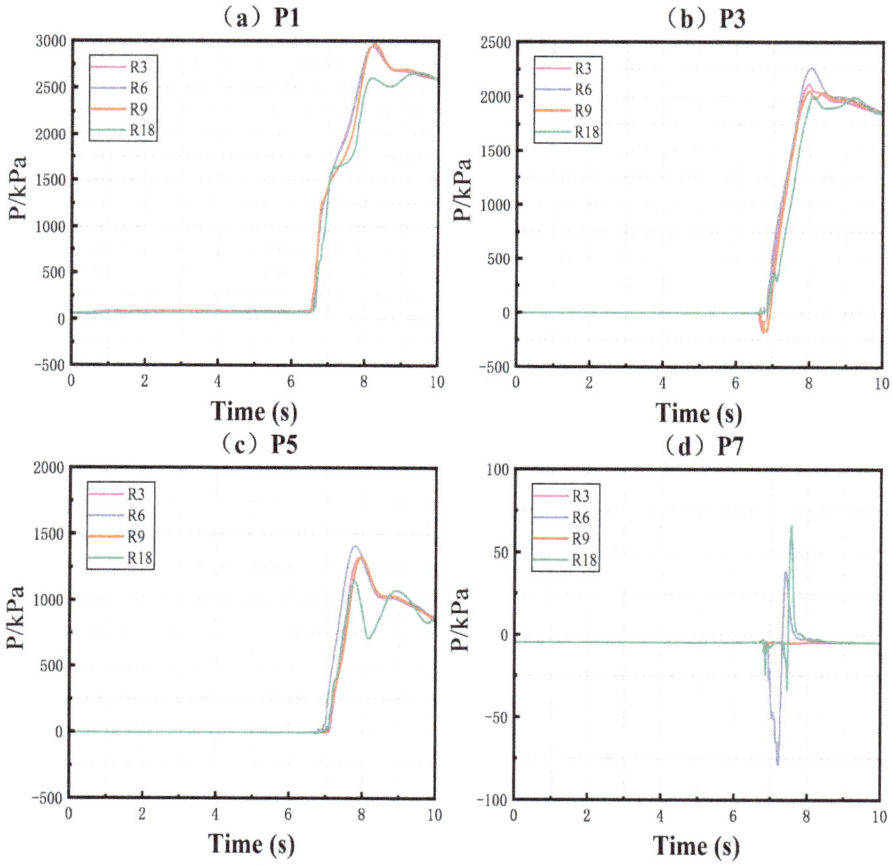

Figure 9. For the impact of $h_s = 0.21$, the different radii of the shell reef affect the pressure at the four measurement points: (**a**) represents point P_1, (**b**) represents point P_3, (**c**) represents point P_5, and (**d**) represents point P_7.

P_1 is located at the toe of the slope, close to the point of wave impact, and thus experiences the greatest impact. Under different reef radii, the pressure trends are similar, with the maximum value occurring around 8 s, peaking near 3000 kPa. Smaller radii (R3, R6) generate slightly higher pressures compared to larger radii (R9, R18), indicating that a larger reef radius can dissipate wave energy slightly better. However, the influence of reef radius at P_1 is relatively limited, as the wave impact remains strong at this point.

P_3 is located in the middle of the slope, where the pressure is significantly lower than at P_1, but still notable. The pressure at P_3 is slightly lower than at P_1, reflecting the attenuation of wave energy during its transmission.

P_5 is located higher up on the slope, where the pressure is significantly lower, with peak values around 1500 kPa. The wave impact at this location is relatively minor, and the effect of the reef radius on wave energy dissipation is not as pronounced as at the toe of the slope.

P_7, positioned at the top of the slope, experiences the least wave impact, with pressure values below 100 kPa.

4. Discussion

4.1. Dimensional Analysis for Broken Wave Pressures on Breakwater

As illustrated in Section 3, the broken wave pressures exerted on breakwater depend on the following parameters: (1) the height of the breakwater h; (2) the broken wave height h_s; (3) the radius of the submerged shell dike R; (4) the density of water ρ; and (5) the gravity acceleration g.

In this study, the key variables are the approaching broken wave characteristics (h_s) and the breakwater and submerged shell dike geometry (h, R), while the water properties (ρ, g) were constant. To normalize the variables, the pressure exerted on the breakwater is divided by the hydrostatic pressure ($\rho g h_s$). In addition, as the pressure exerted on the breakwater decreases with the increase in the radius of the submerged shell dike, the nondimensional parameters (R/h) and (h_s/R) were used.

4.2. Estimating Pressure Coefficient

A best-fit curve has been found for each curve in Figure 10. Taking P_0 as an example, the equations for the best-fit curve for each data set in Figure 10a have the following form:

$$\frac{P_0}{\rho g h_s} = a * \ln\left(\frac{R}{h}\right) + b \tag{4}$$

where a and b are undetermined coefficients. Denoting $c = e^{b/a}$, Equation (4) can be expressed as follows:

$$\frac{P_0}{\rho g h_s} = a * \ln\left(\frac{R}{h} * c\right) \tag{5}$$

where a and c are coefficients, which depend on (R/h). From the experimental data fitting, the following empirical equations were obtained:

$$a = 2248.9\left(\frac{R}{h}\right) + 685.74 \tag{6}$$

$$c = 3.6879\left(\frac{R}{h}\right) + 1.4899 \tag{7}$$

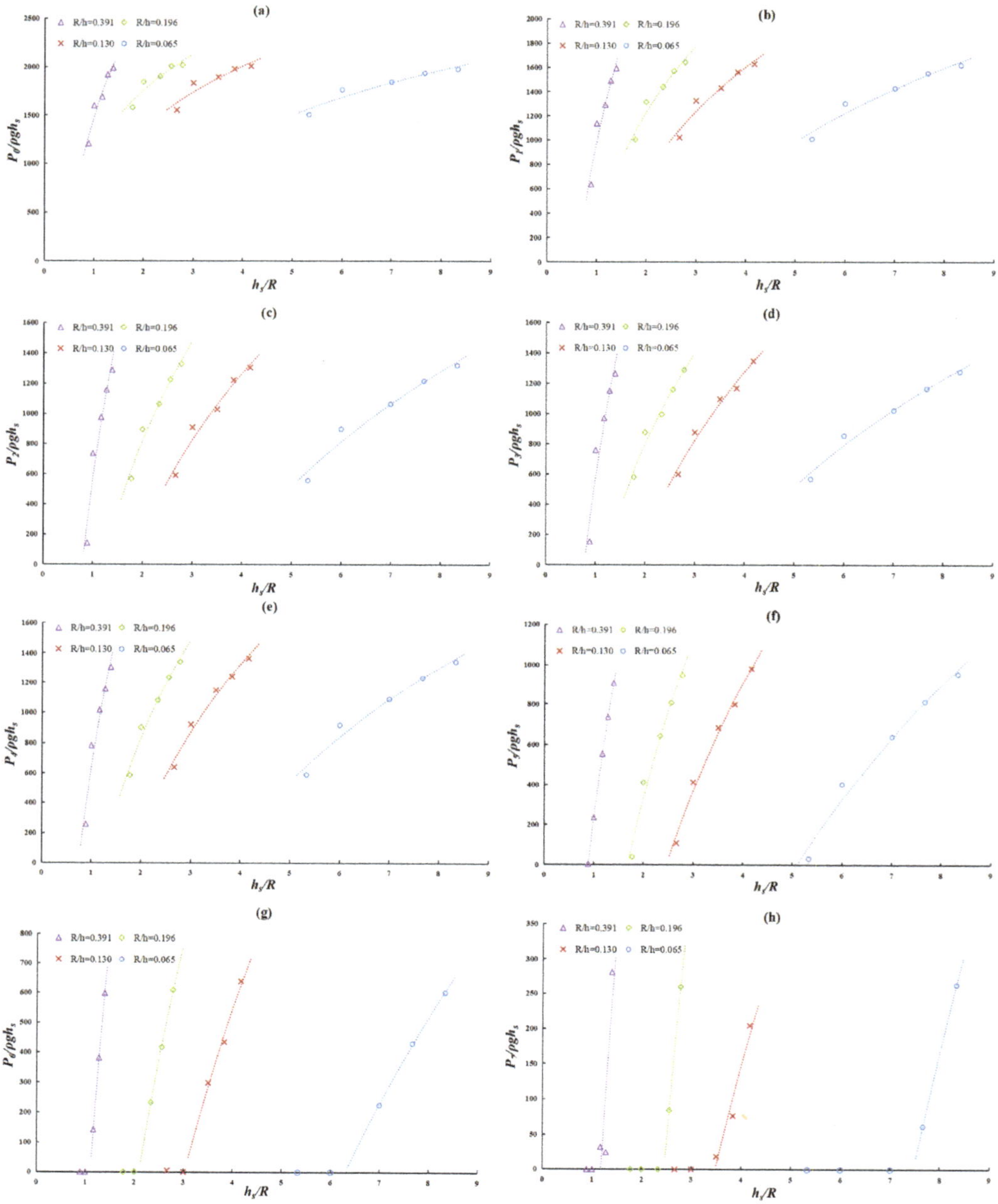

Figure 10. The relationship between the pressure coefficients and normalized broken wave height: (**a**) is the pressure coefficient of P_0, (**b**) is the pressure coefficient of P_1, (**c**) is the pressure coefficient of P_2, (**d**) is the pressure coefficient of P_3, (**e**) is the pressure coefficient of P_4, (**f**) is the pressure coefficient of P_5, (**g**) is the pressure coefficient of P_6, and (**h**) is the pressure coefficient of P_7.

Combining Equations (5)–(7) gives:

$$\frac{P_0}{\rho g h_s} = (2248.9\frac{R}{h} + 685.74)\ln[\frac{R}{h} * (3.6879\frac{R}{h} + 1.4899)] \tag{8}$$

Therefore, pressure can be expressed as follows:

$$P_0 = \rho g h_s(2248.9\frac{R}{h} + 685.74)\ln[\frac{R}{h} * (3.6879\frac{R}{h} + 1.4899)] \quad (9)$$

Similar to Equation (9), other points can be expressed by experimental data as follows:

$$P_1 = \rho g h_s(2340\frac{R}{h} + 1026.1)\ln[\frac{R}{h} * (3.4775\frac{R}{h} + 0.3651)] \quad (10)$$

$$P_2 = \rho g h_s(2676.2\frac{R}{h} + 1274.3)\ln[\frac{R}{h} * (2.9312\frac{R}{h} + 0.1618)] \quad (11)$$

$$P_3 = \rho g h_s(2569.9\frac{R}{h} + 1223.8)\ln[\frac{R}{h} * (2.9855\frac{R}{h} + 0.1624)] \quad (12)$$

$$P_4 = \rho g h_s(1959.6\frac{R}{h} + 1364.1)\ln[\frac{R}{h} * (3.144\frac{R}{h} + 0.1442)] \quad (13)$$

$$P_5 = \rho g h_s(256.87\frac{R}{h} + 1910.7)\ln[\frac{R}{h} * (2.8267\frac{R}{h} + 0.0277)] \quad (14)$$

$$P_6 = \rho g h_s(1712.5\frac{R}{h} + 1886.1)\ln[\frac{R}{h} * (2.265\frac{R}{h} + 0.0251)] \quad (15)$$

$$P_7 = \rho g h_s\left(-1823.4\frac{R}{h} + 2102.5\right)\ln[\frac{R}{h} * (2.1852\frac{R}{h} - 0.0076)] \quad (16)$$

It should be noted that Equations (9)–(16) share the same form, but the coefficients vary depending on the position of the point. In this study, the ranges of the present data are limited to $0.89 \le h_s/R \le 8.3$ and $0.065 \le R/h \le 0.391$.

5. Summaries and Conclusions

Through comparisons with physical experiments, the numerical model accurately reproduces the main characteristics of broken waves, including the rapid increase in wave height and gradual stabilization during wave propagation. The validation error of the numerical model does not exceed 5.6%, confirming its reliability for simulating broken waves in this numerical flume. The temporal pressure profiles of breaking waves at various sections of the breakwater reveal that peak pressures decrease as wave height lessens and the radius of the submerged shell dike increases. Dimensional analysis established relationships between broken wave pressures and dimensionless parameters, such as wave height, breakwater height, and the radius of the submerged shell dike. This research introduces predictive equations for estimating pressures at different points on the breakwater after wave mitigation by the submerged shell dike. These equations account for reduced wave height, the radius of the submerged shell dike, and the height of the breakwater, offering a practical tool for coastal defense planning.

Author Contributions: Conceptualization, N.W.; methodology, G.W.; software, N.W.; validation, H.Z.; formal analysis, X.L.; investigation, X.L.; resources, X.L.; data curation, X.L.; writing—original draft preparation, N.W.; writing—review and editing, X.L.; visualization, G.W.; supervision, H.Z.; project administration, X.L.; funding acquisition, X.L. All authors have read and agreed to the published version of the manuscript.

Funding: This work was supported by the National Key R&D Program of China (2024YFD2400800) and the Basic Research Fund of the Fisheries Engineering Research Institute, Chinese Academy of Fishery Sciences (2024HY-ZC001).

Data Availability Statement: The original contributions presented in the study are included in the article, Further inquiries can be directed to the corresponding author.

Acknowledgments: N.W. acknowledges G.W. for the inspiration.

Conflicts of Interest: The authors declare that there are no conflicts of interests regarding the publication of this article.

References

1. Xu, G.; Wang, J.; Chen, Q.; Xue, S.; Xu, Y. A novel tri-semicircle shaped submerged breakwater for mitigating wave loads on coastal bridges part I: Efficacy. *Ocean Eng.* **2022**, *245*, 110462. [CrossRef]
2. Jiang, L.; Zhang, J.S.; Tong, L.L.; Guo, Y.K.; He, R.; Sun, K. Wave motion and seabed response around a vertical structure sheltered by submerged breakwaters with Fabry-Perot resonance. *J. Mar. Sci. Eng.* **2012**, *10*, 1797. [CrossRef]
3. Hassanpour, N.; Vicinanza, D.; Contestabile, P. Determining wave transmission over rubble-mound breakwaters: Assessment of existing formulae through benchmark testing. *Water* **2023**, *15*, 1111. [CrossRef]
4. Lee, C.; Kim, T.; Hwang, T.; Lee, W.D. Effects of solitary wave breaking on the hydrodynamic performance of a submerged horizontal plate: A physical modeling study. *Ocean Eng.* **2024**, *300*, 117405. [CrossRef]
5. Li, L.; Fu, Z.; Qin, M.; Liu, S.; Zeng, W.; Liu, X. Three-dimensional numerical wave tank containing submerged breakwaters based on the localized method of fundamental solutions. *Math. Comput. Simul.* **2025**, *229*, 273–287. [CrossRef]
6. Xuan, T.L.; Vu, H.T.D.; Oberle, P.; Dang, T.D.; Ba, H.T.; Manh, H.L. Hydrodynamics and wave transmission through a hollow triangle breakwater. *Estuar. Coast. Shelf Sci.* **2024**, *302*, 108765. [CrossRef]
7. Zhao, X.Z.; Zhou, Y.W.; Zong, Y.W.; Yang, Z.L.; Luo, M. A CIP-based numerical simulation of wave interaction with a fluid-filled membrane submerged breakwater. *Ocean Eng.* **2022**, *260*, 111819. [CrossRef]
8. Setyandito, O.; Nizam, N.; Pierre, A.J.; Suputra, G.D.; Wijayanti, Y.; Anda, M. Numerical analysis of velocity magnitude on wave energy converter system in perforated breakwater. *Int. J. Renew. Energy Dev.* **2022**, *11*, 27–33. [CrossRef]
9. Yin, M.; Zhao, X.; Luo, M.; Sun, H. Flow pattern and hydrodynamic parameters of pile breakwater under solitary wave using OpenFOAM. *Ocean Eng.* **2021**, *235*, 109381. [CrossRef]
10. Guler, H.G.; Baykal, C.; Arikawa, T.; Yalciner, A.C. Numerical assessment of tsunami attack on a rubble mound breakwater using OpenFOAM. *Appl. Ocean Res.* **2018**, *72*, 76–91. [CrossRef]
11. Mata, M.I.; van Gent, M.R.A. Numerical modelling of wave overtopping discharges at rubble mound breakwaters using OpenFOAM. *Coast. Eng.* **2023**, *181*, 104274. [CrossRef]
12. Chowdhury, M.S.N.; Walles, B.; Sharifuzzaman, S.; Shahadat Hossain, M.; Ysebaert, T.; Smaal, A.C. Oyster breakwater reefs promote adjacent mudflat stability and salt marsh growth in a monsoon dominated subtropical coast. *Sci. Rep.* **2019**, *9*, 8549. [CrossRef] [PubMed]
13. Zhuang, J.Q.; Wang, Y.W.; Guo, Y.D.; Lin, H.J.; Yu, Q. High-resolution DEMs reveal multi-scale roughness of oyster reefs via smartphone photogrammetry. *Geo-Mar. Lett.* **2024**, *44*, 23. [CrossRef]
14. Michalzik, J.; Paul, M.; Schlurmann, T. Response of grass covers for dikes to coastal stressors during establishment phase. *Ecol. Eng.* **2025**, *212*, 107488. [CrossRef]
15. Shinn, J.P.; Bredes, A.; Bushek, D.; Jerr, L.; Kreeger, K.; McCulloch, D.; Miller, J.; Moody, J.; Rothermel, E.; Zito-Livingston, A. Seven years of monitoring the development of an oyster reef living shoreline. *Estuaries Coasts* **2025**, *48*, 11. [CrossRef]

water

MDPI

Article

Two-Dimensional Numerical Analysis of Submerged Dike Hydrodynamics

Xiaojie Zhang [1,2], Yachao Zhang [1,2,*], Yanfen Deng [1], Xianghuang Li [1] and Bowen Guan [1]

1 East China Sea Survey Center, Ministry of Natural Resources, Shanghai 200137, China
2 Key Laboratory of Marine Ecological Monitoring and Restoration Technologies, MNR, Shanghai 201206, China
* Correspondence: zhychzhych0101@163.com

Abstract

Many studies have been conducted on wave and sediment movement with submerged dikes. However, the effect of a submerged dike's height and orientation on hydrodynamics has not been thoroughly examined from the perspective of the marine ecology impact. This paper employs a two-dimensional numerical model to investigate effects of submerged dike height and orientation on flow, specifically flow velocity and cross-dike flux. The findings indicate that the most significant velocity variation occurs at a distance of approximately one-fifth of the dike length (0.2 L) from the dike head, when the flow is perpendicular to the dike and parallel to the coastline. And this area as the submerged dike's protection zone will have the least impact on the surrounding environment. The change pattern of the flow velocity with the distance apart from the submerged dike varies for different submerged dike heights. A submerged dike height of 0.7 times the water depth (0.7 H) is a dividing value. Additionally, as the orientation angle increases, the cross-dike flux rises. From the perspective of the impact on the marine ecological environment, the design angle of the submerged dike should be as small as possible. The findings establish a theoretical hydrodynamic basis that may support future integrated studies on coastal zone management.

Keywords: submerged dike; height; orientation; flow velocity; flux across submerged dike; property of flow field

1. Introduction

Dikes, including emerged dikes and submerged dikes, are a kind of hydraulic structure that has long been employed to damp waves, redirect flows, safeguard coastlines, adjust water depth, and build local deposition [1]. Consequently, dikes are crucial and extensively implemented coastal engineering projects. Dikes are usually constructed at an angle to the coastline to redirect flow [2].

Many relevant investigations are focusing on emerged dikes. Emerged dike obstruction leads to a multidimensional turbulent flow pattern close to the structure, causing localized regions of steep velocity changes and sediment transport [3]. Weitbrecht et al. indicated that mass and momentum exchange occur in the vicinity of emerged dike heads [4]. J. W. et al. [5] described the analysis of 10 years of measurements together with a recently measured storm that included wave overtopping. Tomohiro Suzuki [6] showed that the vertical wall induces seaward velocity on the dike, which might be an extra risk during extreme events. Beji and Battjes [7] applied nonlinear shallow-water wave theories to investigate the transformation of periodic waves passing over a dike. Previous studies have revealed that the surrounding flow region can be split into three distinct zones [8]. Chen et al. [9] categorized the three zones concerning the dike length, defining them as Y = 2/3 L for the Dike Field Zone (DFZ), Y = 4/3 L for the Momentum Exchange Zone (MEZ), and Y = 3 L for the (Main-Stream Zone) MSZ (where Y is the channel bund width and L is the dike length). Sujantoko et al. [10] present a two-dimensional numerical analysis of the hydrodynamic performance of a submerged dike, focusing on the effects of slope and porosity on wave transmission and reflection. Ikha Magdalena and Owen Nathanael [11] introduce a mathematical model and demonstrate that submerged porous breakwaters effectively mitigate shoaling effects by attenuating wave amplitudes and reducing resonance risks.

Meanwhile, there are some studies on wave and sediment movement with submerged dikes. Harmonic generation past a submerged porous dike has been studied by experiments [12]. Altomare et al. [13] introduced a new "equivalent slope" concept to estimate average wave overtopping discharges on sea dikes with shallow and very shallow foreshores. Tsal et al. [14] studied the wave reflection coefficient of the submerged dike and the attenuation coefficient of wave energy by experiment. Sharifahmandian and Simons [15] published a numerical model that could predict the wave transmission coefficient in submerged dike space under regular waves in shallow water. Based on the RBF method, the model can accurately simulate wave motion near a submerged dike. Chen et al. [16] showed that the wave profile may become more asymmetrical when the wave propagates over the breakwater. Chai et al. [17] employed a 3D numerical model to investigate how dike shape and height affect sediment transport. Their results demonstrated that a narrower hexagonal dike outperforms rectangular designs, and suspended sediment concentration could be reduced by 30% when the height of dike varied from 0.3 m to 0.5 m. In a complementary study, Pang et al. [18] conducted flume experiments examining mud flotation at various dike elevations, fluid mud thickness variations, channel siltation, and sediment deposition upstream of submerged dikes. Their findings revealed that optimal sedimentation effects occur when the submerged dike height-to-depth ratio ranges from 0.2 to 0.5.

Compared with emerged dikes, submerged dikes have the advantages of requiring little engineering work and less investment, and having reduced impacts on the marine ecological environment. With the popularization of the national ecological civilization strategy, more ecologically friendly submerged dikes have attracted more and more attention. It is necessary to analyze the hydrodynamic characteristics of submerged dikes from the perspective of marine ecology impact. Therefore, studying the hydrodynamics near submerged dikes will have engineering and ecological significance. Studies of the current features around submerged dikes are usually based on specific projects. Xu et al. [19] studied the spatial–temporal variation characteristics of current velocity in the south dike of the Yangtze Estuary by using FVCOM, calculated the cross-dike flux, and analyzed the current change past the dike before and after the engineering. Li [20] studied the hydrodynamic environmental impact of the East China Sea Dike on Zhanjiang Bay, showing that the dike reduced current velocity and that the tidal prism decreased by 21% or more in the bay.

A physical model can only be tested under a specific similarity ratio, making it impractical to cover all possible conditions. It is difficult to fit a submerged dike's height in physical models. With the rapid development of computer modeling, numerical simulation has become an effective research method. Using a two-dimensional numerical model, this paper analyzes hydrodynamic characteristics, including the flow velocity and the cross-dike flux, under different heights and orientations of a submerged dike. It can provide a basis for the study of submerged dike hydrodynamics, submerged dike design, and coastal zone ecological restoration.

2. Materials and Methods

This paper uses MIKE21 FM (MIKE ZeroRelease 2014, Service Pack 3), which is developed by DHI Water & Environment, to simulate numerical models. MIKE21 FM uses a cell-centered finite-volume method to fit complicated terrain well and ensure material flux conservation with high speed. It can forecast water level change and flow under different driving forces by solving continuity and momentum equations to calculate the water level at different times and the velocity of each grid point. Related research [21,22] on submerged dikes has proved the applicability of MIKE.

The research focuses on the large-scale, horizontal redistribution of flow and the calculation of bulk cross-dike flux, for which depth-averaged dynamics are dominant. The use of a 2D model allows for a computationally efficient exploration of the parameter space (dike height and orientation) while isolating their first-order effects from the complexities of three-dimensional turbulence. This approach is well-established for investigating coastline-scale hydrodynamics around submerged structures and provides a critical baseline for understanding primary flow patterns.

2.1. Governing Equations

This paper employs two-dimensional incompressible Reynolds-depth-averaged shallow-water equations. The basic equations in this model are as follows.

Mass conservation equation:

$$\frac{\partial \zeta}{\partial t}+\frac{\partial}{\partial x}(hu)+\frac{\partial}{\partial y}(hv)=0 \quad (1)$$

Momentum equation:

$$\frac{\partial u}{\partial t}+u\frac{\partial u}{\partial x}+v\frac{\partial u}{\partial y}-\frac{\partial}{\partial x}\left(\varepsilon_x\frac{\partial u}{\partial x}\right)-\frac{\partial}{\partial y}\left(\varepsilon_x\frac{\partial u}{\partial y}\right)-fv+\frac{gu\sqrt{u^2+v^2}}{C_z^2H}=-g\frac{\partial \zeta}{\partial x} \quad (2)$$

$$\frac{\partial v}{\partial t}+u\frac{\partial v}{\partial x}+v\frac{\partial v}{\partial y}-\frac{\partial}{\partial x}\left(\varepsilon_y\frac{\partial v}{\partial x}\right)-\frac{\partial}{\partial y}\left(\varepsilon_y\frac{\partial v}{\partial y}\right)+fu+\frac{gv\sqrt{u^2+v^2}}{C_z^2H}=-g\frac{\partial \zeta}{\partial y} \quad (3)$$

where ζ indicates the surface elevation; h indicates the depth of water; H indicates the total depth; $H=\zeta+h$; u and v are velocity components in the x and y directions; g is the gravitational acceleration, g = 9.81 m/s^2; f is the parameter of the Carioles force ($f=2\omega\sin\varphi$, ω is rotational angular velocity of the earth, φ is engineering sea latitude); C_z is the Chezy resistance; $C_z=\frac{1}{n}H^{\frac{1}{6}}$; ε_x and ε_y are horizontal eddy viscosity coefficients of components in the x and y directions; and t indicates time. Figure 1 is a schematic diagram of the mass conservation equation.

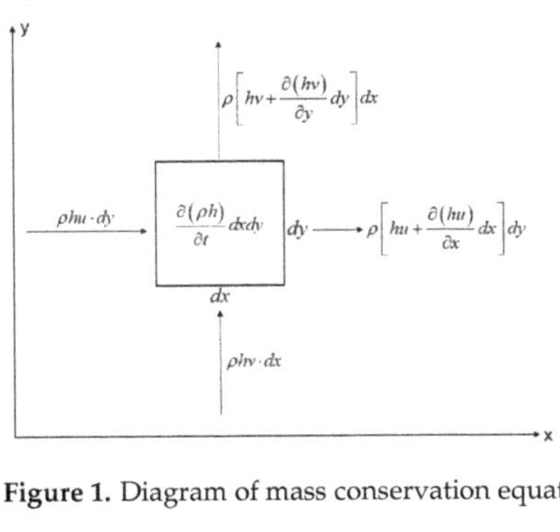

Figure 1. Diagram of mass conservation equation.

The model assumes water is incompressible, which is valid for all practical hydraulic and coastal engineering applications where Mach number Ma << 0.3. The model assumes that the horizontal scales of motion are much greater than the water depth, neglecting vertical acceleration effects.

This 2D approach cannot capture 3D flow structures, particularly strong vertical separation and secondary currents in the dike wake region. The hydrostatic assumption may introduce errors in regions of rapidly varying bathymetry.

2.2. Initial Conditions

$$\begin{cases} \zeta(x,y,t)|_{t=t0} = \zeta(x,y,t_0) = 0 \\ u(x,y,t)|_{t=t0} = v(x,y,t)|_{t=t0} = 0 \end{cases} \tag{4}$$

where t_0 indicates the initial time.

2.3. Boundary Conditions

The closed-boundary normal-direction velocity is zero, $\vec{V} \cdot \vec{n} = 0$.

2.4. Open Boundaries

The open-boundary-condition velocity is 1 m/s constantly. The direction of flow is from left to right.

2.5. Other Conditions

The *Manning* is normally 0.02.

This article employs an idealized case study. To reduce the grid scale effect, the domain is a 20,000 m long and 10,000 m wide rectangular channel with a constant water depth of 10 m. This simulates a large submerged dike about 2500 m long and 700 m wide. The water depth (represented by H) is set as a vertical length unit, and the length of the submerged dike (represented by L) is set as a horizontal length unit.

The model terrain diagram is shown in Figure 2.

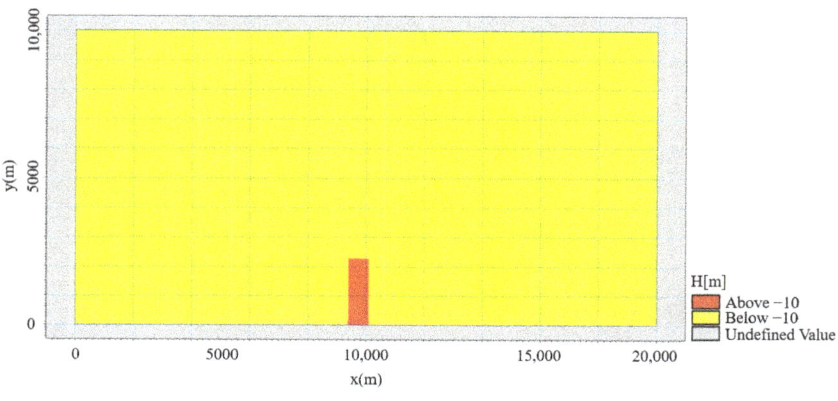

Figure 2. Model terrain schematic.

In this paper, the experimental data presented by Iqbal and Tanaka [23] are used to validate the numerical model. At the same time, the error relative to the measured current value is calculated, and the Brier Skill Score (BSS) is 0.91. BSS is an empirical coefficient that relates to the variance between the measured data and the model results. A BSS close to 1 indicates that the simulation performance of the model is good, and a BSS close to 0 indicates that the fitting result is poor, and the model cannot be used for calculation.

$$\mathrm{BSS} = 1 - \frac{\mathrm{Var}(c - m)}{\mathrm{Var}(m)} \tag{5}$$

where m is the measured value and c is the calculated value of the model.

Although the model does not accurately simulate the variation rule, the overall trend is consistent, and the statistical error analysis results are reasonable (Figure 3).

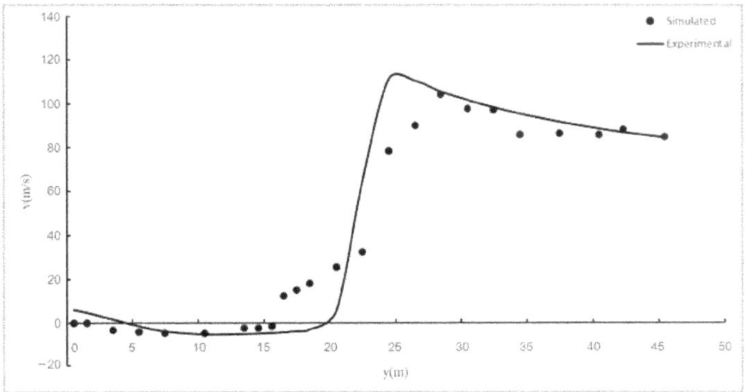

Figure 3. Comparison between numerical and experimental results.

2.6. Schemes Setup

The numerical modeling conditions were designed to be idealized in order to isolate the fundamental effects of dike height and orientation. A rectangular channel with constant depth and a steady, uniform inflow velocity was employed. This approach eliminates confounding factors from complex bathymetry and unsteady flow, allowing for a clear attribution of the observed hydrodynamic patterns to the dike parameters under investigation. While this setup does not replicate a specific field site, it represents a canonical scenario for studying current–structure interaction, similar to methodologies used in foundational studies.

2.6.1. Schemes of Different Heights

To study the characteristics of a submerged dike in terms of dynamics, this paper sets up 10 groups with different heights. The dike heights are set to 0.1 H, 0.2 H, 0.3 H, 0.4 H, 0.5 H, 0.6 H, 0.7 H, 0.8 H, 0.9 H, and 1 H. The upper and lower boundaries of the model are land boundaries, and the left and right boundaries are water boundaries. The boundary conditions are 1 m/s, and the model maintains a constant flow of 1 m/s.

2.6.2. Schemes of Different Orientations

To study the influence of submerged dike orientation on hydrodynamics, this paper sets up 7 groups of different orientations. The length and width of the submerged embankment are the same as those in the aforementioned plan, and the water depth is also the same. The flow direction, which is from due east to west, is defined as a reference. Then, the submerged dikes' orientations—the angles between their center lines and the flow direction— are set at 30°, 45°, 60°, 90°, 120°, 135°, and 150°. The boundary conditions of this plan are also consistent with those of the aforementioned plan.

3. Results

3.1. Hydrodynamic Characteristics for Different Submerged Dike Heights

3.1.1. Flow Velocity Distribution for Different Submerged Dike Heights

1. Distance–velocity distribution

From the flow velocity contour map (Figure 4), it can be seen that the velocity contour map is slightly rounded at the dike head, slightly straight at the tail, and looks quarter-round. With the decrease in the contour value, the contour gathers at the dike head. This indicates that the velocity changes most dramatically near the head of the submerged dike. The velocity at about 0.2 L from the dike head shows the most significant change along the flow direction. Therefore, we analyzed the decay of flow velocity with distance for different submerged dike heights by studying the velocity at a line located about 0.2 L from the dike head, when the orientation angle of submerged dike is 90°.

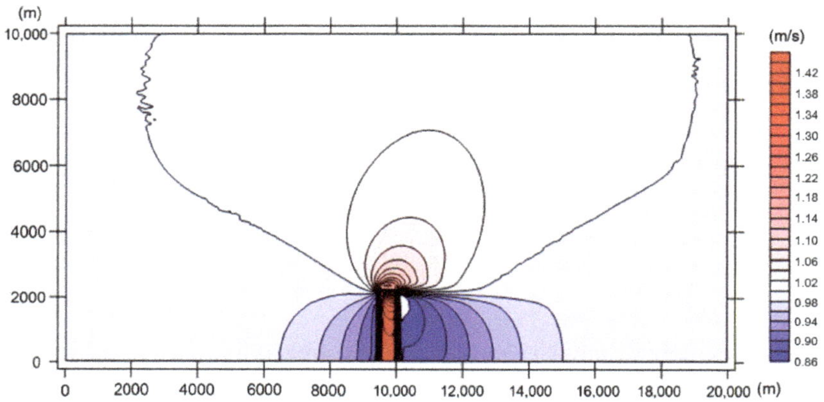

Figure 4. Flow velocity contour map (Flow direction: from left to right).

2. Flow velocity variation with distance for different dike heights

The separate flow zones are different in every case, depending on the height of submerged dike. To reflect the velocity variation with distance for different dike heights, the velocity along the line of 0.2 L distance from the submerged dike was selected to study the velocity law. Figure 5 shows that the flow velocity in general decreases rapidly at a distance of about one-tenth of L. Then, as the distance increases, the flow velocity continues to recover to its original velocity, and finally becomes stabilized. A submerged dike height of 0.7 H is a dividing value. When the submerged dike's height is shorter than or equal to 0.7 H, flow velocity recovery is substantially proportional to distance. While the distance increases, the flow velocity increases. When the height of the submerged dike is higher than 0.7 H, the velocity is invariant or changes little from a distance of about one-tenth of L to half of L. The velocity quickly recovers at distances further than half of L. In all schemes, the velocity recovers to 85 percent of the initial flow velocity at a distance of one dike length, but the velocity values with different dike heights are slightly different.

3.1.2. Flux Characteristics for Different Submerged Dike Heights

To study the hydrodynamics of submerged dikes with different heights, the average velocity across submerged dikes is reflected by the flux across the center line of the submerged dikes. Figure 6 shows that the flux is reduced as the height of the submerged dike increases. When the relative height of the submerged dike is between 0.5 H and 1 H, the relationship between cross-dike flux and the relative height of the submerged dike is linear. When the relative height is less than 0.5 H, the cross-dike flux decreases slowly.

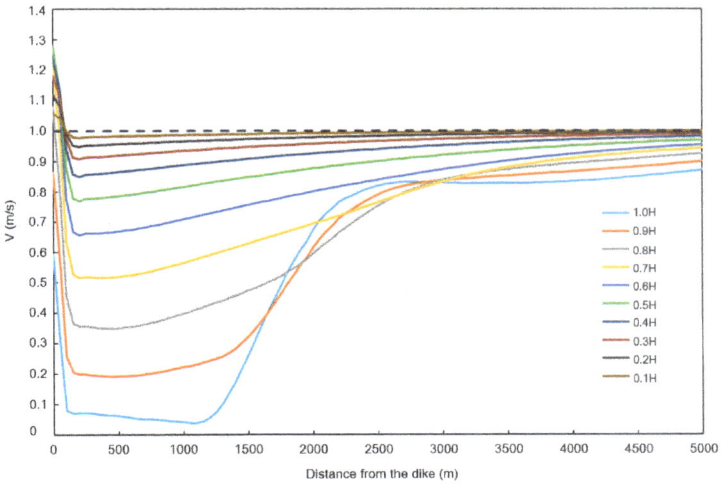

Figure 5. Flow velocity variation with distance for different submerged dike heights.

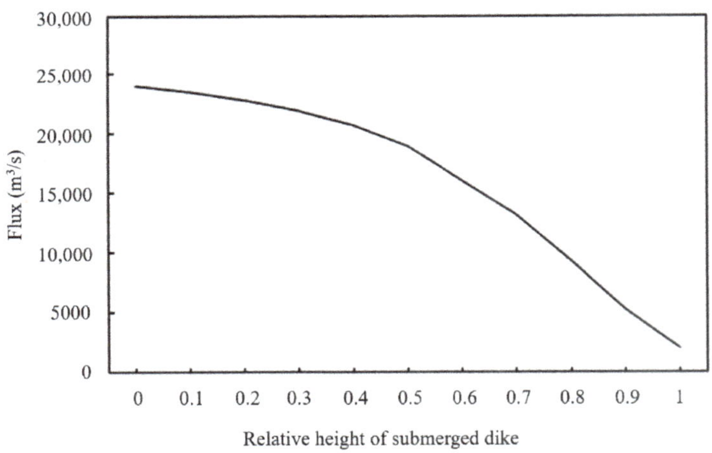

Figure 6. Flux across submerged dike of different heights.

3.2. Hydrodynamic Characteristics Under Different Submerged Dike Orientations

3.2.1. Flow Velocity Distribution Under Different Submerged Dike Orientations

Figure 7 shows that the laws of velocity distribution can be divided into three categories by the angle between the submerged dike and the flow direction. The first category is acute angles. The flow velocity decreases on the backside of the submerged dike and the front side of the dike foot, and increases on the face of the submerged dike and the front of the dike head. As the angle increases, the area of flow velocity decrease becomes larger, and the area of flow velocity increase becomes smaller. The second category is a right angle. In this circumstance, the flow velocity decreases on the backside and front side of the submerged dike, and increases on the face of the submerged dike and the front of the dike head. Thirdly are obtuse angles. In these cases, the flow velocity decreases on the frontier of the submerged dike head and the backside of the submerged dike foot, and increases on the dike, the front of the dike foot, and the backside of the dike head. With an increase in angle, the area of flow velocity decrease becomes smaller. In addition, it is easily discovered that with a decrease in angle, the range and the degree of flow velocity decrease become smaller and lower. This means that submerged dikes with smaller angles have lower influence on flow fields.

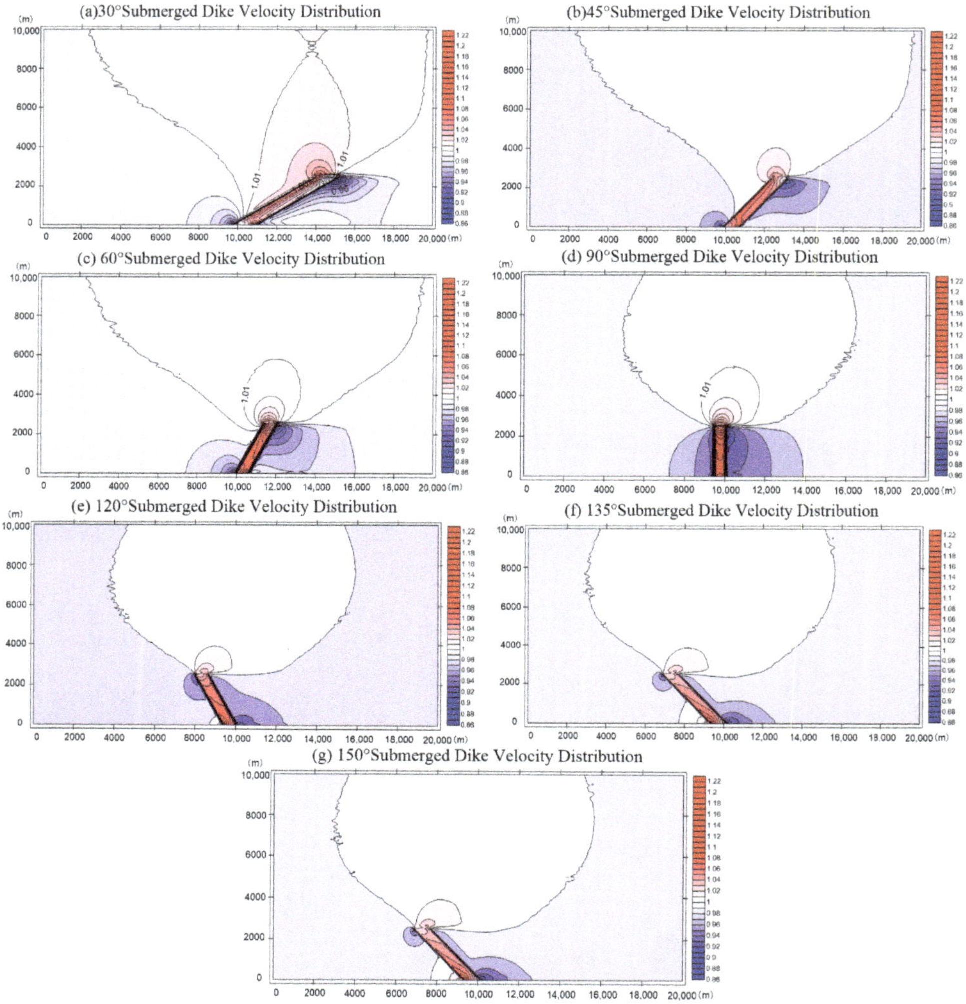

Figure 7. Flow velocity distribution for different submerged dike orientations (flow direction: from left to right).

3.2.2. Flux Characteristics Under Different Submerged Dike Orientations

As depicted in Figure 8, the cross-dike flux increases with angle. Because the flow direction follows the submerged dike orientation, the blocking effect of the submerged dike is tiny. In orientations with acute angles, the cross-dike flux changes little. A smooth linear relationship is present when the angle is less than 60°. If the angle is more than 60°, the flux increases sharply with angle.

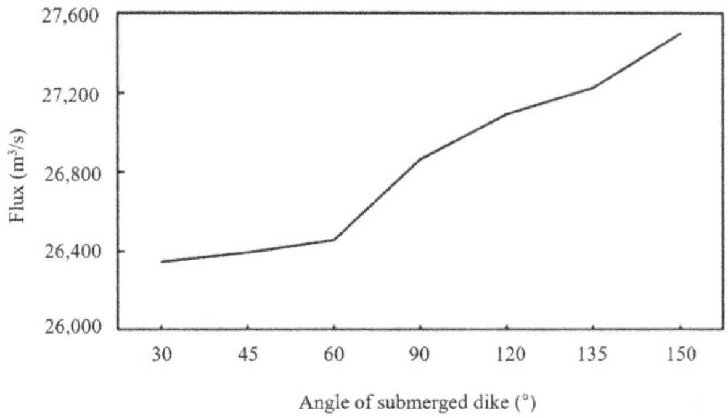

Figure 8. Flux under different submerged dike orientations.

4. Discussion

The fundamental ecological premise of this study is that structures which minimize alterations to natural flow patterns generally have lower environmental impacts. Reduced velocity changes and maintained cross-dike flux help preserve natural sediment transport regimes, nutrient distribution, and habitat connectivity—all critical factors for ecosystem health. Macroinvertebrate abundance and diversity both generally declined in response to alteration inflow magnitude, whether an increase or a decline [24].

In general, when a submerged dike is perpendicular to the coastline and the flow direction is parallel to the coastline, owing to the presence of the submerged dike, the flow velocity increases on the dike's head side and decreases on the dike's tail side. The velocity at about 0.2 L from the dike head shows the most significant change along the flow direction. The velocity on the dike's head side increases because the water section along the dike reduces, and the velocity of all the water tries to keep its original state. The velocity on the dike's tail side decreases because the water along the dike is hindered by the dike. The range of the flow velocity variation is concentrated in the area of 2 L.

4.1. *Influence of Submerged Dike Height on Flow Velocity and Flux*

The velocity distribution law for different dike heights can be used to adjust the marine hydrodynamic impact. According to the result shown in Figure 4, if the height is below 0.7 H, a location velocity can be predicted by the empirical formula: $y = ax + b$ (a is the slope). Thus, the initial height of the dike can be confirmed from the perspective of the flow velocity. In this paper, the relationship between relative height and slope is summarized in Table 1.

Table 1. The relationship of relative height and slope.

Relative Height	0.6 H	0.5 H	0.4 H	0.3 H	0.2 H	0.1 H
Slope (a)	4.36	2.64	1.58	0.90	0.47	0.18

When the dike orientation is perpendicular to the coastline, the flux is reduced as the height of the submerged dike increases. The changing of the cross-dike flux of different dike heights can be divided into two types: basically unchanged (dike height < 0.5 H) and linear change (dike height > 0.5 H). It may be the following reasons. When the relative height of the submerged dike is small (less than 0.5 H), as the height of the submerged dike increases, the area of water across the submerged dike is reduced, but the velocity is increased. These two changes cause the flux to decrease slowly. However, when the relative height of the submerged dike is high (greater than 0.5 H), as the height of the submerged dike increases, the area of the water across the submerged dike is reduced. It becomes a decisive factor and causes the cross-dike flux to linearly decrease as the height increases.

To preserve natural flow patterns, considering the effect of the submerged dike height, the submerged dike height should be designed below the critical relative height of 0.7 H. This ensures the flow regime remains in a "low-disturbance" state, which is fundamental for protecting in situ benthic communities and natural sediment dynamics. Considering the influence on cross-dike flux, it is recommended to limit the submerged dike height to below 0.5 H.

4.2. Influence of Submerged Dike Orientation on Flow Velocity and Flux

The laws of velocity distribution can be divided into three categories: the acute angle, the right angle, and the obtuse angle law. For the first category, the submerged dike orientation is nearly the same as the flow direction. The flow velocity on the frontier of the dike head increases because of the dike's deflecting flow effect. The flow velocity on the submerged dike increases because the area of water across the submerged dike becomes small. The flow velocity on the backside of the submerged dike and the front side of the dike foot decreases because of the blocking effect of the submerged dike. For the second category, the submerged dike is perpendicular to the flow direction. The flow velocity on the backside and front side of the submerged dike decreases because of the blocking effect of the submerged dike. The flow velocity on the submerged dike also increases because the area of water across the submerged dike becomes small. For the third category, the orientation of the submerged dike is opposite to the flow direction, and they are shaped like a horn mouth. The flow is adjusted by the orientation and gathers in front of the dike's foot. This makes the velocity increase. The flow velocity on the submerged dike increases because the area of water flowing across the submerged dike becomes small. The flow velocity on the backside of the submerged dike decreases because of the blocking effect of the submerged dike.

To achieve an optimal compromise between moderating flow velocity variations and preserving ecological connectivity, the orientation of the submerged dike should be constrained to angles under 60° relative to the primary current direction. If maximizing cross-dike flux is a project objective, an orientation angle greater than 90° is recommended.

5. Conclusions

A numerical model is set up using MIKE 21 FM to analyze the hydrodynamical behaviors of a submerged dike. A series of model tests have been conducted to study the variation characteristics of the flow velocity and the cross-dike flux under certain conditions. The flow patterns are different in each case, depending on the height and orientation of the dike. Therefore, the height and orientation must be considered during the design of a submerged dike. According to our simulated results, the following conclusions can be obtained.

The velocity at a distance of about 0.2 L from the head of the dike shows the most significant change along the flow direction, which is perpendicular to the dike and parallel to the coastline. The range of flow velocity variation is concentrated in the area of 2 L.

In every different height condition, the flow velocity decreases rapidly at a distance of 0.1 L. Then, as distance increases, the flow velocity continues to recover. The relative dike height of 0.7 H marks a clear division beyond which the flow velocity exhibits a distinct decay pattern with distance compared to all other heights.

The cross-dike flux is reduced as the height of the submerged dike increases. When the relative height of the submerged dike varies from 0.5 H to 1 H, the relationship between the flux and the relative height of the submerged dike is linear. When the relative height is less than 0.5 H, the flux across the submerged decreases very slowly.

Under different orientation conditions, the laws of velocity distribution can be divided into three categories by the angle between the dike and the flow direction. And cross-dike flux increases as the angle increases.

For ecological restoration projects that aim to minimize hydrodynamic disturbance, a relative dike height lower than 0.7 H and an orientation angle smaller than 60° are preferable under steady current conditions similar to this study.

While recognizing that real-world submerged structures exhibit complex geometries and porosity, this study intentionally employs a simplified, impermeable, rectangular dike section. This fundamental approach is adopted to isolate the first-order hydrodynamic effects of dike height and orientation, free from the confounding influences of other parameters. Establishing this theoretical baseline is a critical first step in developing a comprehensive understanding. It provides a foundation for the future systematic evaluation of additional complexities, such as slope, porosity, and surface roughness.

Author Contributions: Conceptualization, Y.Z. and X.Z.; methodology, Y.Z.; software, Y.Z. and X.L.; validation, Y.D. and X.L.; formal analysis, X.Z.; investigation, X.Z.; resources, Y.Z. and X.Z.; data curation, X.Z., Y.Z., and B.G.; writing—original draft preparation, Y.Z. and X.Z.; writing—review and editing, X.Z. and Y.Z.; visualization, Y.D. and X.Z.; supervision, Y.Z.; project administration, Y.Z.; funding acquisition, X.Z. All authors have read and agreed to the published version of the manuscript.

Funding: This work was funded by the Youth Marine Science Fund of East China Sea Bureau, Ministry of Natural Resources, grant number 2023180504.

Data Availability Statement: The datasets generated and analyzed during the study are available. Please contact the corresponding author to obtain them.

Conflicts of Interest: The authors declare no conflicts of interest.

References

1. Li, G.; Lang, L.; Ning, J. 3D Numerical Simulation of Flow and Local Scour Around a Spur Dike. In Proceedings of the IAHR World Congress, Chengdu, China, 13 September 2013; pp. 1–9.
2. Dai, W.H.; Ding, W. Hydrodynamic improvement of a goose-head pattern braided reach in lower Yangtze River. *J. Hydrodyn.* **2019**, *31*, 614–621. [CrossRef]
3. Ma, D.; Zhou, J.; Wang, Z. Simulation of scour and deposition in the flow field and adaptability of fishes around eco-friendly notched groins. *J. Hydrol.* **2022**, *609*, 127754. [CrossRef]
4. Weitbrecht, V.; Socolofsky, S.A.; Jirka, G.H.; Asce, F. Experiments on mass exchange between groin fields and main stream in rivers. *J. Hydraul. Eng.* **2008**, *134*, 173–183. [CrossRef]
5. Van der Meer, J.W.; Nieuwenhuis, J.W.; Steendam, G.J.; Reneerkens, M.; Steetzel, H.; van Vledder, G.P. Wave overtopping measurements at a real dike. *Coast. Struct.* **2019**, *2019*, 1107–1117.
6. Suzuki, T.; Altomare, C.; Yasuda, T.; Verwaest, T. Characterization of overtopping waves on sea dikes with gentle and shallow foreshores. *J. Mar. Sci. Eng.* **2020**, *8*, 752. [CrossRef]
7. Beji, S.; Battjes, J.A. Numerical simulation of nonlinear wave propagation over a bar. *Coast. Eng.* **1994**, *23*, 1–16. [CrossRef]
8. Iqbal, S.; Tanaka, N. Analysis of hydrodynamic behavior in response to diverse pile arrangements adjacent to an impermeable dike. *Water Sci.* **2024**, *38*, 587–603. [CrossRef]
9. Chen, Y.; Lu, Y.; Yang, S.; Mao, J.; Gong, Y.; Muhammad, W.I.; Yin, S. Numerical investigation of flow structure and turbulence characteristic around a spur dike using large-eddy simulation. *Water* **2022**, *14*, 3158. [CrossRef]
10. Sujantoko, S.; Djatmiko, E.B.; Wardhana, W.; Firmasyah, I.N.G. Numerical Modeling of Hydrodynamic Performance on Porous Slope Type Floating Breakwater. *Nase More* **2023**, *70*, 77–87. [CrossRef]
11. Magdalena, I.; Nathanael, O. Wave resonance reduction over a linear transition bottom with a submerged porous breakwater. *Results Eng.* **2025**, *26*, 104580. [CrossRef]
12. Losada, I.J.; Losada, M.A.; Martin, F.L. Experimental study of wave-induced flow in a porous structure. *Coast. Eng.* **1995**, *26*, 77–98. [CrossRef]

13. Altomare, C.; Suzuki, T.; Chen, X.; Verwaest, T.; Kortenhaus, A. Wave overtopping of sea dikes with very shallow foreshores. *Coast. Eng.* **2016**, *116*, 236–257.
14. Tsal, C.P.; Chen, H.B.; Jeng, D.S.; Chen, K.H. Wave Transformation and Soil Response due to Submerged Permeable Breakwater. In Proceedings of the OMAE 2006, Hamburg, Germany, 4–9 June 2006; pp. 113–120.
15. Sharifahmandian, A.; Simons, R.R. A 3D numerical model of nearshore wave field behind submerged breakwaters. *Coast. Eng.* **2014**, *83*, 190–204. [CrossRef]
16. Chen, J.; Jiang, C.B.; Hu, S.X.; Huang, W.W. Numerical study on the characteristics of flow field and wave propagation near submerged breakwater on slope. *Acta Oceanol. Sin.* **2010**, *29*, 88–99. [CrossRef]
17. Chai, J.C.; Hayashi, S.; Yamanishi, H. Effect of the shape of submerged dike/mound on mud transport. *Sediment Ecohydraulics INTERCON* **2005**, *9*, 329–544.
18. Pang, Q.X.; Zhang, R.B.; Yang, H. Flume experimental study on the heights of submerged dike to diminish siltation. *Ocean Eng.* **2012**, *30*, 66–71.
19. Xu, P.; Ge, J.Z.; Ding, P.X.; Fu, G. Numerical simulation about temporal and spatial variations of overtopping flow flux at the south leading jetty in the deep waterway project of the Changjiang Estuary. *J. East China Normal Univ. (Nat. Sci.)* **2016**, *2*, 50–78.
20. Li, X.B. Research on the Effects of Donghai Dam on the Hydrodynamic Environment of the Zhanjiang Bay. Master's Thesis, Ocean University of China, Qingdao, China, 2008; pp. 112–127.
21. Johnson, H.K.; Karambas, T.V.; Avgeris IZanuttigh, B.; Gonzalez-Marco, D.; Caceres, I. Modelling of waves and currents around submerged breakwaters. *Coast. Eng.* **2005**, *52*, 949–969. [CrossRef]
22. Zanuttigh, B.; Lamberti, A. Experimental analysis and numerical simulations of waves and current flow around low-crested rubble-mound structures. *J. Waterw. Port Coast. Ocean Eng.* **2006**, *132*, 10–27. [CrossRef]
23. Iqbal, S.; Tanaka, N. Experimental Study on Flow Characteristics and Energy Reduction Around a Hybrid Dike. *Int. J. Civ. Eng.* **2023**, *21*, 1045–1059. [CrossRef]
24. Poff, N.L.; Zimmerman, J.K.H. Ecological responses to altered flow regimes: A literature review to inform the science and management of environmental flows. *Freshw. Biol.* **2010**, *55*, 194–205. [CrossRef]

water

MDPI

Article

A Numerical Study on the Pullback Process of a Submarine Cable Based on Trenchless Directional Drilling Technology

Gang Qian [1], Wei Kang [1], Yun Cong [1] and Zhen Liu [2,*]

[1] Zhoushan Power Supply Company, State Grid Zhejiang Electric Power Co., Ltd., Zhoushan 316000, China; zsxm2021@126.com (G.Q.); zsxm2017@126.com (W.K.); congyun198708@gmail.com (Y.C.)

[2] Shandong Provincial Key Laboratory of Ocean Engineering, Ocean University of China, Qingdao 266100, China

* Correspondence: liuzhen@ouc.edu.cn

Abstract: Horizontal directional drilling (HDD) can be utilized in a submarine cable landing operation to solve the problems of a deficient buried depth and a limited route. In this study, a numerical model of the pullback process of a submarine cable using HDD technology is established based on the commercial finite element method platform OrcaFlex 11.3, which is validated using the in situ measured data of an HDD operation project for a pipeline. The effects of the crossing length, incident angle, and pullback velocity of the cable on the effective tension in the cable are investigated and analyzed. The results indicate that an increase in the crossing length and incident angle can significantly enhance the tension in the cable. Under the specific conditions in the Zhoushan islands, the maximum crossing length and incident angle are 1700 m and 35°, respectively. The pullback velocity has a minor influence on the tension in the cable, and an extremely large velocity might lock the cable during its pullback operation. The permissible values derived in this study can provide valuable information to similar engineering cases and projects.

Keywords: submarine cable; laying operation; horizontal directional drilling; pullback; tension in the cable

1. Introduction

During the landing processes of submarine cable laying constructions, the traditional methodology is to trench the underwater groove on the seabed, locate the cable routing, release the supporting floaters, and bury the cable in the groove [1,2]. The contradictions between the increased construction volume during cable laying and the space limitations at the landing points have recently become more significant [3]. The cable landing route plan and construction are more difficult because space resources are becoming more limited, and the ecological conservation redline proposes more requirements [4]. On the other hand, the trenching construction and protection methodology during cable landing always causes an insufficient burying depth and inadequate protection, resulting in the exposure of buried cables on the shoaling beach [5–7]. In addition, submarine cables are always damaged by external impacts, such as ship anchors, in shallow water [8,9]. Therefore, the landing section has been the weak point of the entire submarine cable.

Horizontal directional drilling (HDD) is a popular non-trenching methodology for cable laying, which can drill and expand the pipeline following the planned route, connect the drawing head at the endpoint, and drag the cable to the starting point to accomplish the cable laying [10]. Compared to the traditional trenching methodology, HDD has significant advantages, including fewer environmental impacts [11], a large construction range, good adaptability, and a sufficient burying depth and length [12,13]. HDD has been employed in different projects, such as the pipeline landing construction project in Yacheng [14], the pipeline landing project in the South China Sea [15], the cable landing project in San Nicolas Island in California [16], and the cable landing project in Biscayne Bay [17].

During the landing process, the submarine cables can be affected by many factors, resulting in different varying trends of the cable profile and tension distribution. These trends might cause the drags to exceed the permissible stress and possible mechanical damage, which will affect the serviceability of the cable [18]. Yoshizawa and Yabuta studied the cable variation during the laying process [19] and found that the bottom tension of the cable was determined mainly by the laying vessel and the water depth. The bottom tension increases as the water depth decreases and the vessel speed increases. Nagatomi et al. used the concentrated mass method to calculate the motion and force of the submarine cable and compared the computational results with the on-site measuring data [20]. Prpić-Oršić and Nabergoj reported a numerical model for predicting the motion and tension of the cable during the laying process under rough sea conditions [21] and found that significant tension variations could be observed at the deployment point under the conditions of higher wave frequencies, heavier cables, and deeper waters. Zhang and Hu proposed that the tension should be the control variable during the laying operation of the submarine cable [22].

Yang et al. investigated tensions in the cable during the laying operation and found that the tension varies linearly with the water depth and reaches its peak at the laying wheel point [23]. Feng et al. established a three-dimensional (3D) finite element method (FEM) cable model and studied the dynamic properties of the submarine cable under the wave impacts [24]. Zhang et al. simplified the submarine cable and employed a lumped mass method to analyze the dynamic characteristics of the cable during its laying process, validated using experimental data [25]. Chang et al. conducted numerical simulations on the dynamic responses of a submarine cable under coupled tensile, torsional, and compressive loads and found that coupled loads acting on the cable could be reduced to 40% of the total when the water depth was greater than 4000 m [26]. Sun et al. reported that the moving velocity of the cable-laying vessel, the releasing velocity of the cable, and the cross-sectional diameter of the cable significantly influenced the tension distributions in the cable [27]. Okkerstrøm established a numerical model using the commercial software OrcaFlex based on the finite element method and found that the cable weight had a significant effect on the peak tension, which should be considered during the design of the cable armor [28]. Kuang et al. studied the influences of the cable length, current velocity, incident waves, and wind direction on the tensions in a cable during the laying process and reported that the floating cable with floaters for smaller lengths could be considered a mooring line for the vessel, which could significantly enhance the tension in the cable [29]. Shen et al. analyzed the effects of the burial depth on the tension in a submarine cable [30]. The results indicated that the peak tension in the submarine cable, appearing at the laying wheel point of the cable-laying vessel, increases linearly with an increase in the burial depth. Peres et al. analyzed the influences of the cable mass, laying length, and environmental conditions on the kinematic and dynamic responses of a submarine cable and found that the cable mass plays a dominant role [31]. In addition, a strong correlation between the chute vertical downward velocity and compression on the touchdown point can be observed.

For HDD applications in cable installation, Li et al. proposed that the trade-off between the friction resistance and allowable traction force remains a key constraint in long-distance laying [32]. Khasanov et al. reported a modified calculation method to analyze the tension in the pipeline during HDD operations and enhanced the calculation precision for practical applications [33]. Faghih et al. found the height variation of the downhole has significant influences on the stress and strain in the pipeline by analyzing the field-measured tensions in the pipeline during HDD operation [34]. Guo and Liang numerically investigated cable tension dynamics during pullback operations [35], identifying linear tension variation patterns along the curved borehole trajectory and peak tension occurrences during the final pullback phases. Based on the domestic codes and previous engineering cases, Tang et al. systematically evaluated cable pull force safety parameters through regulatory compliance analysis and empirical case studies [36], demonstrating that an enhanced conductor cross-sectional area correlates with elevated safety factors under equivalent weight conditions. Utilizing ANSYS Workbench simulations, Vasilescu and Dinu conducted a parametric analysis of pipeline pullback mechanics [37], establishing a parametric sensitivity hierarchy (descending order) for pipeline stress: diameter, wall thickness, internal pressure, drilling fluid density, and burial depth. Through multi-phase analysis of stepped borehole crossing projects, Liang et al. characterized tension evolution patterns during various installation stages, subsequently refining the computational model for the HDD pullback force [38]. Zheng numerically examined force distributions in HDD pipelines using field monitoring data, verifying an inverse proportionality between the bend section tensions and curvature radii [39].

The literature review indicates that previous studies predominantly concentrated on cable tension mechanics and cable–vessel interactions in conventional laying configurations. For HDD-assisted cable installations, the existing findings primarily stem from analytical and numerical approaches assuming elevation parity between pullback termini. In contrast, HDD-enabled submarine cable landing operations involve complicated engineering challenges during extended pullback processes with significant elevation differentials, including pull force optimization, cable–pipe friction management, and tension distribution across variable curved segments. These operational complexities underscore persistent challenges in ensuring cable integrity during pipeline transitions. This study employed a representative Chinese HDD case study to establish numerical simulation parameters for cable pullback dynamics with particular emphasis on tension fluctuations at critical connection nodes. A comparative analysis of the curvature gradients, crossing distances, and pullback velocities against submarine cable specifications provides a systematic evaluation of the tension development mechanisms, delivering crucial insights for engineering design optimization.

2. Problem Description

This investigation derives from an actual offshore construction project implemented in the Zhoushan Archipelago, Zhejiang, China. The submarine cable installation was completed using HDD methodology. The pullback operation of the cable through the HDD pipeline is depicted in Figure 1. The analysis presumes that the HDD pipeline was successfully constructed for submarine cable landing, with its geometric configuration remaining unchanged. The entry point was situated onshore at 120.0 m from the initiation point of the pipeline horizontal segment, elevated 10.0 m above the mean water level. The seabed topography exhibited a planar configuration with a constant water depth of 10.0 m. The horizontal span between the entry and exit points is designated as L_H, termed the crossing length. The vertical elevation and horizontal span of the burial pipeline segment are defined as H_B and L_{HB}, respectively. The entry angle of the pipeline at the terrestrial interface is denoted by φ.

The guidance wire entered the pre-drilled pipeline from the entry point, traversed the pipeline, and emerged at the exit point for integration with the cable through a dedicated coupling mechanism. The synchronized retrieval of both the guidance wire and cable maintained a constant velocity throughout the pipeline. During initial deployment, the cable-laying vessel dispensed the cable via the sheave assembly at a velocity marginally exceeding the pulling velocity v_{pull}, ensuring the controlled accumulation of cable slack beyond the exit point. In subsequent phases, the dispensing rate achieved parity with the pulling velocity. The ambient hydrodynamic conditions in the operational zone exhibited minimal wave–current activity, exerting negligible influence on the cable installation. Given sufficient residual cable length beyond the exit point, vessel maneuvering dynamics also demonstrated an insignificant impact on the traction process. Consequently, this study excluded analytical consideration of environmental hydrodynamic forces and vessel operational kinematics.

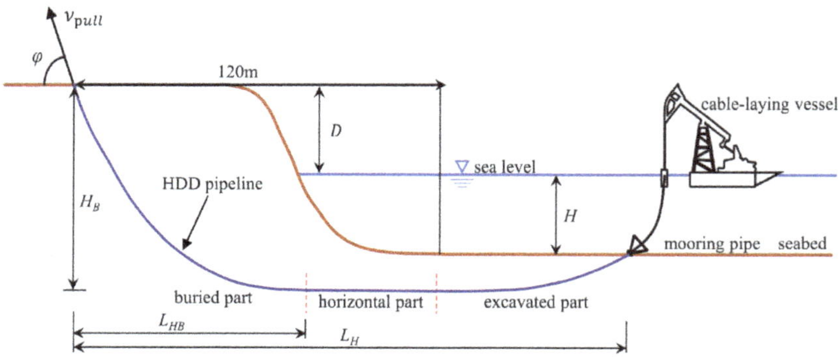

Figure 1. Schematic of the submarine cable in the HDD pipeline.

The submarine transmission line implemented in this investigation constitutes a standard tri-axial optoelectronic composite cable, with material specifications as systematically tabulated in Table 1. The predominant materials for protective conduit systems comprise seamless steel tubing and high-density polyethylene (HDPE). Steel conduits exhibit superior structural rigidity and deep-water operational capability, albeit necessitating advanced welding methodologies. HDPE systems provide enhanced hermetic sealing and corrosion resilience, combined with cost-efficient installation and accelerated project timelines. Notable constraints of HDPE include the reduced tensile capacity and compromised load-bearing stability. Following cost–benefit analysis, the HDPE configuration was selected as the optimal solution. The material characteristics compliant with the domestic national code are comprehensively detailed in Table 2 [40].

Table 1. Detailed parameters of the submarine cable.

Item	Value (Unit)	Item	Value (Unit)
Outer diameter	199.2 mm	Mass in the water	47.9 kg/m
Mass in the air	79.8 kg/m	Permissible tension	171.7 kN
Axial stiffness	7.0×10^5 N/mm	Bending stiffness	1.0×10^8 N/mm

Table 2. Detailed parameters of the PE pipe.

Item	Value (Unit)	Item	Value (Unit)
Classification	PE 100	Density	910.0 kg/m^3
Outer diameter	355.0 mm	Single section length	12.0 m
Thickness	13.6 mm	Friction coefficient with the cable	0.35
Axial stiffness	1.3×10^4 N/mm	Bending stiffness	2.6×10^{11} $N \cdot mm^2$

3. Numerical Model

3.1. Fundamental Theory of FEM Cable Modeling

The frictional force computation in the numerical model was systematically formulated using the modified Coulomb friction model [41], as graphically represented in Figure 2. A comparative analysis, as shown in Figure 2a, reveals that the standard Coulomb friction model exhibits discontinuity in friction transition, whereas the modified formulation, shown in Figure 2b, demonstrates a linear friction variation from $-\mu R$ to $+\mu R$ within the critical displacement range [$-D_{crit}$, $+D_{crit}$]. The critical displacement threshold D_{crit} is mathematically expressed as follows:

$$D_{crit} = \frac{\mu R}{k_s a}, \quad (1)$$

where μ denotes the friction coefficient and R the normal contact force; k_s and a correspond to the shear strength and contact surface area, respectively.

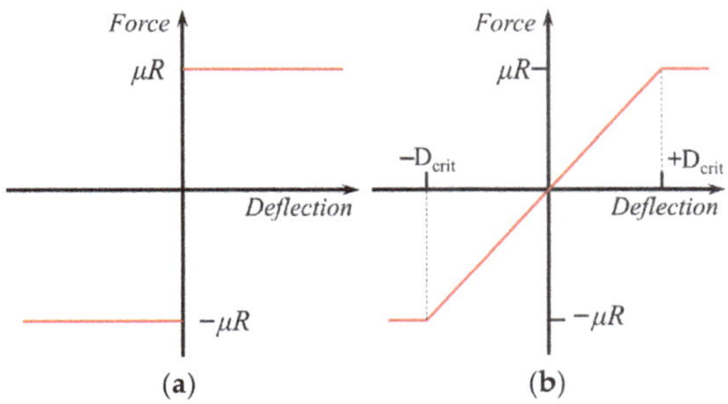

Figure 2. Variation mode change in the modified Coulomb friction model [36]. (**a**) Sudden change mode; (**b**) linear change mode.

In the finite element formulation, the submarine cable is discretized into a series of massless linear elements, as schematically illustrated in Figure 3. Each element configuration comprises dual nodal components at its extremities, with dynamic properties (mass, gravitational force, buoyancy, etc.) lumped at nodal points. This discretization scheme effectively represents the cable as interconnected nodal masses bridged by quasi-static beam segments. The geometric configuration features half-beam segments adjoining each nodal element, with terminal segments positioned at cable extremities. Mechanical connectivity is achieved through coaxial elastic bars integrated with axial damping mechanisms and torsional spring–damper assemblies, thereby enabling comprehensive characterization of the torsional behavior.

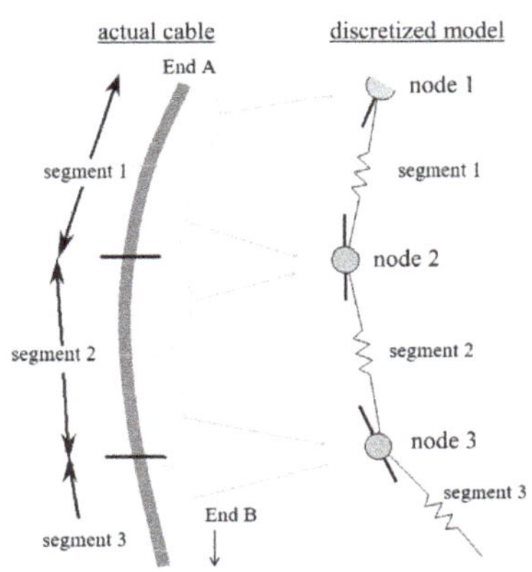

Figure 3. Submarine cable in the FEM model [36].

Stress distributions across discrete elements require systematic computation to derive the global cable tension. Fundamental kinematic parameters must be preliminarily determined, including the inter-nodal displacement within linear elements, their temporal derivatives, and the orientation unit vector S_z along the inter-nodal axis. The mid-element tensile force manifests as vectorial stress aligned with S_z, quantified by the effective tension T_e, mathematically defined as follows:

$$T_e = T_w + (p_0 a_0 - p_i a_i), \tag{2}$$

where p_0 and p_i denote the external and internal hydrostatic pressures, respectively; a_0 and a_i represent the external and internal stress-bearing areas, respectively; and T_w signifies the circumferential wall tension. Given the linear axial stiffness characteristics, T_w is formulated as follows:

$$T_w = EA \cdot \varepsilon - 2\nu(p_0 a_0 - p_i a_i) + k_{tt}\frac{\tau}{l_0} + EA \cdot c\frac{dl}{dt}\frac{1}{l_0}, \tag{3}$$

where E indicates the effective Young's modulus and A the cross-sectional area, constituting the axial rigidity EA; l_0 denotes the undeformed segment length; ν is the Poisson's ratio; k_{tt} characterizes the tension–torsion coupling; τ quantifies the segment twist angle (in radians); c embodies the viscous damping coefficient (s); dl/dt signifies the change rate of the length; and ε expresses the total mean axial strain, operationally defined as follows:

$$\varepsilon = \frac{l - \lambda l_0}{\lambda l_0}, \tag{4}$$

where l corresponds to the instantaneous segment length; λ indicates the expansion factor of the segment.

In addition, the pullback force acting on the cable T is governed by the following:

$$T = T_1 + Wk_0 L \tag{5}$$

where T is subject to the mechanical constraint: $T \leq \min\{F_q, F_t\}/\gamma_0$, where F_q denotes the maximum tensile capacity at the cable termination, F_t the permissible operational stress, and γ_0 the safety factor; W represents the volumetric weight; k_0 signifies the cable–PE conduit friction coefficient; L indicates the safe pullback distance; and T_1 corresponds to the initial pre-tension in the cable. Given the presence of cable redundancy in this study, T_1 was excluded from the pullback force computation.

3.2. Numerical Model Setup

The numerical model was established using the commercial FEM software OrcaFlex 11.3. The seabed profile was configured through the Seabed module, with the free sea surface positioned at z = 0.0 m in the global coordinate system. Within this coordinate framework, the onshore terrain and seabed elevation were, respectively, defined at z = +10.0 m and z = −10.0 m. The initial positioning of the model components on the seabed interface employed global coordinates, while the rope assemblies, connectors, and articulated cable system were parameterized using local coordinate systems. The pipeline geometry was specifically defined within the nodal references.

The pipeline system was modeled through the Homogeneous Pipeline Type module, with PE material characteristics implemented via the dimensional parameters (diameter), mass density, and stiffness properties. The Mid-line Connections module incorporates a lumped mass formulation to specify critical physical attributes at discrete pipeline nodes, including spatial coordinates, angular orientation, and arc-length parameters. These definitions collectively characterize the burial depth, curvature profile, and horizontal span, ensuring numerical representation fidelity to actual pipeline behavior. The rope, connector, and cable elements were simulated using the General Line Type module with appropriate diameter specifications, stiffness parameters, and material definitions. Hinged joint elements model the mechanical interaction during cable-pulling operations. The section division and mesh generation were primarily determined by the curved section of the pipeline. The section division should ensure the curved section is as smooth as possible to avoid any penetration and non-physical oscillation during simulation. On the other hand, the straight section can be set as long as possible to accelerate the simulation process. Following the criterion ruled by the OrcaFlex, the straight section length was set at 10.0 m and the curved section length at 5.0 m, ranging from 1/10 to 1/5 of the curvature radius. From the trial simulations, it was found that the present section length setup balanced calculation precision and cost.

For the contact mechanics between the PE pipeline and cable during the pullback phases, the Penetration Contact algorithm within the Line Contact module was implemented. The computational efficiency was optimized by abstracting the soil–pipeline friction and mud viscous resistance into equivalent friction coefficients. The pullback velocity was parameterized through the Payout Rate specification in the Feeding module. A ramp function generator (Apply Ramp module) ensured smooth transition from static equilibrium to dynamic operation, preventing numerical discontinuities during the initial motion phases.

3.3. Numerical Model Validation

The Lanzhou–Zhengzhou–Changsha pipeline crossing project served as an engineering case study to validate the numerical model's capability in predicting the cable tension during pullback operations [42]. The simulation results of the pipeline forces were benchmarked against field measurements at representative monitoring locations. Key engineering parameters for this validation are documented in Table 3.

Table 3. Detailed parameters of the engineering [42].

Item	Value (Unit)	Item	Value (Unit)
Crossing length	1970.0 m	Pipe specification	$\Phi610 \times 12.7$ mm
Pipe length	2129.0 m	Pipe material density	7850 kg/m^3
Crossing depth	73.5 m	Mud density	1200 kg/m^3
Incident angle	16.0°	Unearthed angle	14°
Friction coefficient	0.3	Pipeline radius	0.65 m

Figure 4 presents a comparative analysis between the numerically predicted pullback forces and the field-measured values. The data demonstrate satisfactory agreement between the simulations and measurements, with a maximum deviation of 9.8% observed at monitoring point C. Notably, significant under-prediction occurs at point A near the entry location. This discrepancy primarily stems from the geometric simplifications in the numerical representation: the simulation modeled the pipeline as an idealized Line module directly pulled through the channel, whereas the field conditions involved non-negligible pipe–floor friction from seabed deployment. Despite these simplifications, the developed model successfully replicated the directional drilling channel configurations and achieved reasonable friction prediction accuracy during pullback simulation. This validation confirmed the model's applicability for simulating submarine cable retrieval operations in HDD channels.

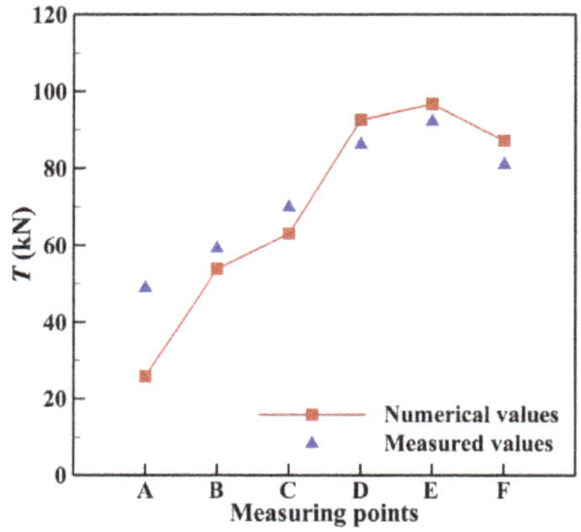

Figure 4. Comparison of the predicted and measured forces in the engineering project [42].

4. Results and Discussion

4.1. Effects of the Crossing Length on the Tension in the Cable

During pullback operations, the progressive increase in the crossing length L_H extends the contact interface between the cable and pipeline, consequently amplifying the frictional resistance and modulating the tension dynamics. To systematically evaluate this relationship, six L_H values (600, 800, 1000, 1200, 1400, and 1600 m) were simulated under fixed parameters: $\varphi = 35°$ and $v_{\text{pull}} = 0.1$ m/s. Figure 5 presents the temporal evolution of the effective cable tension across varying L_H. For any given L_H, the pullback tension exhibits a triphasic evolution.

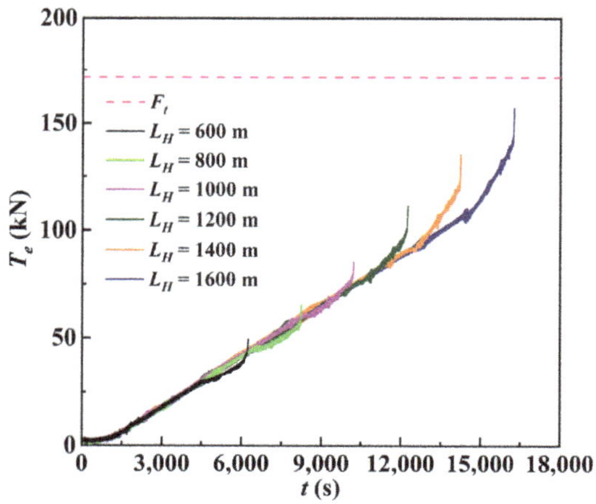

Figure 5. Time histories of the effective tension in the cable for various crossing lengths.

Taking L_H = 1600 m as an example, a non-zero tension manifests during the initial stage (0–1000 s) due to pre-existing stress in the semi-submerged cable prior to pipeline entry. This baseline tension undergoes gradual accumulation as the cable transitions from a vessel-deployed suspended configuration to a seabed-contacting relaxed state via the chock, a process accompanied by continuous marine deployment, as shown in Figure 6. The tension increment arises from the geometric reconfiguration during this transitional phase.

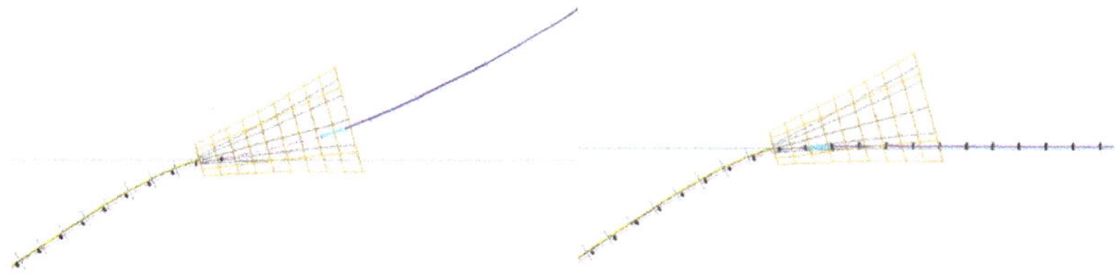

(**a**) (**b**)

Figure 6. Different states of the cable during its pullback process. (**a**) Suspension state; (**b**) relaxed state.

During the intermediate stage (1000–14,000 s), the effective tension exhibits linear escalation proportional to the cumulative frictional resistance along the straight pipeline segment, arising from the uniformly distributed sliding contact between the cable and conduit. The terminal phase (>14,000 s) demonstrates accelerated tension amplification attributable to intensified curvature-derived friction, as the cable navigates the ascending bend. Here, enhanced geometric constraints induce circumferential cable–pipe wall interactions governed by material stiffness, superimposing the bending resistance onto the baseline sliding friction. A tensile discontinuity occurs upon extraction completion, reflecting the instantaneous constraint of removal at disengagement. Crucially, passage through the curved segment introduces two supplementary mechanisms: (1) the gravitational alignment effects modify the interfacial pressure distribution under non-planar geometries, and (2) the localized armor-layer deformation induces hysteresis through stiffness-mediated contact with the conduit wall. Engineering analysis confirms that pipeline diameter upscaling reduces the contact stress concentration, effectively mitigating the armor fatigue risks during high-curvature transits. Furthermore, empirical validation reveals a strict linear proportionality between the crossing length and duration (ΔL_H =200 m/Δt =2000 s), enabling precise schedule optimization through parametric extrapolation.

Figure 7 illustrates the effective tension versus crossing length. The maximum effective tension demonstrates monotonic escalation with increasing L_H, rising from 49.4 kN to 156.8 kN (217% amplification ratio), governed by progressive frictional accumulation between the submarine cable and pipeline during axial pullback. A linear proportionality between the peak tension and L_H is established, with an average incremental rate of 21.5 kN of 200 m. This quantized growth coefficient serves as a deterministic parameter for selecting a tension control system and defining the safety margins in an engineering operation.

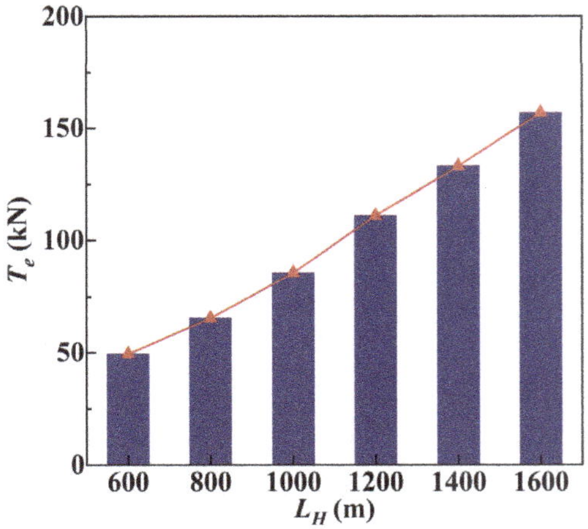

Figure 7. The maximum values of the effective tension for different crossing lengths.

To evaluate the permissible crossing length under current construction constraints, a supplementary simulation was executed for L_H = 1700 m, as shown in Figure 8. The tension evolution mirrors the triphasic profile observed at L_H = 1600 m in Figure 5, with a peak effective tension of 169.7 kN, approaching the cable's permissible threshold. Consequently, the critical crossing length under operational limits was validated as 1700 m, as exceeding this induces closeness to the structural tolerance boundary.

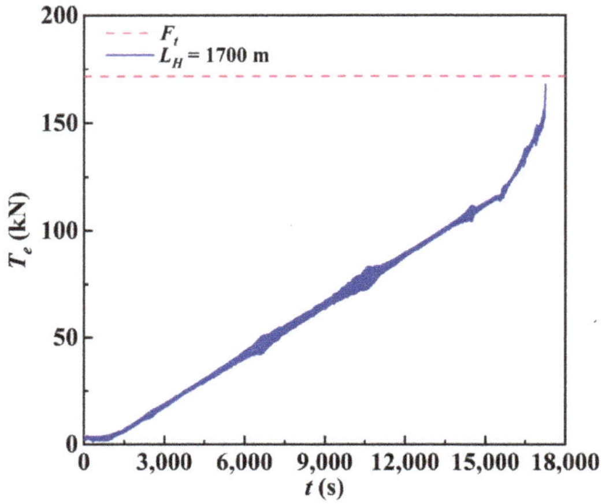

Figure 8. Time histories of the effective tension in the cable for L_H = 1700 m.

4.2. *Effects of the Incident Angle on the Tension in the Cable*

Diverging from terrestrial horizontal HDD, submarine cable installations operate under spatially constrained marine environments, necessitating strategic incident angle adjustments to accommodate infrastructural boundaries. The frictional resistance at the cable–pipeline interface is governed by the incident angle variation, a critical operational parameter examined here under the fixed crossing length (L_H = 1700 m) and the pullback velocity (v_{pull} = 0.1 m/s). Five values of the incident angle were parametrically examined: $\varphi = 10°$, $\varphi = 20°$, $\varphi = 30°$, $\varphi = 40°$, and $\varphi = 50°$.

To isolate the φ-dependent effects, the numerical framework preserved the following geometric invariants: the horizontal and excavated pipeline segment remained static, with terminal points rigidly anchored. This protocol enabled automated trajectory generation for the buried pipeline section as φ varied while maintaining kinematic consistency. Figure 9 quantifies the resultant mechanical responses through the comparative analysis of the horizontal bending displacement (H_B) and the lateral horizontal bending length (L_{HB}), elucidating their parametric dependencies on φ.

Figure 10 delineates the temporal evolution of the effective tension T_e in the cable under distinct pipeline incident angles. While T_e exhibits phase-dependent convergence during the initial and intermediate pullback stages, divergent dynamics emerge in the terminal phase, governed by the acute sensitivity to φ-modulated geometrical constraints. This triphasic behavior arises from invariant pipeline routing in excavated/horizontal segments (nullifying φ-dependent frictional variations) versus critical geometry, the tribology coupling in the buried region. During final pullback, curvature-induced contact transitions to low-φ regimes ($\varphi \leq 30°$), sustaining unidirectional sliding friction, whereas $\varphi > 30°$ instigates multi-axial contact regimes due to helix-like cable–pipeline interactions. Such geometric locking amplifies the frictional resistance by synchronizing the contact area enlargement and static friction coefficient elevation, directly manifested as steepened T_e: temporal gradients and heightened peak tensions. Crucially, the threshold at

$\varphi = 30°$ demarcates the friction regimes dominated by kinematic linearity versus nonlinear geometric coupling.

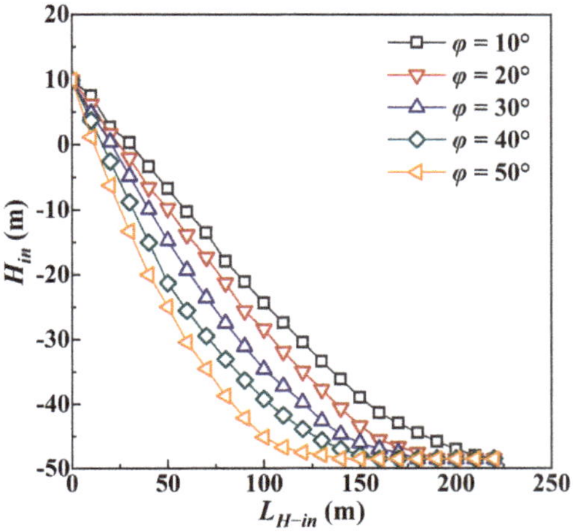

Figure 9. Shape variations in the buried part of the pipeline for various incident angles.

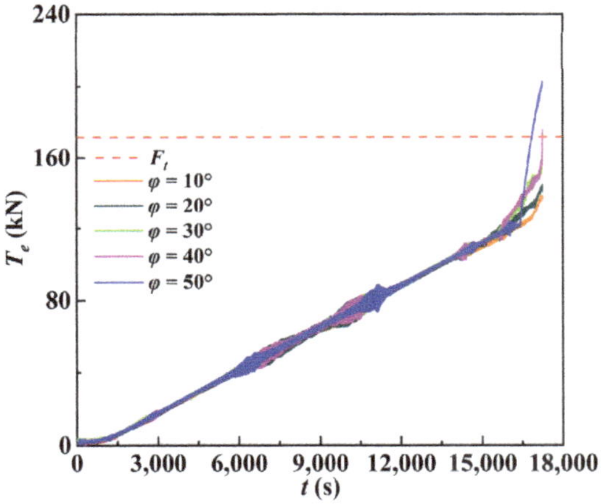

Figure 10. Time histories of the effective tension in the cable for various incident angles.

Meanwhile, as the incident angle increases from 10° to 40°, the crossing period remains constant. However, at $\varphi = 50°$, the cable tension surges abruptly at t = 16,400 s and exceeds the permissible threshold by 171.7 kN at t = 16,858 s. This nonlinear escalation stems from the intensified geometric constraints at elevated angles, which amplify the cable–pipeline contact pressure and shift the interfacial friction from dynamic to static dominance. The resultant kinematic lock forces the pulling system to generate excessive driving power, inevitably breaching the safe tension limits. For fields where tension constraints cannot be mitigated, establishing an operational platform to reduce φ becomes imperative to ensure safety compliance.

Figure 11 illustrates the maximum effective cable tensions across varying incident angles φ. The amplification ratio exhibits a linear proportionality to φ, confirming enhanced tensile forces at larger angles. Upon pullback completion, the peak tensions measure 137.6 kN, 143.2 kN, 175.7 kN, and 201.8 kN for the respective φ values. For $\varphi < 50°$, the average tension amplification (12.7 kN) remains marginal compared to the crossing length effects, indicating limited φ-dependency in cable–pipeline contact mechanics. Notable tension escalation occurs exclusively under geometric locking at $\varphi = 50°$. At $\varphi = 35°$, the peak tension of 169.7 kN suggests that the optimal incident angle for 1700 m crossings should be maintained within the 35–40° range to balance the operational feasibility and safety thresholds.

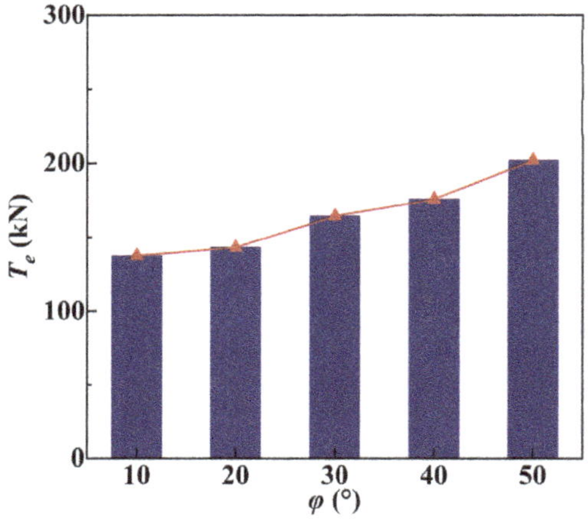

Figure 11. Maximum effective tensions in the cable for various incident angles.

4.3. *Effects of the Pullback Velocity on the Tension in the Cable*

The pullback velocity of the submarine cable is another critical factor for the tension in the cable. Enhancing the velocity requires higher power with higher costs and causes tension variation in the cable. Therefore, it is necessary to investigate the effects of the pullback velocity on the tension in the cable. Here, we set the crossing length and the incident angle at L_H = 1700 m and $\varphi = 35°$. Four pullback velocities ν_{pull} were employed in the comparison: 0.1 m/s, 0.2 m/s, 0.3 m/s, and 0.4 m/s.

Figure 12 illustrates the temporal evolutions of the effective cable tensions for various pullback velocities. The three varying phases still exist as the pullback velocity increases. Furthermore, the pulling out period decreases significantly as the pullback velocity increases. As ν_{pull} increases from 0.1 m/s to 0.4 m/s, the pulling out period shortens nonlinearly. For ν_{pull} = 0.1 m/s, the period measures 17, 250 s, decreasing to 4, 309 s at 0.4 m/s, while the extreme case (0.4 m/s) shortens the period to <25% of the baseline (0.1 m/s). This inverse proportionality between the velocity and duration is attributed to the kinematic friction stabilization at elevated speeds, which attenuates the rate of temporal reduction. The results suggest a saturation effect in period reduction at higher velocities, aligning with quasi-steady friction behavior under dynamic cable–pipeline interactions.

Figure 13 illustrates the peak effective cable tensions under varying pullback velocities. As the pullback velocity ν_{pull} increases from 0.1 m/s to 0.3 m/s, the peak effective tension fluctuates between 169.7 kN, 169.4 kN, and 170.2 kN, with deviations remaining below 0.5%, and all values are under the permissible threshold of 171.7 kN. However, at ν_{pull} = 0.4 m/s, the peak effective cable tension rises to 172.5 kN, exceeding the permissible limit. Compared to the crossing length and incident angle, the pullback velocity has a limited influence on the peak effective cable tension. The peak effective tension exceeds the permissible value only when ν_{pull} = 0.4 m/s. As the crossing length and incident angle are fixed, the tension in the cable is primarily determined by the cable mass and the friction between the cable and pipeline, with limited influence from the pullback velocity. Consequently, the peak effective cable tension remains small. The cable stiffness, however, reduces the deformation period at the curvature zones as v_{pull} increases, enlarging the contact area with the upper pipeline wall and amplifying the frictional resistance, thereby elevating the tension. To mitigate the risks of cable locking and structural failure, v_{pull} must be maintained below 0.4 m/s.

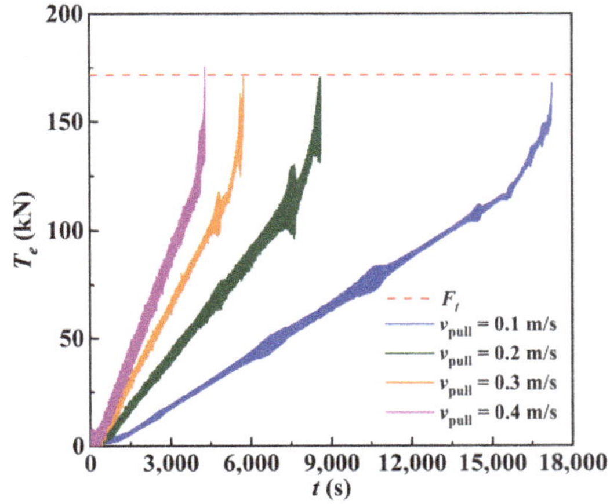

Figure 12. Time histories of the effective tension in the cable for various pullback velocities.

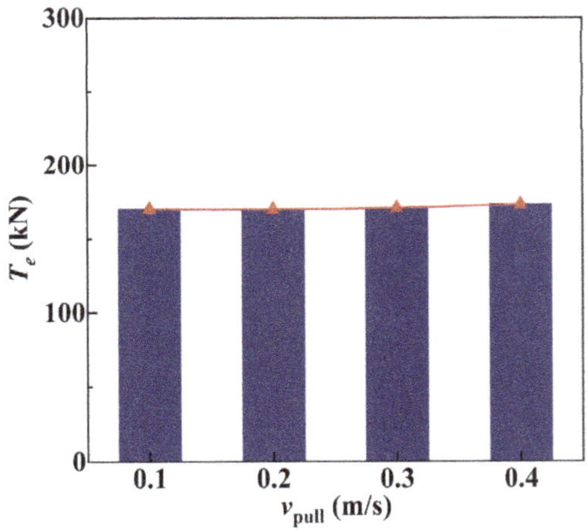

Figure 13. Maximum effective tensions in the cable for various pullback velocities.

5. Conclusions

Based on the HDD project for submarine cable laying in the Zhoushan islands, China, a numerical model using the commercial FEM platform OrcaFlex 11.3 was established to simulate the pullback process of the submarine cable in the pipeline, which was validated using the in situ measured data. The effects of the crossing length, incident angle, and pullback velocity on the effective tension in the cable were investigated and analyzed.

The numerical predictions indicate that the maximum tension in the cable increases gradually as the crossing length of the pipeline increases. Under the investigated project conditions, the peak tension in the cable was close to the permissible value of 171.7 kN when the crossing length was 1700 m. As the incident angle increased, the maximum tension in the cable also increased gradually. When the incident angle was too large, the cable contacted the pipe, causing the tension to be close to the permissible value. The numerical results suggest that the incident angle of the pipeline should be less than 35°. The pullback velocity of the system had a minor influence on the effective tension in the cable, and a gradual increase in the pullback velocity might lock the cable in the pipeline, resulting in damage to the cable.

Future work should address the cable burial depth and emergence angle to expand the tension mitigation strategies. Additionally, while this study assumed a uniform friction coefficient, practical engineering requires sectional distinctions (e.g., excavated vs. seawater-submerged pipeline segments). Buoyancy effects, which reduce the submerged cable weight and alter the friction dynamics, should also be incorporated in simulations for enhanced field application. The environmental loadings, ignored in this study, will be included in the related numerical simulations, and their influences on the pullback process of the submarine cable will be evaluated.

Author Contributions: Conceptualization, Z.L. and G.Q.; methodology, W.K.; software, Y.C.; validation, Y.C.; formal analysis, G.Q. and W.K.; writing—original draft preparation, Z.L.; writing—review and editing, Z.L. and G.Q.; visualization, Y.C. All authors have read and agreed to the published version of the manuscript.

Funding: The research comes from the science and technology project fully supported by the State Grid Zhejiang electric power Co., Ltd. of China (grant no.: 5211ZS230003).

Data Availability Statement: The original contributions presented in this study are included in the article. Further inquiries can be directed to the corresponding author.

Conflicts of Interest: Authors Gang Qian, Wei Kang and Yun Cong were employed by the Zhoushan Power Supply Company, State Grid Zhejiang Electric Power Co., Ltd. The remaining authors declare that the research was conducted in the absence of any commercial or financial relationships that could be construed as a potential conflict of interest.

Nomenclature

Symbol	Definition
a	Contact area
a_0	External stress area
a_i	Internal stress area
A	Cross-sectional area of the cable
c	Damping coefficient
dl/dt	Change rate of the length
D	Height from the excavation point to the water surface
D_{crit}	Deflection
E	Effective Young's modulus
EA	Axial stiffness of the cable
F_q	Maximum pulling force in the drawing head of the cable
F_t	Permissible tension in the cable
H	Water depth
H_B	Height of the buried part of the pipeline
k_0	Friction coefficient between the PE pipe and the cable
k_s	Shear strength
k_{tt}	Tension/torque coupling
l	Instantaneous length of the segment
l_0	Unstretched length of the segment
L	Pulling safety distance of the cable
L_H	Horizontal distance between the entry and exit points
L_{HB}	Horizontal length of the buried part of the pipeline
p_0	External pressure

p_i	Internal pressure
R	Contact reaction force
S_z	Unit vector
t	Time
T	Pullback force
T_1	Initial tension in the cable
T_e	Effective tension
T_w	Wall tension
ν	Poisson ratio
ν_{pull}	Pullback velocity
W	Volume weight of the cable
φ	Incident angle
τ	Twist angle of the segment
ε	Total mean axial strain
λ	Expansion factor of the segment
γ_0	Safety coefficient of the cable load
μ	Friction coefficient

References

1. Worzyk, T. *Submarine Power Cables: Design, Installation, Repair, Environmental Aspects*; Springer Science & Business Media: Berlin/Heidelberg, Germany, 2009; pp. 136–139.
2. Bao, S.; Zhang, Z.; Yuan, Z.; Wang, J.; Yu, Q.; Liu, Z. Effects of Floating Airbag on Cable Hydrodynamic Behaviors: An Experimental Study. *J. Mar. Sci. Eng.* **2022**, *10*, 402. [CrossRef]
3. Taormina, B.; Bald, J.; Want, A.; Thouzeau, G.; Lejart, M.; Desroy, N.; Carlier, A. A review of potential impacts of submarine power cables on the marine environment: Knowledge gaps, recommendations and future directions. *Renew. Sustain. Energy Rev.* **2018**, *96*, 380–391. [CrossRef]
4. Tian, Z. The Policy About Submarine Cable Pipeline Project in China. Master's Thesis, Harbin Engineering University, Harbin, China, 2021. (In Chinese)
5. Wu, A.; Yuan, Z.; Gong, Y. Overview of construction technology for deep-buried laying of submarine cables in China. *Zhejiang Electr. Power* **2015**, *3*, 57–62. (In Chinese) [CrossRef]
6. Zou, D.; Hu, J. Study on local sediment scour and stress state of submarine cables in offshore wind farms. In Proceedings of the IOP Conference Series: Earth and Environmental Science, 2021 8th International Conference on Coastal and Ocean Engineering, Tokyo, Japan, 3–5 April 2021. [CrossRef]
7. The Cables Buried in the Seabed Run Out for "Breathability"! Consultation with All Parties to Ensure Winter Electricity Consumption on Chongming Island. Available online: https://baijiahao.baidu.com/s?id=1654575915573352837&wfr=spider&for=pc (accessed on 1 April 2024).
8. Osthoff, D.; Heins, E.; Grabe, J. Impact on submarine cables due to ship anchor—Soil interaction. *Geotechnik* **2017**, *40*, 265–270. [CrossRef]
9. Vandals Cut San Francisco Area Fiber Optic Lines for 11th Time in a Year. Available online: https://edition.cnn.com/2015/07/01/tech/california-fiber-optic-cable-cuts/index.html (accessed on 10 April 2024).
10. Yan, X.; Ariaratnam, S.T.; Dong, S.; Zeng, C. Horizontal directional drilling: State-of-the-art review of theory and applications. *Tunn. Undergr. Space Technol.* **2018**, *72*, 162–173. [CrossRef]
11. The Distinct Advantages of Horizontal Directional Drilling. Available online: https://www.powrmole.com/blog/the-distinct-advantages-of-horizontal-directional-drilling/ (accessed on 12 April 2024).
12. Syms, E.I.B. *Horizontal Directional Drilling*; Butterworth-Heinemann: Oxford, UK, 2004; pp. 297–326.
13. Sun, P.; Liu, W.; Yang, H.; Wei, B.; Xia, Y. Progress in research and applications of trenchless horizontal directional drilling equipment and technology in China. *Chin. J. Eng.* **2022**, *44*, 122–130. (In Chinese) [CrossRef]
14. Li, D.; Wang, W.; Lin, Y.; Zhang, Z.; Zhang, X. Exploration and Practice of Land to Sea Directional Drilling and Crossing Technology in Marine Engineering. *J. Ocean Technol.* **2020**, *39*, 95–100. (In Chinese)
15. Tang, P.; Chen, C.; Huang, S.; Huang, D.; Yang, X.; Wei, J.; Jin, L. Research of Land-to-sea Horizontal Directional Drilling Crossing Technology. *Pipeline Tech. Equip.* **2022**, *2*, 31–34. (In Chinese)
16. Black, S.; Cable, B.; Fredrickson, B.; Warren, D. Horizontal directional drilling cable shorelanding at san nicolas island, California. In Proceedings of the Oceans' 04 MTS/IEEE Techno-Ocean'04 (IEEE Cat. No. 04CH37600), Kobe, Japan, 9–12 November 2004; pp. 2104–2111. [CrossRef]

17. Bascom, E.C.R.; Rezutko, J. Novel installation of a 138kv pipe-type cable system under water using horizontal directional drilling. In Proceedings of the 2014 IEEE PES T&D Conference and Exposition, Chicago, IL, USA, 14–17 April 2014. [CrossRef]
18. Jiao, Y.; Meng, J.; Pan, L.; Zhang, Q.; Ku, M. Laying Construction Techniques of 35 kV Submarine Cable in Offshore Wind Farm. *Hydropower New Energy* **2022**, *36*, 44–47. (In Chinese) [CrossRef]
19. Yoshizawa, N.; Yabuta, T. Study on submarine cable tension during laying. *IEEE J. Ocean. Eng.* **1983**, *8*, 293–299. [CrossRef]
20. Nagatomi, O.; Nakamura, M.; Wataru, K. Dynamic simulation and field experiment of submarine cable during laying and recovery. In Proceedings of the International Ocean and Polar Engineering Conference, Kitakyushu, Japan, 26–31 May 2002.
21. Prpić-Oršić, J.; Nabergoj, R. Nonlinear dynamics of an elastic cable during laying operations in rough sea. *Appl. Ocean Res.* **2005**, *27*, 255–264. [CrossRef]
22. Zhang, T.; Hu, X. Tension analysis in submarine cable laying. *Ship Boat* **2009**, *20*, 15–20. (In Chinese)
23. Yang, N. Tension Analysis of Submarine Cables During Laying Operations. Master's Thesis, Shanghai Jiao Tong University, Shanghai, China, 2015. (In Chinese)
24. Feng, Y.; Shang, Q.; Lv, A. Dynamic Finite Element Analysis of Fiber-Optic Composite Submarine Cable Under Wave. *Study On. Opt. Commun.* **2017**, *4*, 30–34. (In Chinese) [CrossRef]
25. Zhang, D.; Bai, Y.; Zhao, W.; Zhu, K. Dynamic analysis of submarine cable during the process of laying back to the seabed. *Ships Offshore Struct.* **2019**, *15*, 153–161. [CrossRef]
26. Chang, H.-C.; Chen, B.-F. Mechanical behavior of submarine cable under coupled tension, torsion and compressive loads. *Ocean Eng.* **2019**, *189*, 106272. [CrossRef]
27. Sun, Y.; Shu, C.; Gao, Z. Analysis of Influencing Factors in the Process of Laying Submarine Optical Cable. In Proceedings of the 2020 IEEE 9th Joint International Information Technology and Artificial Intelligence Conference (ITAIC), Chongqing, China, 11–13 December 2020. [CrossRef]
28. Okkerstrøm, L. Sensitivity Analysis of Peak Tension Loads on Subsea Power Cables During Laying Operations. Master's Thesis, Western Norway University of Applied Sciences, Haugesund, Norway, 2021.
29. Kuang, J.; Chen, G.; Yuan, Z.; Qi, X.; Yu, Q.; Liu, Z. Dynamic Interactions of a Cable-Laying Vessel with a Submarine Cable during Its Landing Process. *J. Mar. Sci. Eng.* **2022**, *10*, 774. [CrossRef]
30. Shen, J.; Liang, Y.; Hong, H.; Chen, J. Numerical Investigation of Burial Depth Effects on Tension of Submarine Power Cables. *J. Mar. Sci. Eng.* **2024**, *12*, 1972. [CrossRef]
31. Peres, M.; Wang, S.; Soares, C.G. Sensitivity analysis of loads on subsea power cables during installation. In *Advances in Maritime Technology and Engineering*; CRC Press: Boca Raton, FL, USA, 2024; pp. 477–487.
32. Li, Y.; Jiang, L.; Xie, M.; Yu, J.; Qian, L.; Xu, K.; Chen, M.; Wang, Y. Advancements and Challenges in Power Cable Laying. *Energies* **2024**, *17*, 2905. [CrossRef]
33. Khasanov, R.R.; Pascal, M.A.; Sultanmagomedov, T.; Magomedtagirovich, S.S. Improving the methodology for calculating the thrust force during underwater crossing by directional drilling method. *J. Integr. Sci. Technol.* **2024**, *12*, 823. [CrossRef]
34. Faghih, A.; Goerz, B.; Bayat, A. Stress analysis of steel pipe installation in horizontal directional drilling based on strain monitoring. *Tunn. Undergr. Space Technol.* **2024**, *147*, 105673. [CrossRef]
35. Guo, Z.; Liang, Z. Analysis of the pipeline pull-back load for horizontal directional drilling. *KSCE J. Civ. Eng.* **2018**, *22*, 5133–5142. [CrossRef]
36. Tang, J.; Qin, T.; Zhang, H. Research on influencing factors of cable traction force in horizontal directional drilling. In Proceedings of the IOP Conference Series: Earth and Environmental Science, Sanya, China, 16–18 November 2021. [CrossRef]
37. Vasilescu, V.-F.; Dinu, D. Installation of submarine cables and pipes in shallow waters and connecting them to the shore by the method of horizontal directional drilling. In Proceedings of the 10th International Conference on Thermal Equipments, Renewable Energy and Rural Development, Bucharest, Romania, 26–27 June 2020. [CrossRef]
38. Liang, X.-Q.; Hu, D.; Li, Y.-S.; Yang, X.; Jiang, L.; Zhang, Y.-Y. An Improved Model to Calculate Pullback Force of Trenchless Horizontal Directional Drilling Pipeline. *Adv. Civ. Eng.* **2022**, *2022*, 1–13. [CrossRef]
39. Zheng, H. Mechanical response analysis on pipeline during horizontal directional drilling construction and service period. *I* **2025**, *56*, 198–204. (In Chinese) [CrossRef]
40. *GB/T 13663.2-2018*; Polyethylene (PE) Piping Systems for Water Supple—Part 2 Pipes. General Administration of Quality Supervision, Inspection and Quarantine of the People's Republic of China: Beijing, China, 2018.
41. Orcaflex11.3b User's Manual. Available online: https://www.orcina.com/author/heff/page/2/ (accessed on 22 May 2024).
42. Liu, X. Mechanics Analysis of Pipeline in Back-Dragging Process of Horizontal Directional Drilling. Doctoral Thesis, Southwest Petroleum University, Chengdu, China, 2017. (In Chinese)

water

MDPI

Article

Hydraulic Response of Dam-Break Flood Waves to Converging Channel Geometries: A Numerical Investigation

Amir Ghaderi [1,*], Hooman Shahini [1], Hossein Mohammadnezhad [1], Hossein Hamidifar [2] and Jaan H. Pu [3,*]

1 Department of Civil Engineering, Faculty of Engineering, Urmia University, Urmia 57561-15311, Iran; st_h.shahini@urmia.ac.ir (H.S.); mohammadnezhad.hossein@gmail.com (H.M.)
2 Institute of Geophysics, Polish Academy of Sciences, 01-452 Warsaw, Poland; hhamidifar@igf.edu.pl
3 Faculty of Engineering and Digital Technologies, University of Braford, Bradford BD7 1DP, UK
* Correspondence: a.ghaderi@urmia.ac.ir (A.G.); j.h.pu1@bradford.ac.uk (J.H.P.)

Abstract

The topography of the flood path significantly influences the hydraulic characteristics of flood events, necessitating in-depth analysis to better understand the continuous dynamics during dam failure scenarios. These analyses are useful for the hydraulic evaluation of infrastructures downstream of a dam site. This study examined the effects of four distinct converging configurations of guide-banks on the propagation of unsteady flow in a rectangular channel. The configurations studied included trapezoidal and crescent side contractions, as well as trapezoidal and crescent barriers located at the channel's center, each with varying lengths and widths. Numerical simulations using computational fluid dynamics (CFD) simulation were validated against experimental data from the literature. The results reveal that the flow experienced a depth increase upon encountering converging geometries, leading to the formation of a hydraulic jump and the subsequent upstream progression of the resulting wave. The width of the obstacles and contractions had a marked influence on the flow profile. Increased channel contraction led to a more pronounced initial water elevation rise when the flood flow encountered the topography, resulting in a deeper reflected wave that propagated upstream at less time. The reflected wave increased the water elevations up to 0.64, 0.72, and 0.80 times the initial reservoir level (0.25 m), respectively, for cases with 33%, 50%, and 66% contraction ratios to the channel width (0.3 m). For the same cases at a certain time of t = 5.0 s, the reflected wave reached 1.1 m downstream, 0.5 m downstream, and 0.1 m upstream of the initial dam location. Waves generated by the trapezoidal configuration affected the upstream in less time than those formed by the crescent contraction. The length of the transitions or their placement (middle of/across the channel) did not significantly affect the flow profile upstream; however, within the converging zone, longer configurations resulted in a wider increased water elevation. Overall, the intensity of the hydraulic response can be related to one factor in all cases, namely, the convergence intensity of the flow lines as they entered the contractions.

Keywords: dam-break; transitional flow; flood risk management; CFD

1. Introduction

Floods are among the most catastrophic hydrological events, capable of causing severe damage to infrastructure and significant loss of life [1,2]. While in many cases single structures, such as dams, or hybrid structures that combine various flood control measures, are typically among the best options to manage floods [3,4], their failure can also serve as a significant source of flooding in downstream areas. Among the various types of floods, dam-break floods stand out as particularly devastating due to their rapid onset, immense destructive potential, and widespread impact on both infrastructure and human populations [5]. It is expected that the frequency and intensity of extreme weather events be exacerbated, potentially increasing the likelihood of conditions that could lead to dam failures and subsequent floods [6,7]. Hence, understanding the hydraulic response of dam-break waves to different channel geometries is crucial for improving flood risk assessments and designing effective mitigation strategies.

Several factors influence the characteristics and impact of dam-break floods, including the reservoir volume [8], breach formation dynamics [9], downstream topography [10], and channel boundary conditions [11]. The breach development determines the release rate and peak discharge of the flood wave [12]. Hydrodynamic factors such as flow resistance, bed roughness, and sediment transport further modulate wave attenuation and energy dissipation [13,14]. Additionally, external influences such as vegetation [15], infrastructure [16], and tributary inflows [17] can alter flood propagation patterns. Among these factors, channel morphology plays a particularly critical role in shaping the evolution of dam-break waves. Variations in channel slope [18], width [19], sinuosity [20], and cross-sectional geometry [10] influence wave speed, depth, and flood extent. Constrictions or expansions in the channel can also cause wave reflection, energy concentration, or dispersion, significantly altering the flood's destructive potential [21,22]. Understanding these morphological controls is essential for accurate flood hazard predictions and the development of effective mitigation strategies. To comprehensively assess the implications of such an event and to provide a reliable estimate of the associated flood risk, it is essential to investigate various dam failure scenarios. These scenarios can generally be categorized into two groups. The first category involves case studies and numerical simulations aimed at fully modeling the routing of the dam breach flow and examining their propagation through rivers, canals, and urban infrastructure downstream [23–28]. The second category does not relate to complete routing of a dam-break flood but encompasses studies that investigate the dam-break flow dynamics in a closed domain, typically conducted at a laboratory scale, and identify the influence of various factors on the flood. The present study falls into the second category and primarily examines the interaction between floods and structures during a dam-break event.

The water–structure interaction is particularly significant downstream of a failed dam [23]. Unlike river floods, dam-break floods occur much more abruptly and with higher flow velocities, making their impact on downstream. In the event triggered by a dam failure, the flow typically traverses a heterogeneous topography at downstream, influencing its hydraulic characteristics. Various factors such as the presence of bridge piers in the river path, flow-facing structures, changes in channel width, steps or protrusions in the channel invert, and vegetation can significantly alter the flood's flow profiles. Additionally, infrastructures such as roads, bridges, and municipal facilities serves as barriers to the downstream propagation of flood waves following a dam failure [29]. These variations lead to the formation of secondary waves and increasing water turbulence, thereby heightening the potential for more extensive damage. Kocaman et al. [19] investigated the impact of three trapezoidal transitions with varying geometries onto a dam-breach flow in a rectangular channel. Their research focused on flow profiles during the sudden emptying process of a reservoir due to the abrupt removal of a gate. The study found that the flow depth increased as it passed through the transitions, and this increase propagated upstream as a form of a wave. Additionally, they numerically modeled their experiments and demonstrated that the set of Reynolds-Averaged Navier–Stokes (RANS) equations using the standard k-ε turbulence model successfully reproduce the flow profile of a dam-break unsteady flow in similar geometries. Kocaman et al. [30] experimentally and numerically explored the propagation of flood waves resulting from a dam breach in a region with closed boundaries. In their scenario, waves propagate not only downstream but also upstream, as the flow interacts with the boundaries. Their study specifically focused on the effect of tailwater depth on the resulting wave profile. The reliability of RANS equations with a standard k-ε turbulence model coupled with VOF method in simulating wave propagation is reported in their work. Di Cristo et al. [31] introduced two morphodynamic models to analyze the interaction between a dam-break wave and a rigid obstacle in the presence of an erodible bed. Their study emphasized the importance of developing effective predictive tools for assessing impact forces on structures caused by floods, which are crucial for designing appropriate risk mitigation strategies and protective measures. Khoshkonesh et al. [29] conducted a numerical study on two configurations of individual obstacles placed downstream of a dam-break wave. In their model, the obstacles were arranged in a triangular formation within a rectangular channel, allowing the flow to pass through them. Their findings highlighted the significant influence of obstacle arrangement on key hydraulic parameters, including the flow profile, reservoir discharge rate, three-dimensional velocity fields, and flow regimes around the obstacles. Maghsoodi et al. [32] investigated three different scenarios of obstacle arrangement facing a dam-break wave, namely, a rectangular obstacle, a trapezoidal sill, and triangle sill in a channel path. A set of RANS equations with a standard k-ω turbulence model along with a VOF tracking method were used in their study to analyze flow profiles and pressure field with sufficient accuracy. Lee and Nguyen [33] performed numerical simulations using an LES turbulence model coupled with a VOF tracking method to investigate the hydraulic characteristics of a dam-break flood in a laboratory-scale urban area. The variables of their problem included initial water stage, breach size, and position of dam gate. They analyzed flow depth contours, the velocity hydrograph, streamlines, vorticity, and the q-criterion incorporated with these variables. Focusing on wave–structure interactions, Oodi et al. [34] investigated the transient flow characteristics of dam-break waves interacting with downstream obstacles and contractions at varying distances in a channel site. They developed a high-accuracy numerical code by incorporating an LES turbulence model along with an air entrainment model and the VOF method to analyze flow profiles, turbulence structures, air entrainment around obstacles, flow regime variations, and the flood hydrograph in the downstream channel. Their findings contribute to a deeper understanding of the wave–structure interactions during dam-break events by elucidating how obstacles influence wave dynamics at different distances from the dam. Beteille et al. [35] experimentally and numerically studied the dam-break flood in an idealized city layout with reduced scale. They analyzed the impact of different configurations of cubic obstacles on the flow field characteristics including flow depth history and instantaneous velocity around the obstacles. The VOF-based RANS model in their study was able to accurately reproduce wake zones around the obstacles and the velocity field.

While numerous scenarios of wave–structure interactions during a dam-break include the presence of obstacles in an idealized urban area, the intention of the present work is to study the transient flood flow across a river site. The presence of hydraulic structures in a river site is common, and the failure of them can lead to rapid propagation of flood waves across a river and colliding with infrastructure located in or across the river, therefore making it practical to study. In present study, the interacting infrastructures were specifically defined based on practical river control structures. Among them, the presence of "guide-banks" in a river site were evaluated during a dam-break event, which has not been investigated with a focused manner in previous research. Guide-banks are built for guiding and regulating the river flow for various reasons. They can be used for preventing erosion and ensuring safe passage of water around bridges, weirs, and hydraulic structures. They can be constructed in pairs, upstream and downstream of the supported structure, and can be parallel, convergent, or divergent depending on the specific needs of the river and the structure [36,37]. In present work, four different guide-bank placements along a river path were evaluated facing a dam-break flood in a rectangular channel—this includes lateral and mid-width contractions with either trapezoidal or crescent shape. In addition, for each case, various configurations of length and contraction width for guide-banks were assessed. The evaluated configurations covered a reasonable range of guide-bank designs, representing practical results for engineering applications [37]. This study assessed key hydraulic parameters, including flow profiles and velocity fields of the dam-break wave, influenced by various downstream contractions. The findings contribute to a deeper understanding of flood dynamics in river systems following different configurations of guide-banks. The findings will be useful for flood risk management and also can be incorporated for designing and evaluating river guide-banks. Furthermore, this study aims to develop a high-accuracy numerical model capable of predicting flood characteristics resulting from dam failures in channels interacting with downstream constrictions.

2. Materials and Methods

2.1. Numerical Model Description

A Computational Fluid Dynamics (CFD) model was applied to simulate the three-dimensional flow dynamics of the problem. The reference model of this study was adapted from the experiment work of Kocaman et al. [19]. As illustrated in Figure 1, the setup consists of a rectangular channel with a width of 30 cm and a zero-bed slope. At t = 0 (t is time), a still water region with a height of 25 cm exists upstream of the gate's location. The gate has not been modeled. The simulation begins with the sudden releasing of initial water, initiating an unsteady flow downstream while the water upstream gradually drained over time. Throughout this drainage process, no additional inflow is introduced into the computational domain. Downstream of the gate's initial position (set as the origin coordinates system), two converging transitions with a trapezoidal shape in plan are present, significantly influencing the characteristics of the unsteady flow.

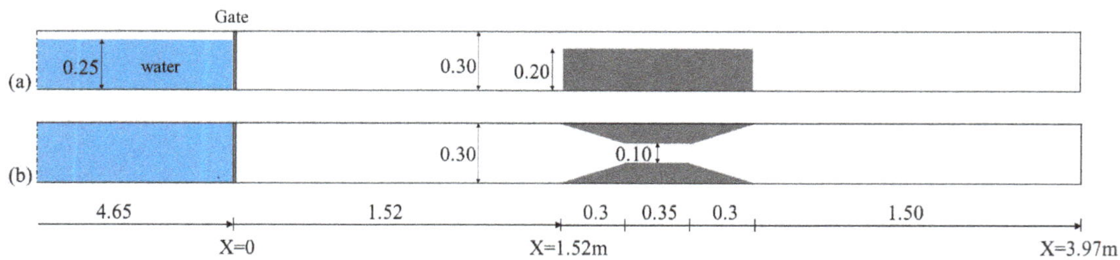

Figure 1. Descriptions of the reference model [19]: (**a**) side view; (**b**) plan view.

In this study, following the simulation of the reference model based on the geometric specifications outlined in Figure 1 and the validation of the results using the experimental data of Kocaman et al. [19] (as detailed in Section 2.4), new geometries in the channel plan were assembled in the numerical model. The tested configurations fell into four categories as shown in Figure 2, namely, parallel lateral contraction with trapezoidal shape (Tr), parallel lateral contraction with crescent shape (Cr), parallel mid-width contraction with trapezoidal shape (Tr-C), and parallel mid-width contraction with crescent shape (Cr-C). In each category, the transition was considered with three contraction ratios along with three various lengths. The channel widths at the contraction (L_C) were selected as 0.1,

0.15, and 0.2 m, which corresponded to contraction ratios of 0.66, 0.5, and 0.33 compared to the channel width (0.3 m). The overall lengths of transitions were selected as 0.6, 0.9, and 1.2 m in which the sloping sides of trapezoidal contractions were constant. These values corresponded to 2W, 3W, and 4W (W = channel width). Consequently, the tested cases provide a conclusion and cover a wide insight about the effect of the placement of guide-banks (lateral versus mid-width), shape of them (crescent versus trapezoidal), contraction ratios, and length of the transitions on the dam-break response in a channel. Overall, 36 distinct guide-bank geometries were evaluated in present study, all of which had a vertical height of 20 cm in numerical simulations. The descriptions of the numerical runs are represented as Table 1.

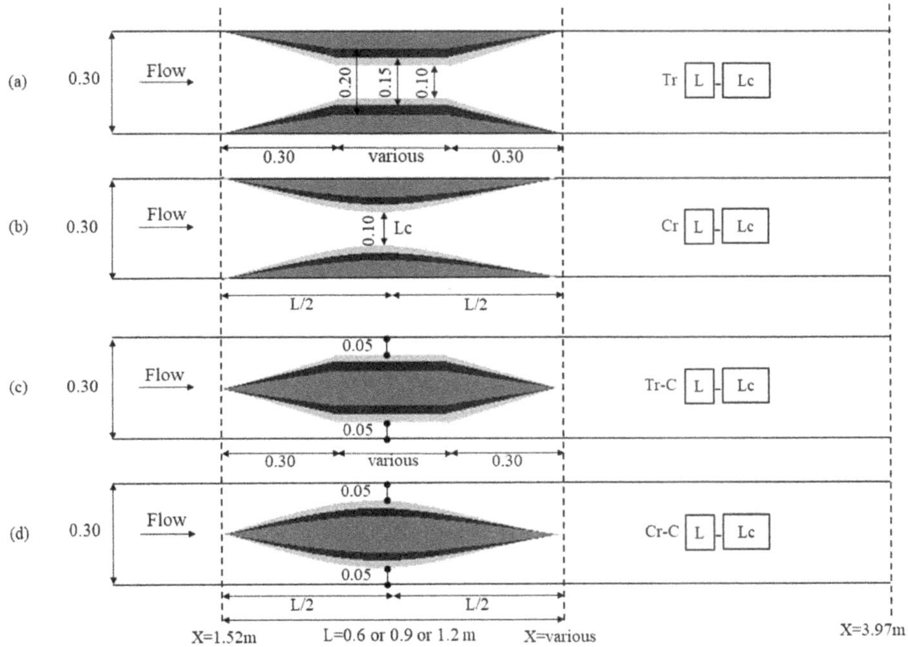

Figure 2. Plan view of the channel contractions used in the numerical simulations of the present study (all dimensions are in meters): (**a**) parallel lateral contraction with trapezoidal shape, (**b**) parallel lateral contraction with crescent shape, (**c**) parallel mid-width contraction with trapezoidal shape, (**d**) parallel mid-width contraction with crescent shape.

Table 1. Definition of numerical cases.

Shape (a) *	Shape (b)	Shape (c)	Shape (d)	L (m)	L_C (m)
Tr 0.6–0.1	Cr 0.6–0.1	Tr-C 0.6–0.1	Cr-C 0.6–0.1	0.6	0.1
Tr 0.6–0.15	Cr 0.6–0.15	Tr-C 0.6–0.15	Cr-C 0.6–0.15	0.6	0.15
Tr 0.6–0.2	Cr 0.6–0.2	Tr-C 0.6–0.2	Cr-C 0.6–0.2	0.6	0.2
Tr 0.9–0.1	Cr 0.9–0.1	Tr-C 0.9–0.1	Cr-C 0.9–0.1	0.9	0.1
Tr 0.9–0.15	Cr 0.9–0.15	Tr-C 0.9–0.15	Cr-C 0.9–0.15	0.9	0.15
Tr 0.9–0.2	Cr 0.9–0.2	Tr-C 0.9–0.2	Cr-C 0.9–0.2	0.9	0.2
Tr 1.2–0.1	Cr 1.2–0.1	Tr-C 1.2–0.1	Cr-C 1.2–0.1	1.2	0.1
Tr 1.2–0.15	Cr 1.2–0.15	Tr-C 1.2–0.15	Cr-C 1.2–0.15	1.2	0.15
Tr 1.2–0.2	Cr 1.2–0.2	Tr-C 1.2–0.2	Cr-C 1.2–0.2	1.2	0.15

Note: * Shapes are described in Figure 2.

2.2. CFD Code

This CFD simulation demonstrates satisfactory performance in solving hydraulic problems involving both steady and unsteady flow in open channels, as validated in previous studies [19,30,38–41]. In this study, Reynolds-Averaged Navier–Stokes (RANS) equations were incorporated to solve the three-dimensional turbulent flow of an incompressible Newtonian fluid. In previous dam-break problems, the RANS approach provides a more accurate results compared to Shallow-Water Equations (SWE), especially in terms of the initial times of dam-break, and moreover provides better insights of the three-dimensional flow field around the obstacles [19,35,40,42,43]. On the other hand, the RANS approach is considered cost-effective in comparison with the Large Eddy Simulation (LES) method in terms of modeling turbulence. The conservation of mass and momentum equations are represented in three-orthogonal cartesian coordinates as Equations (1) and (2), respectively, in which they are discretized by the Finite Volume Method (FVM):

$$\frac{\partial}{\partial x_i}(u_i A_i) = 0 \tag{1}$$

$$\frac{\partial u_i}{\partial t} + \frac{1}{V_F}(u_j A_j \frac{\partial u_i}{\partial x_j}) = -\frac{1}{\rho}\frac{\partial p}{\partial x_i} + g_i + f_i \tag{2}$$

In these equations, x represents the coordinate along the three directions (denoted each time by the subscripts i and j), t is time, V_F is the fractional volume open to flow, p is the pressure, ρ is the fluid density, u_i is the mean velocity, A_i is the fractional area open to flow, g_i is the body acceleration, and f_i is the viscous acceleration in the subscript direction.

Three closure models, namely, the standard k-ε [44], the RNG k-ε [45], and the standard k-ω [46], were adopted to complete solving the RANS equations, and accuracies of them are compared in Section 2.4.2. All of these two-equation models have widespread use in dam-break problems in channels [19,32,35,40–42,47,48]. And of course, these are the available two-equation closure models for RANS equations in FLOW-3D software. The numerical model employs the Volume of Fluid (VOF) method, introduced by Hirt and Nichols [49], to track the free surface of the flow. Additionally, the Fractional Area/Volume Obstacle Representation (FAVOR) algorithm, proposed by Hirt and Sicilian [50], was implemented to approximate solid boundaries within the computational domain.

2.3. Model Setup

The solution domain was meshed in Cartesian coordinates using two distinct mesh blocks. As illustrated in Figure 3, the first block covered the upstream region of the gate, while the second block was located downstream of the gate. The cubic cells in the second block were chosen to be smaller than those in the first block. This is because the interaction between dam-break wave and the transition was included in the second block area, which was set to be investigated. Since the water elevation variation was the main focused parameter, there was no need for setting up finer grids near boundaries. After performing a mesh sensitivity analysis (Section 2.4.1), the optimal cell sizes were determined to be 1.6 cm for the first block and 1.2 cm for the second block. Figure 4 provides a detailed definition of boundary conditions for the present problem. The boundary condition at the entrance of the solution domain was set as symmetry, meaning there was no change across the boundary, and the flow conditions were assumed to be identical on both sides. The boundary condition at the exit of the solution domain was defined as outlet discharge, where the water flow exited the domain. The bottom of the channel and the channel walls were assigned wall-type boundary conditions, which prevented the flow from crossing these boundaries, causing it to return to the solution domain upon contact. The top boundary of the domain was specified as atmospheric pressure, and the interface between the two mesh blocks was defined as symmetry. No physical symmetry boundary was actually imposed there, ensuring proper flow transport across the blocks. Additionally, various probe points were defined along the channel, as shown in Figure 5, to capture the time-history of the flow parameters, including water elevation. These points were located at the center of the channel width.

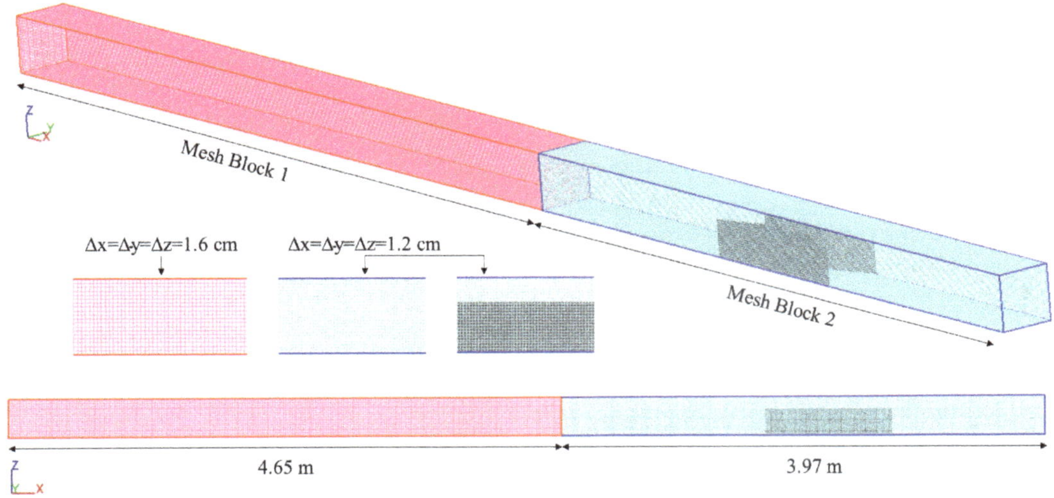

Figure 3. Meshing configuration within the problem domain.

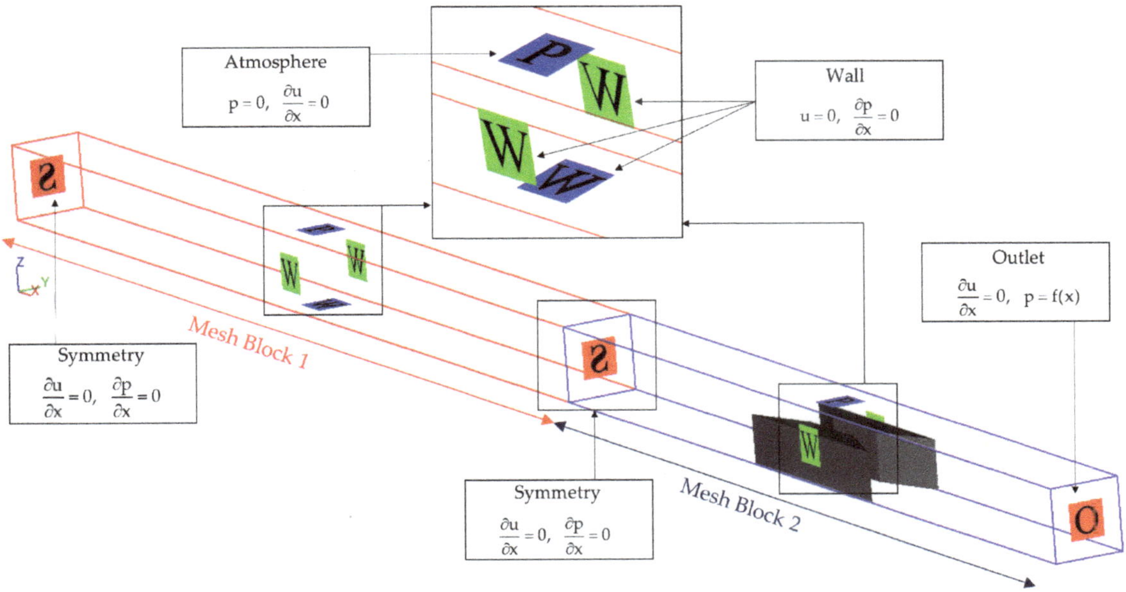

Figure 4. Boundary conditions setup.

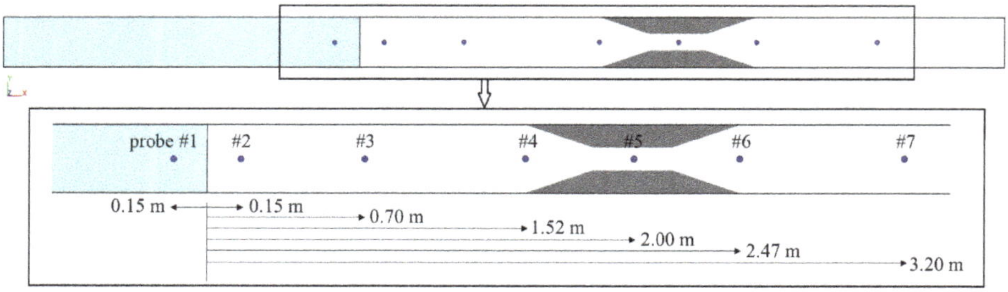

Figure 5. Definitions of probe locations in the numerical model (where # indicates the probe number).

2.4. *Verification of the Numerical Model*

The validation and verification of the numerical model results were performed using water elevation data from the experiment conducted by Kocaman et al. [19]. The reason for choosing this parameter was the quantitative outputs that are needed for the current work. The meshing and accuracy of the numerical model should be sufficient to calculate the water level. In this process, the flow profiles were extracted within the range of 0 to 2.5 m downstream of the gate position at times 2.40 s, 4 s, and 5.50 s. Additionally, time-history water elevations at points p_1, p_3, and p_5 were obtained. The verification process was carried out in three stages, as detailed in Sections 2.4.1–2.4.3.

2.4.1. Mesh Sensitivity Analysis

To analyze the meshing sensitivity, three numerical simulations with different mesh sizes, as detailed in Table 2, were conducted using RNG turbulence model. The mesh sizes had a refinement factor of 0.8 comparing two consecutive cases. The finest grid led to the generation of 848×10^3 cells, which increased the cell counts more than that result in cost-effective and relatively uneconomical numerical calculations. On the other hand, increasing the cell sizes more than Case 1 may result in unreliable outcomes. Therefore, the selected mesh sizes, as shown in Table 2, were considered reasonable and effective for determining the hydraulic parameters of this study. Flow profiles along the central plane of the channel and time-history water elevations were calculated, as shown in Figures 6 and 7, respectively. The results indicate that the water elevation data for the three numerical models showed minimal variation (less than 1%), with the graphs nearly overlapping. This suggests that this parameter was not significantly sensitive to the selected mesh sizes. As seen in Table 2, reducing the mesh size from Case 2 to Case 3 increased the computational volume by 236% based on the number of cells, yet the accuracy of the water depth calculations remained largely unaffected. Conversely, despite the negligible difference in water elevation errors between Cases 1 and 2, the large cell size in Case 1 could cause substantial error in the calculation of other hydraulic parameters, such as the velocity field. Therefore, Case 2 is considered more reliable than Case 1. Given these findings, and in light of the expected results of this study, which includes both qualitative and quantitative water elevation data as well as velocity field observations, Case 2 was selected as the optimal mesh for this problem.

Table 2. Specification of meshing used for the grid convergence analysis.

Case	Cell Counts ($\times 10^3$)	Cell Size of Mesh Block 1 (cm)	Cell Size of Mesh Block 2 (cm)	Refinement Ratio Relative to the Former Case	Increase in Calculations Time
1	206	2.00	1.50	-	-
2	358	1.60	1.20	0.80	174%
3	848	1.20	0.90	0.75	237%

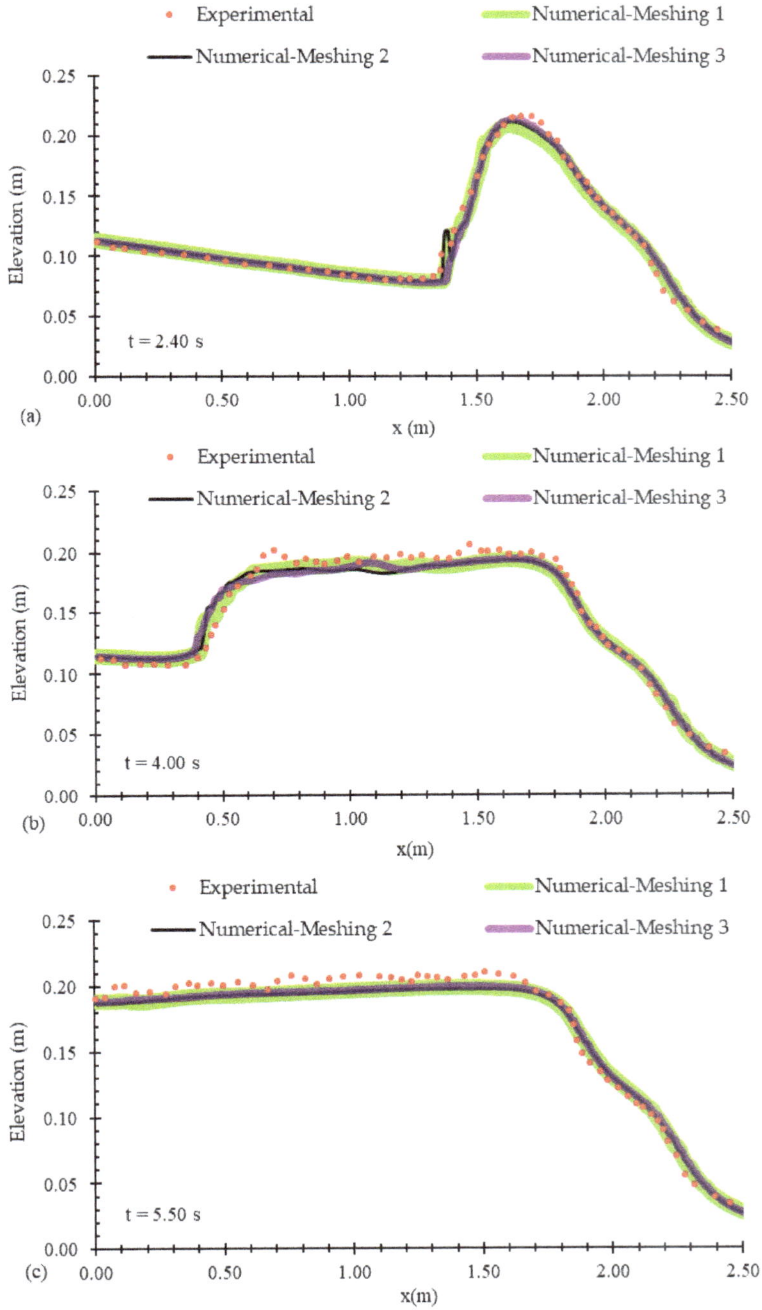

Figure 6. Flow profiles calculated by numerical models with different mesh sizes, compared to the experimental data of Kocaman et al. [19], at times of (**a**) t = 2.40 s, (**b**) t = 4.00 s, and (**c**) t = 5.50 s after the gate removal.

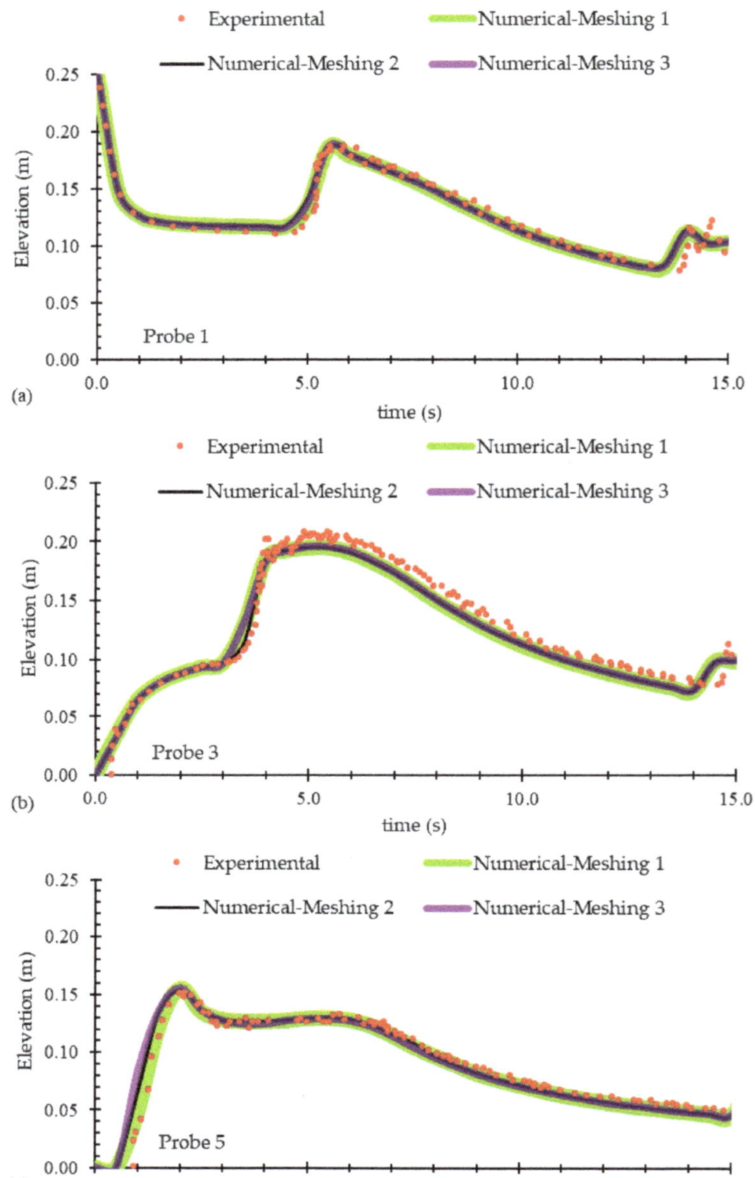

Figure 7. Time-history water elevations calculated by numerical models with different mesh sizes, compared to the experimental data of Kocaman et al. [19], at three locations of (**a**) Probe 1, (**b**) Probe 3, and (**c**) Probe 5, as described in Figure 5.

2.4.2. Accuracy Comparison Between Turbulence Models

After adopting the optimal mesh, the reference model was applied to the three turbulence models: RNG, standard k-ε, and k-ω, to determine the most suitable model. To assess the accuracy of each one, flow profiles were evaluated at three time points: 2.40 s, 4.00 s, and 5.50 s. Figure 8 presents the result of the flow profile analysis at 4 s. The graphs of all three turbulence models show negligible difference, indicating that water elevation parameter was not sensitive to the choice of turbulence model in this problem. Considering the satisfactory performance of the RNG turbulence model in estimating a wide range of hydrodynamic parameters in hydraulic structure problems, specifically in dam-break simulations containing water–structure interactions [23,38,51–57], this model was selected for the numerical simulations in the present study.

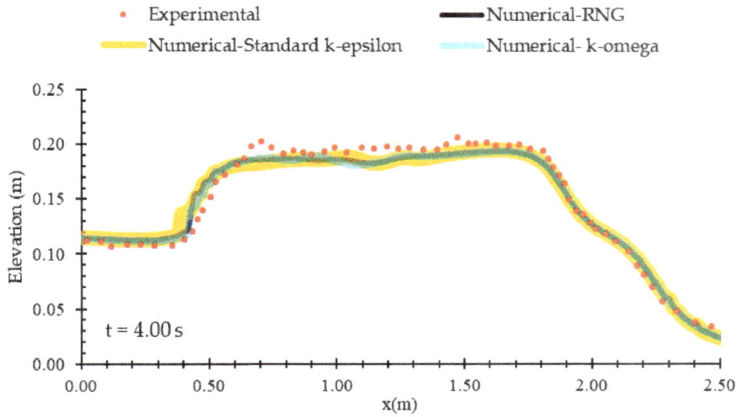

Figure 8. Flow profiles with incorporation of three turbulence models in numerical calculations, compared to the experimental data of Kocaman et al. [19].

2.4.3. Accuracy Assessment of the Optimal Numerical Model

The accuracy of the optimal numerical model in comparison to the laboratory data was determined by calculating the Mean Absolute Percentage Error (MAPE), as defined by Equation (3):

$$MAPE\% = 100 \times \frac{1}{n}\sum_{1}^{n}\left|\frac{X_{exp} - X_{num}}{X_{exp}}\right| \quad (3)$$

In this equation, X_{exp} represents the experimental value, X_{num} represents the numerical value, and n is the number of data points.

To calculate the error in the flow profiles (Figure 6), 15 points along the channel length on the central axis were selected for comparison. Water elevations at those points were compared between the numerical model outputs and the experimental data. This comparison was repeated for the three selected times in Figure 6. Additionally, to determine the percentage of error in the time-history elevations (Figure 7), water elevations at three selected points (p_1, p_2, and p_3) were compared at 15 different times intervals, ranging from 1 s to 15 s after the event began.

3. Results

3.1. General Observations

The resulting average errors for the determination of flow profiles and time-history water elevations are represented in Table 3. Overall, the accuracy of the reference numerical model was calculated as 5.27% for the flow profiles and 5.59% for time-history water elevations, which is considered acceptable. Furthermore, the general observations of the numerical model results in Section 3.1, which indicate the satisfactory performance of the numerical model in describing the flow field, further confirm the accuracy of the outputs.

Table 3. Mean Absolute Percentage Error (MAPE) of the numerical output compared to the experimental data of Kocaman et al. [19].

Flow Profile		Time-History Water Elevations	
t (s)	MAPE (%)	Probe No.	MAPE (%)
2.40	3.80	1	4.07
4.00	6.21	3	5.49
5.50	5.82	5	7.23
Average	5.27	Average	5.59

The flow profiles for the reference numerical run (as described in Figure 1) are plotted in Figure 9. As time progressed from the start of the event (the sudden removal of the gate), the dam-break wave reached the transition location, causing the water elevation increase at the converging point, as seen in Figure 9a. This increase in water elevation propagated upstream over time as a reflected wave, as observed in Figure 9b–d. It can be inferred that a hydraulic jump occurred at the transition point due to the collision between the high-velocity main flow and the slower-moving fluid. The formation of this initial hydraulic jump reduced the velocity and increased the water elevation upstream of the transition. Frequently, as the flow encountered upstream currents with low-velocity zones, hydraulic jumps were repeated. The position of this hydraulic jump shifted over time and moved upstream. The occurrence of a moving hydraulic jump was also evident in the Froude number contours presented in Figure 10. The Froude number in a vertical axis is defined as $Fr = u. (g.h)^{-0.5}$, where u is the flow depth-averaged velocity at x direction, h is the average flow depth at corresponding axis, and g is gravitational acceleration. The flow coming from upstream initially had a Froude number greater than 1.0 and is considered supercritical. In open channels, for an approaching flow having a certain specific energy ($h + v^2/2g$, where h is the flow depth and u is the flow velocity), the maximum discharge is achieved when the flow experiences the critical depth. When a supercritical flow encounters a narrower section, the flow depth across the transition increases, unless it reaches the critical depth. In that case, chocking occurs, and the water is repelled to gain the specific energy required to pass through the constriction at the critical depth. As a result, a hydraulic jump occurs, and the inflow changes from a supercritical state to a subcritical state [58]. This fact justifies the formation of the hydraulic jump during the initial times of collision between the incoming dam-break wave and the contractions shown in Figure 9, which can also be concluded in Figure 10 by observing the Froude number variations. Overall, the observation of the hydraulic response of a dam-break wave crossing the contraction align well with the experimental results of Kocaman et al. [19]. The response factors include time of flow choking and formation of hydraulic jump, water elevation rise at choking time, the reflected wave elevation, and the upstream propagation pattern of the reflected wave, which are obvious in Figure 9.

The changes in the flow velocity at various times are demonstrated in the contours of Figure 11. The propagation state of the main flow as well as the hydraulic jump have reliable conformity with experimental results of Kocaman et al. [19]. At times 1.5, 1.8, and 2.1 s, the flow velocity at the transition site decreased. After 2.4 s, in the regions affected by the hydraulic jump, flow bifurcation became apparent, where the deeper parts of the flow exhibited a positive velocity (downstream direction) while the upper parts showed a negative velocity (upstream direction). Over time, the upper wave propagated upstream, and its speed gradually decreased. However, in the area affected by the reflected wave, the velocity of the main flow downstream decreased compared to the approaching flow due to the hydraulic jump. Upstream of the hydraulic jump, the velocity remained greater than 1.0 m/s in all cases. However, downstream of the hydraulic jump, the main flow velocity reduced to approximately 0.5 m/s. Meanwhile, the propagation velocity of the hydraulic jump (the upper wave velocity) initially reached a maximum of about −1 m/s and averaged around −0.5 m/s over time.

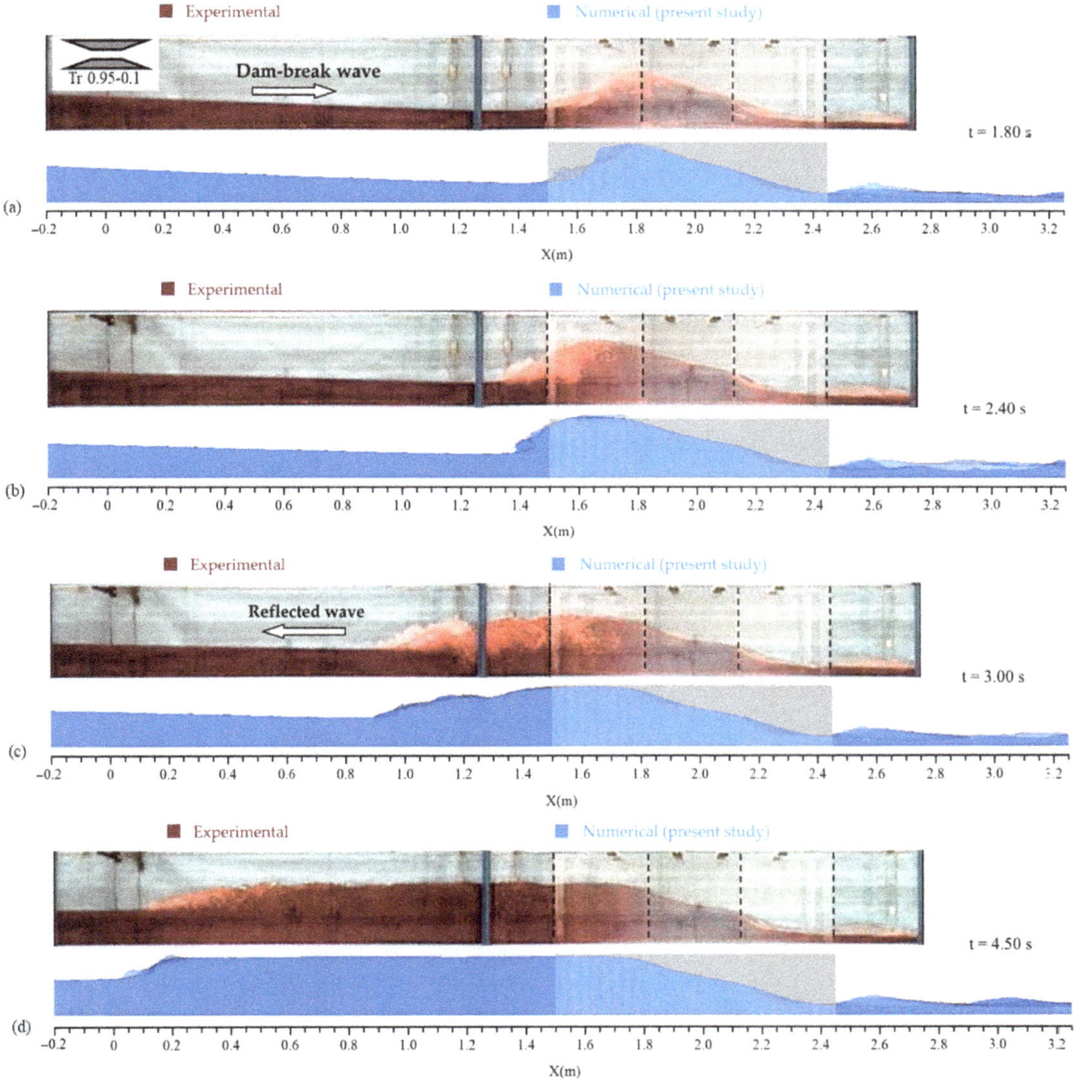

Figure 9. Flow profiles for the reference case (present study) and experimental data [19] at times of (**a**) t = 1.80 s, (**b**) t = 2.40 s, (**c**) t = 3.00 s, and (**d**) t = 4.50 s after the gate removal.

3.2. *Effect of Transition Shape on Flow Hydraulics*

According to Figure 2, four types of contraction shapes were tested along the channel path. To compare the effect of each transformation on the dam-break flow characteristics, the flow profiles were drawn at five different times, as shown in Figure 12. The profiles were extracted for the central plane (y = 0.15 m) and a lateral plane (y = 0.078 m) along the channel. For the cases tested in this section, the length of the transformation and the contraction width of the channel remained constant (L = 0.6 m and L_c = 0.1 m). It is worth noting that the cases of Tr 0.6–0.1 and Tr-C 0.6–0.1 had a triangle shape. As observed in Figure 12, the flow profiles for the central plane of the channel (y = 0.15 m) show that the water elevations in the range of x = 1.52 m to x = 2.12 m for both Tr-C and Cr-C states were zero, which is attributed to the presence of an obstacle in the flow path. Similarly, in the water surface profiles for the lateral plane (y = 0.078 m), the water levels were zero in the region where the flow encountered obstacles. Examining the flow profiles at times t = 1.0 s and t = 1.6 s (Figure 12a,b), it is clear that the profiles for the lateral transitions (Tr and Cr) followed a similar pattern, while the profiles for the obstacles in the center of the channel (Tr-C and Cr-C) followed a different pattern. At t = 1.6 s (Figure 12b), as the flow passed through the central obstacles, the water depth increased significantly along the central axis. However, over time, it became evident from the profiles at t = 2.0, 3.0, and 5.0 s (Figure 12c–e) that the reflected waves generated by the interaction of the flow with the trapezoidal obstacles (Tr and Tr-C) exhibited a consistent upstream propagation pattern. Similarly, the reflected waves associated with the crescent-shaped obstacles (Cr and Cr-C) followed a comparable pattern. In other words, from t = 2.0 s onward, the positioning of the obstacles no longer affected the reflected wave profile; instead, the wave profile was primarily determined by the shape of the obstacle. At t = 5.0 s (Figure 12e), the upstream wave elevation for the crescent-shaped transitions was approximately 1 cm (5.5%) greater than that of the trapezoidal obstacles. Moreover, the crescent-shaped contractions outperformed the trapezoidal ones in upstream wave propagation. Notably, in the four tested conditions shown in Figure 12, the maximum flow elevation at t = 5.0 s ranged between 19 and 20 cm, nearly filling the upstream distance up to the gate location (x = 0). This indicates that the reflected wave can elevate water elevations close to the initial reservoir level (25 cm) within 5 s of water release.

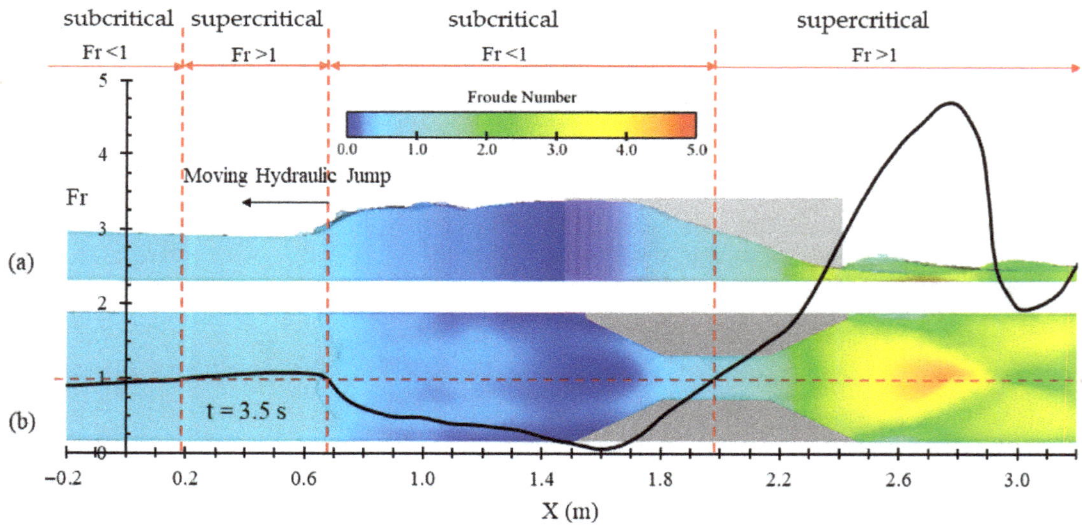

Figure 10. Variation of the Froude number along the channel at the time of 3.5 s after the gate removal for the reference case: (**a**) sideview (across the channel's centerline); (**b**) plan view.

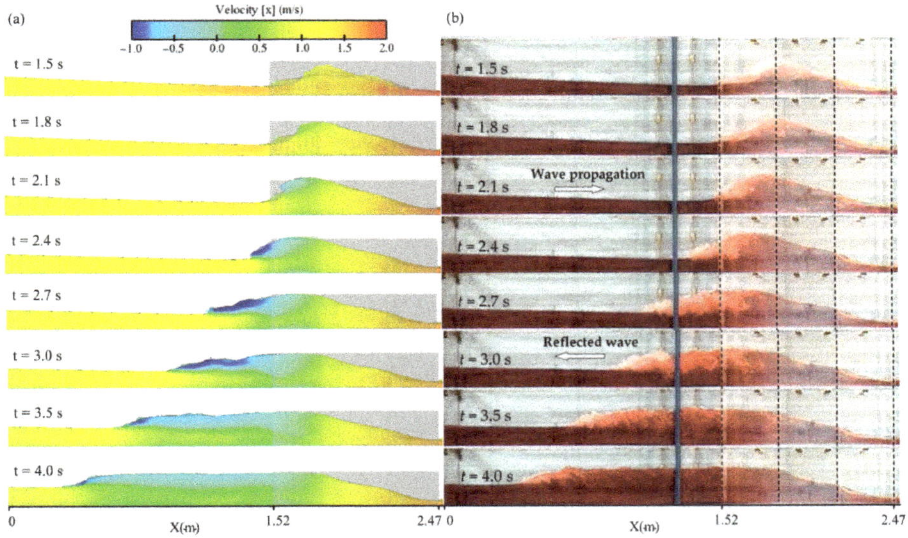

Figure 11. (**a**) Contours of velocity field along the channel for the reference case (numerical results of present study); (**b**) corresponding flow profiles (experimental data [19]).

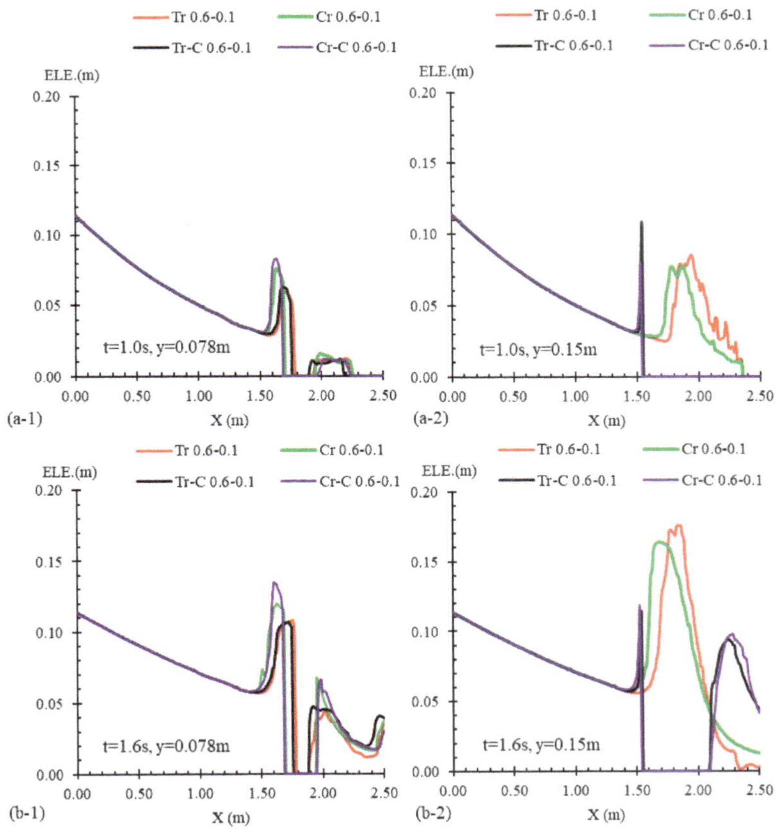

Figure 12. *Cont.*

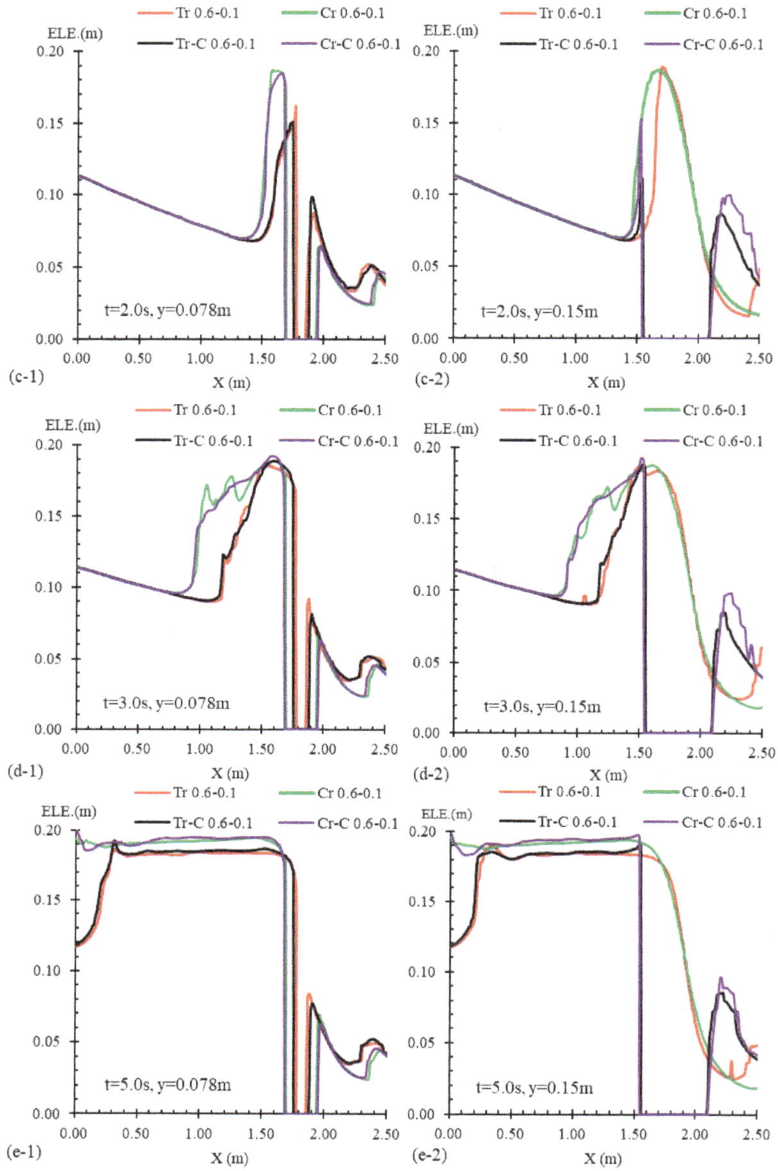

Figure 12. Flow profiles for cases with different transition shapes, including Tr 0.6–0.1, Cr 0.6–0.1, Tr-C 0.6–0.1, and Cr-C 0.6–0.1, alongside the channel's mid-width (y = 0.15 m) and quarter-width (y = 0.078 m) and at times of (**a**) t = 1.0 s, (**b**) t = 1.6 s, (**c**) t = 2.0 s, (**d**) t = 3.0 s, and (**e**) t = 5.0 s after the gate removal.

For a more detailed analysis of the effect of contraction shape on water elevations, their contours along the channel were extracted, as shown in Figure 13, for the four tested transformations with L = 0.9 m and L_c = 0.1 m. These configurations were longer than those in Figure 12, where L = 0.6 m. At first glance, the contours confirmed the three-dimensional nature of the flow, with water elevation varying across the transverse direction of the channel. At initial times (t = 2.0 s), the maximum flow depth was nearly 20 cm. In the cases with trapezoidal constrictions (Figure 13a,c), the water elevation extended over a larger area at the entrance of the contraction compared to the crescent ones (Figure 13b,d). This was due to the reduced disruption of streamlines in the crescent contractions, leading to less water accumulation. From t = 4.0 s onwards, the patterns of the wave for the crescent obstacle (Cr-C 0.9–0.1) and crescent contraction (Cr 0.9–0.1) became visually similar, as did those for the trapezoidal obstacle (Tr-C 0.9–0.1) and trapezoidal contraction (Tr 0.9–0.1), as noted in the water elevations. At t = 5.0 s, the reflected wave related to trapezoidal constrictions (Figure 13a,c) progressed upstream by about 0.2 m farther than the waves in the crescent configurations (Figure 13b,d), and, to some extent, the water elevations were also larger. The sharp edges of the trapezoidal contraction caused a more significant increase in water level during the initial stages of the dam-break flow collapse on the constriction, leading to a more progressive reflected wave. However, for the shorter transitions examined in Figure 12, the trapezoidal transformation with a length of 0.6 m formed a triangle in the flow path, which caused less convergence of flow lines compared to a trapezoidal alternative. As a result, the propagation pattern of the corresponding hydraulic jump differed between a triangle transition and a trapezoidal one.

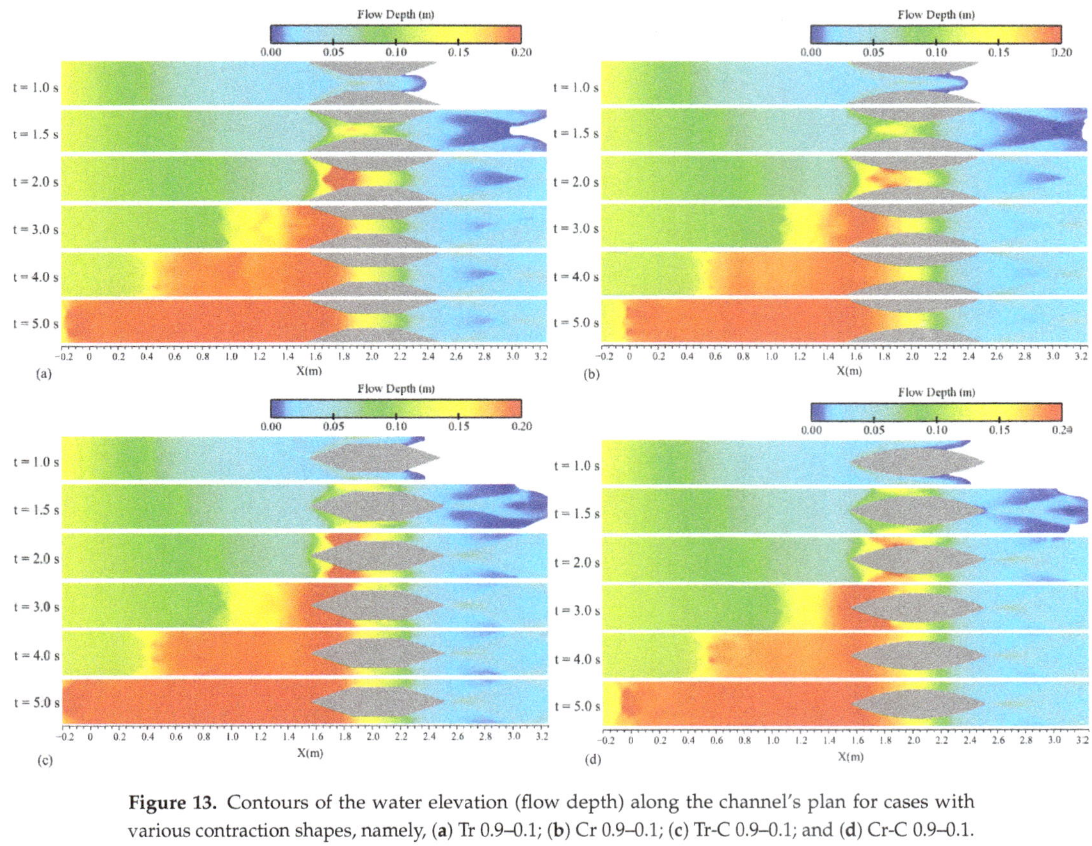

Figure 13. Contours of the water elevation (flow depth) along the channel's plan for cases with various contraction shapes, namely, (**a**) Tr 0.9–0.1; (**b**) Cr 0.9–0.1; (**c**) Tr-C 0.9–0.1; and (**d**) Cr-C 0.9–0.1.

In addition, the propagation of the wave alongside the velocity field in the central plane of the channel are presented in Figure 14 for four tested shapes of contraction with L = 0.95 m and L_c = 0.1 m. The propagation patterns complied with observations of the flow profiles and water elevation contours. In the area affected by the moving hydraulic jump, the flow bifurcation was observed for all tested conditions, where the upper flow layer had negative velocity and the lower part had positive velocity with small values, but there was not a significant pattern in comparing the concentrations of the two velocity zones between the four examined cases. The quantitative velocity interval of the reflected wave and also the main wave in the bottom in all subfigures of Figure 14 were similar for the four examined contractions, and the only difference was the propagation pattern. Faster propagation of the upstream wave in the cases of Tr and Tr-C was related to higher water elevations in t = 2.0 s (initial times as Figure 14a), as mentioned before. It is noticeable that at t = 2.0 s (Figure 14a), the flow exiting the contractions Tr and Cr had relatively larger velocity than the cases of Tr-C and Cr-C, because the water flowed downstream from the middle of the channel in the cases of Tr and Cr, not being obstructed by contractions.

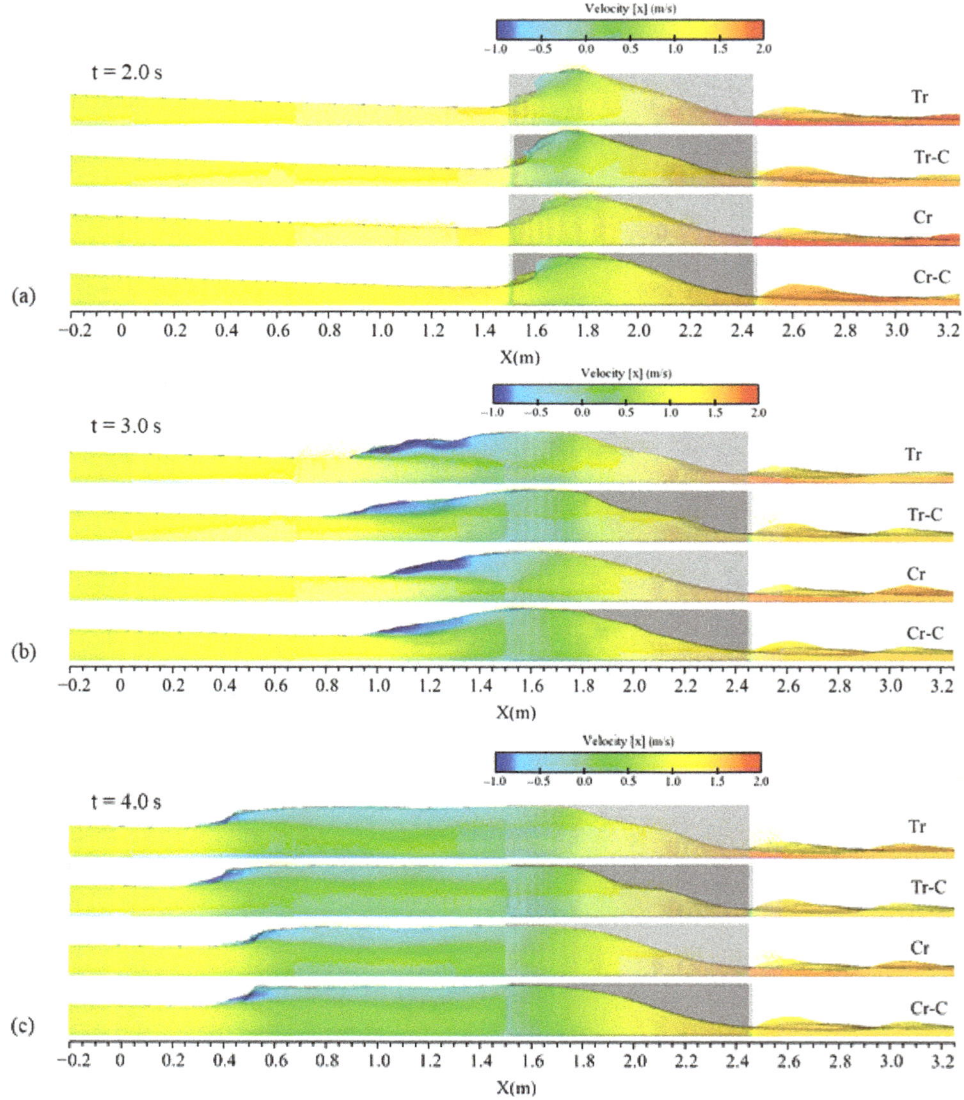

Figure 14. *Cont.*

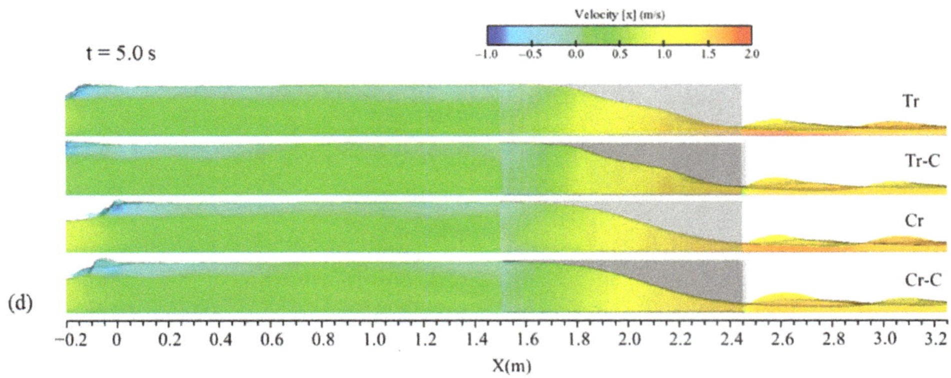

Figure 14. Velocity contours alongside the flow profiles (side view) for cases with various contraction shapes, including Tr 0.95–0.1, Tr-C 0.95–0.1, Cr 0.95–0.1, and Cr-C 0.95–0.1 at times of (**a**) t = 2.0 s, (**b**) t = 3.0 s, (**c**) t = 4.0 s, and (**d**) t = 5.0 s after the gate removal.

3.3. Effect of Channel Contraction Width on Flow Hydraulics

Flow profiles for three cases including trapezoidal transitions of equal length (L = 0.9 m) but varying widths are presented in Figure 15. These profiles were observed along the central axis of the channel (y = 0.15 m) and a lateral axis at a quarter of the channel width (y = 0.078 m). The transitions were located between x = 1.52 m and x = 2.42 m. For the conditions tested, the channel widths at the transition point (L_c) were 0.1 m, 0.15 m, and 0.2 m corresponding to 33%, 50%, and 66% of the total channel width. The flow lines passing through the central axis (y = 0.15 m) did not encounter any of the transitions, while those passing through the lateral axis (y = 0.078 m (interacted with the Tr 0.9–0.1 transition. As a result, in a section of the channel that overlapped with the contraction, the water elevation was zero. From the initial seconds (t = 1.0 s and t = 2.0 s in Figure 15a,b), it was evident that greater contractions resulted in an increase in water elevation. The wave's peak generated by the flow interacting with the transitions varied significantly across the different contraction conditions. At t = 2.0 s (Figure 15a), the wave's peak at the central axis for the Tr 0.9–0.1 case was 20 cm. In contrast, for the Tr 0.9–0.15 and Tr 0.9–0.2 cases, these values decreased to 15 cm and 11 cm, respectively. This indicates that the transition with a larger contraction had a more significant effect on the water elevation. The streamlines passing through the larger contraction underwent a more substantial change in direction, resulting in greater energy dissipation and a higher accumulation of water entering the contraction. As a result, the water elevation increased. At t = 5.0 s (Figure 15c), the reflected waves in all three scenarios propagated upstream in a unidirectional pattern. The upstream water level in the case of Tr 0.9–0.1 reached 20 cm, whereas for the Tr 0.9–0.15 and Tr 0.9–0.2 cases, it decreased to 18 cm and 16 cm, respectively, at the central axis. In addition to the differences in the water elevation, the position of the wave front also varied among the three tested scenarios. At t = 5.0 s (Figure 15c), the wave fronts for the Tr 0.9–0.1, Tr 0.9–0.15, and Tr 0.9–0.2 cases were positioned approximately before the gate, 0.5 m downstream, and 1.1 m downstream of the gate, respectively. This indicates that the reflected wave formed by the dam-break flow encountering a more compact contraction travelled faster upstream, which resulted from larger water elevations at the entrance of the contraction at the initial times (t = 2.0 s). On the downstream side of the channel contraction (x > 2.47 m), the flow profiles followed a distinct pattern, as illustrated in Figure 15. In the initial seconds (t = 1.0 s and t = 2.0 s in Figure 15a,b), the flow profiles showed minimal variation, but over time, at t = 5.0 s, as shown in Figure 15c, the profiles diverged. At this point, the average water elevation downstream of the Tr 0.9–0.1 transition was 4 cm at axis y = 0.078 m, while downstream of Tr 0.9–0.15 and Tr 0.9–0.2 transitions, the average water elevations were 5 cm and 6 cm, respectively. Therefore, in contrast to the upstream reflected wave profile, the water elevation downstream of the transitions decreased as the contraction ratio increased.

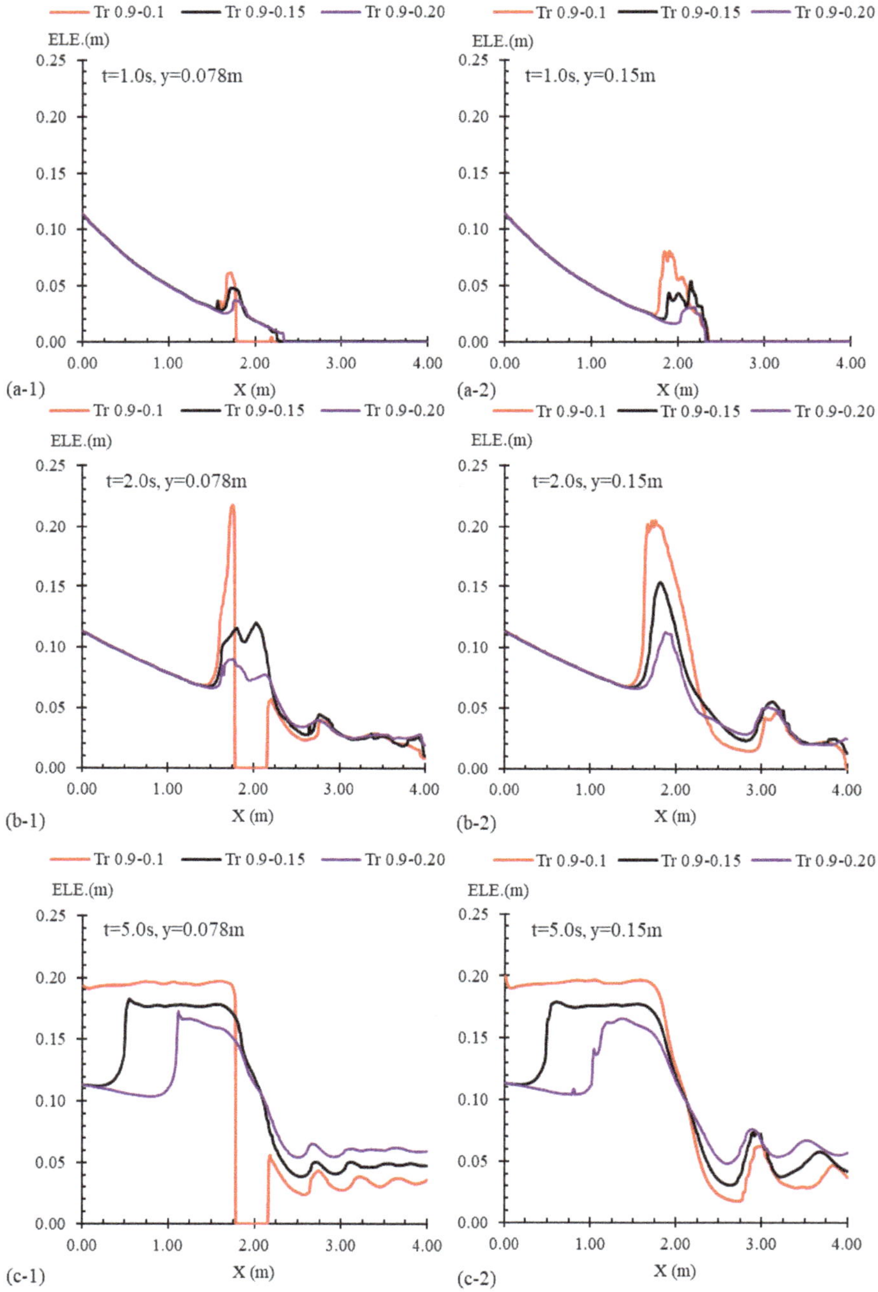

Figure 15. Flow profiles for three cases with various contraction widths including Tr 0.9–0.1, Tr 0.9–0.15, and Tr 0.9–0.2 alongside the channel's mid-width (y = 0.15 m) and quarter-width (y = 0.078 m) and at times of (**a**) t = 1.0 s, (**b**) t = 2.0 s, and (**c**) t = 5.0 s after the gate removal.

Furthermore, an analysis of the water elevation contours for the three cases is presented in Figure 16. It is observed that for all three Tr 0.9–0.1, Tr 0.9–0.15, and Tr 0.9–0.2 cases, in which the channel experienced 66%, 50%, and 33% compaction, hydraulic jump was present. In other words, in all cases, the approaching flow could not pass the transition without accumulating specific energy that led to the formation of hydraulic jump and a reflected progressive wave. For the case of Tr 0.9–0.1 (Figure 16a), the formation of the reflected wave was observed at t = 2.0 s, after the collision between the supercritical approaching flow with the channel contraction. The formation of the reflected wave was observed at larger times for the less compacted transitions. In the case of Tr 0.9–0.15 (Figure 16b) and Tr 0.9–0.2 (Figure 16c), the upstream wave started to propagate at t = 3.0 s and t = 4.0 s, respectively. This was because, in earlier times, the specific energy of the approaching supercritical flow could be large enough for passing the water during the contraction with higher water elevations without changing the flow regime to subcritical. However, due to unsteady water inflow, this equation could be changed, and the hydraulic jump would present at later times. Overall, Figure 16 reveals two key observations that comply with the flow profiles. Firstly, at similar times, the elevation of the hydraulic jump was greater when the flow interacted with a more converging transformation. Secondly, the hydraulic jump extended over a larger area upstream as the contraction increased.

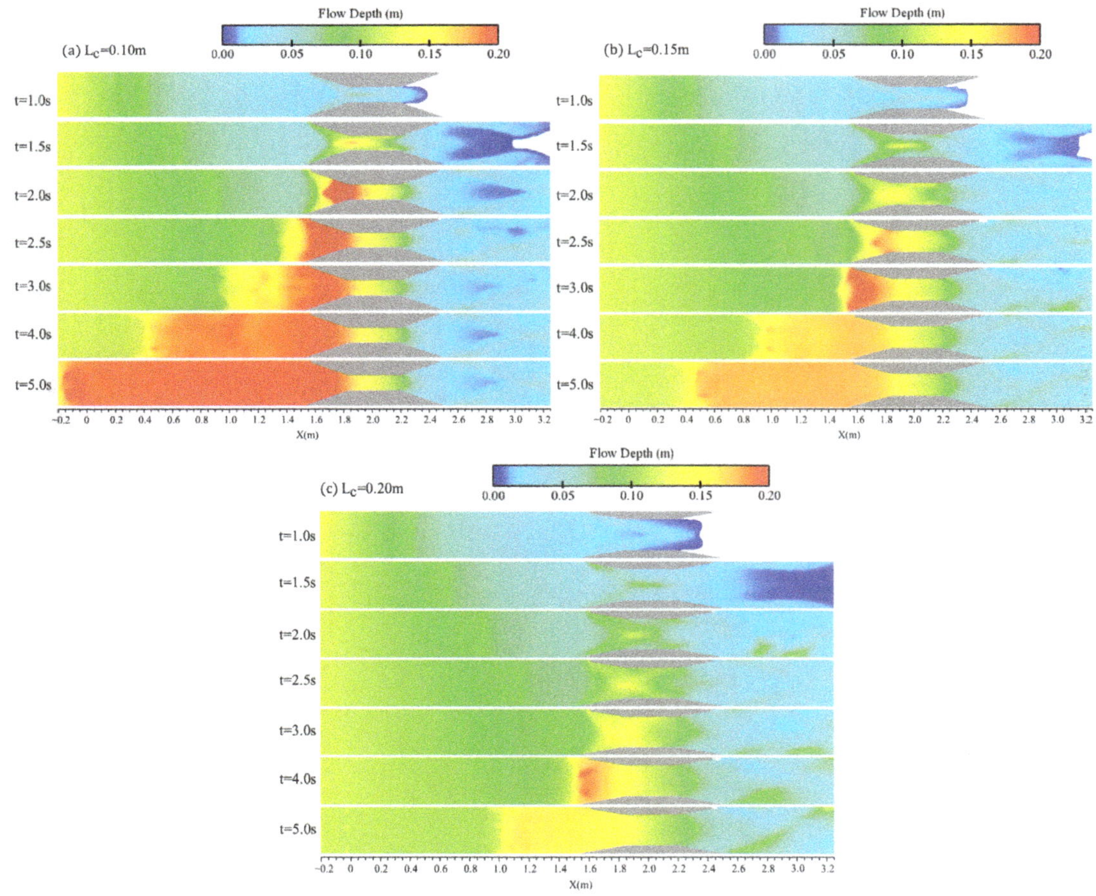

Figure 16. Plan view of the water elevation (flow depth) contours for the three tested transitions with varying contraction widths: (**a**) Tr 0.9–0.1; (**b**) Tr 0.9–0.15; (**c**) Tr 0.9–0.2.

Figure 17 presents a comparison of velocity contours at four different time intervals for the three transition cases: Tr 0.9–0.1, Tr 0.9–0.15, and Tr 0.9–0.2. At t = 2.0 s (Figure 17a), an increase in water elevation was observed at the channel constriction for all three cases. However, the formation of a reflected wave was initially evident only in the Tr 0.95–0.1 case. Over time, flow bifurcation became noticeable upstream of the channel constriction, where velocity of the upstream wave, as shown in Figure 17b, approached approximately −1 m/s in the Tr 0.95–0.1 case. In the case of the transition with greater constriction, the upstream wave velocity was greater. The position of the upstream wave front varied among the three cases. At t = 4.0 s and t = 5.0 s (Figure 17c,d), the positions of the wave front differed by approximately 0.6 m, with the reflected wave in the most constricted transition advancing further upstream. The larger increase in water elevation at initial times (t = 2.0 s), along with the broader low-velocity zones at t = 5.0 s (Figure 17c,d), contributed to the intensification of the reflected wave originating from the more compact transition.

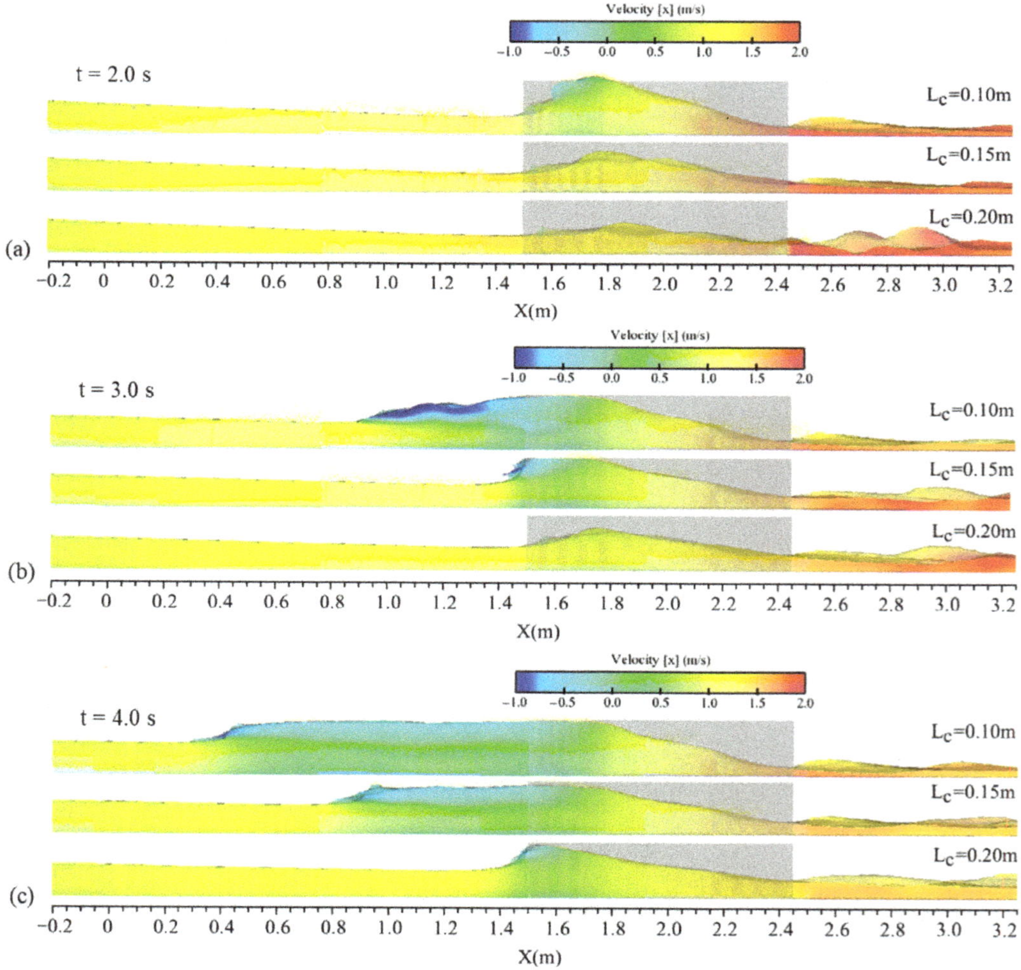

Figure 17. *Cont.*

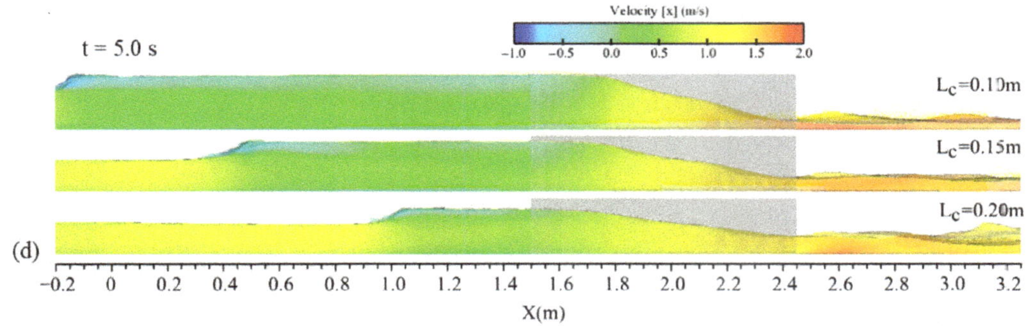

Figure 17. Contours of longitudinal velocity along the channel for cases with different contraction ratios including Tr 0.9–0.1, Tr 0.9–0.15, and Tr 0.9–0.2 at times of (**a**) t = 2.0 s, (**b**) t = 3.0 s, (**c**) t = 4.0 s, and (**d**) t = 5.0 s after the gate removal.

3.4. Effect of the Transition Length on Flow Hydraulics

Flow profiles for three crescent-shaped transitions of varying lengths are depicted in Figure 18. In all cases, the channel width at the contraction section (L_c) remained constant at 0.15 m, while the transition lengths (L) were 0.6 m, 0.9 m, and 1.2 m. The flow lines passing through these transitions, both along the center axis of channel width (y = 0.15 m) and quarter-channel width (y = 0.078 m), did not directly interact with the transitions but rather passed in close proximity. Since all three transitions originated from the same position within the channel, the reflected waves within the first 2 s following gate failure (Figure 18a,b) had small differences in elevations, which was quite a bit higher for shorter transitions in the central axis. However, variations in the water elevation along the transition and downstream sections became more pronounced over time. By t = 5.0 s, as shown in Figure 18c, the reflected wave associated with Cr 0.6–0.15 advanced further upstream compared to Cr 0.9–0.15, showing a notable difference. Similarly, a comparable trend was observed for Cr 0.9–0.15 relative to Cr 1.2–0.15. The propagation of the hydraulic jump was linked to larger water elevations at the contraction site, which formed at the initial times. In the case of shorter contractions, the water elevations were slightly larger, contributing to a more pronounced reflection of the wave. Nonetheless, the differences in the reflected wave's front remained subtle, and the water elevations remained nearly identical across all three cases. It is worth mentioning that at the contraction site, a notable increase in the water level was observed for the longer transition, which also influenced a greater longitudinal extent of the channel.

Contours of the water elevation for the same examined cases with varying contraction lengths are compared in Figure 19. At initial times, up to t = 2.0 s, the water elevation experienced a larger increase when the dam-break wave encountered a shorter transition (Figure 19a). At t = 2.0 s, the region of increased water level at the entrance of the Cr 0.6–0.15 case was wider than the Cr 0.9–0.15 case, and similarly, this area was wider for Cr 0.9–0.15 compared to Cr 1.2–0.15. The longer transitions featured softer edges that directed the flow lines downstream with a gentler slope, leading to less energy dissipation and a smaller velocity reduction at the contraction entrance, as well as a more modest increase in water elevation. In all three cases, the formation of the reflected wave caused by an initial hydraulic jump was evident, as shown in Figure 19. However, the start time of the upstream propagation differed by around 0.5 s between the cases. In the case of Cr 0.6–0.15 (Figure 19a), flow accumulation was observed at t = 2.0 s that led to an upstream wave observation at t = 2.5 s. Meanwhile, in the other two cases (Figure 19b,c), the flow chocking started at t = 2.5 s. This difference contributed to the lower convergence of the flow lines passing through the cases Cr 0.9–0.15 and Cr 1.2–0.15 than the case Cr 0.6–0.15. From t = 3.0 s onwards, differences in the location of the reflected wave became more apparent, with the position at t = 5.0 s differing by approximately 0.1 m between the three cases. These observations align with the flow profiles in Figure 18.

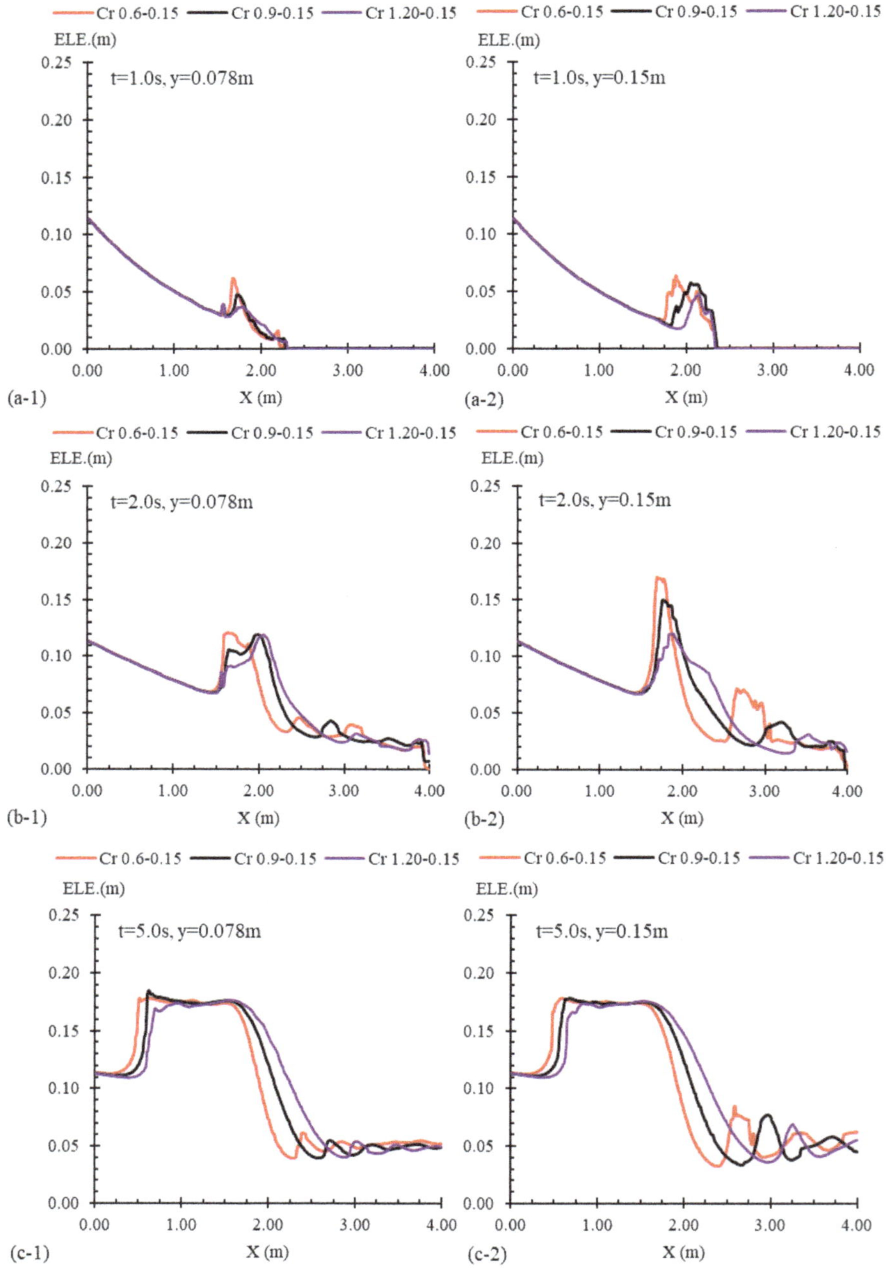

Figure 18. Flow profiles for three cases with varied transition lengths, namely, Cr 0.6–0.15, Cr 0.9–0.15, and Cr 1.2–0.15, alongside the channel's mid-width (y = 0.15 m) and quarter-width (y = 0.078 m) and at times of (**a**) t = 1.0 s, (**b**) t = 2.0 s, and (**c**) t = 5.0 s after the gate removal.

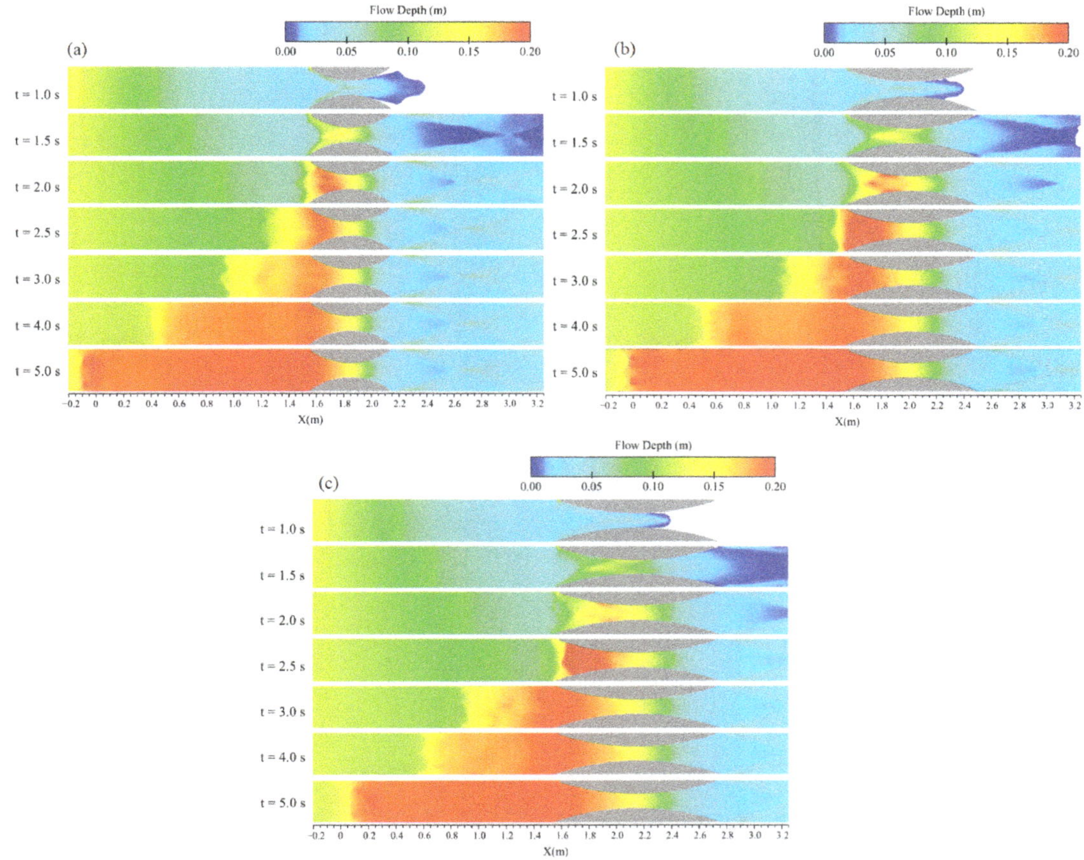

Figure 19. Plan view of water elevation (flow depth) contours for three tested transitions with varying lengths: (**a**) Cr 0.6–0.15; (**b**) Cr 0.9–0.15; (**c**) Cr 1.2–0.15.

In addition, the propagation of the wave alongside the velocity field in the central plane of the channel is presented in Figure 20 for the same contractions with different lengths. The observed propagation patterns align with the results from the flow profiles and water elevation contours. In the region affected by the moving hydraulic jump, flow bifurcation was observed in all tested conditions, where the upper flow layer exhibited negative velocity and the lower part showed positive velocity with small magnitudes. However, there was no significant pattern when comparing the concentrations of the two velocity zones across the three examined cases. The quantitative velocity interval of the reflected wave and the main wave in the bottom in all subfigures of Figure 20 were almost similar for the three examined contractions, and the only difference was the propagation pattern. The faster propagation of the upstream wave in cases with shorter contractions was related to higher water elevations at t = 2.0 s as Figure 20a (initial times). This indicates that the shorter transitions resulted in a more rapid advance of the reflected wave due to the larger increase in water elevations in the earlier stages, which is justified in Figure 19.

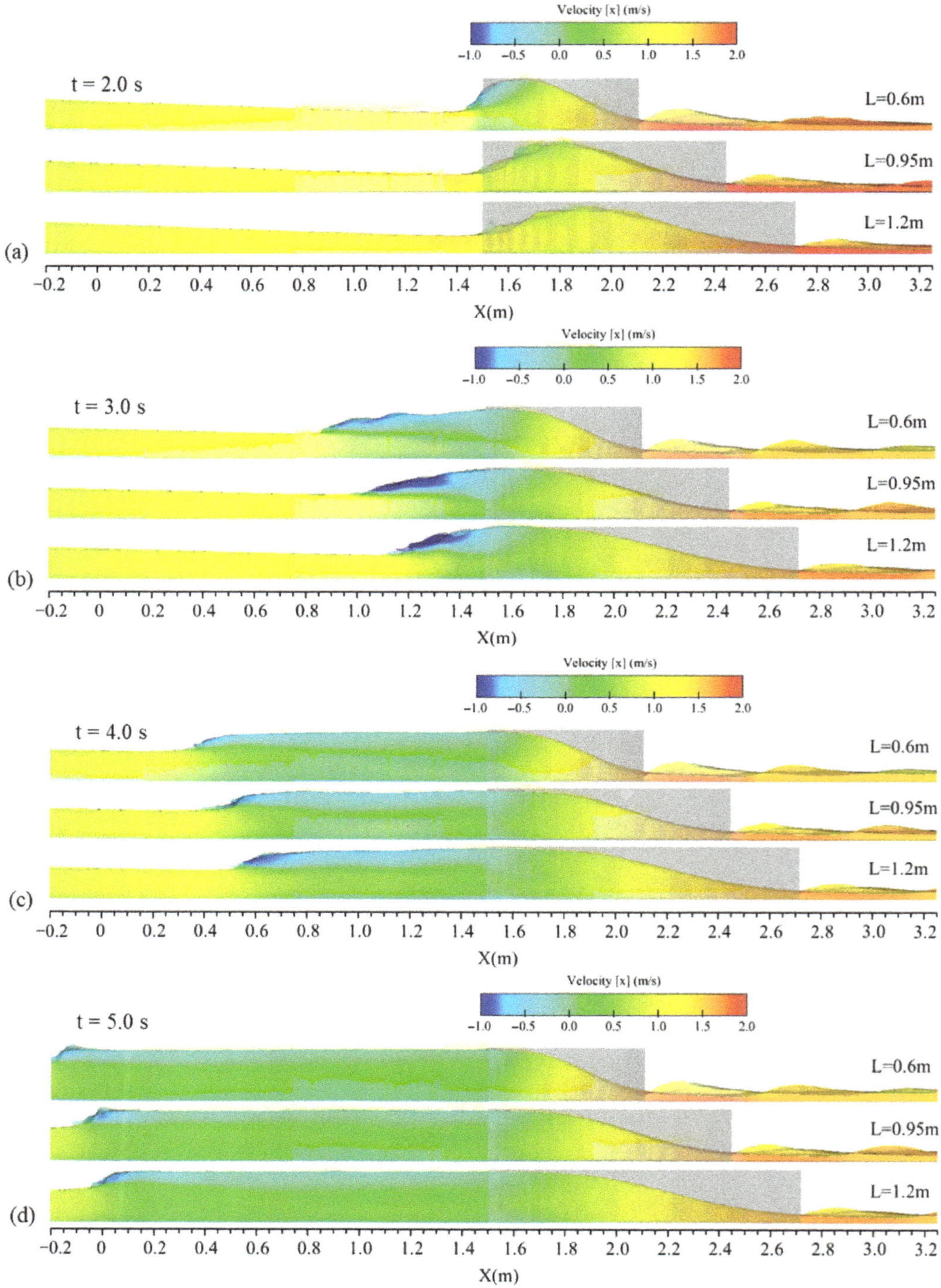

Figure 20. Contours of longitudinal velocity distribution along the channel for cases with different contraction lengths including Cr 0.6–0.15, Cr 0.9–0.15, and Cr 1.2–0.15 at times of (**a**) t = 2.0 s, (**b**) t = 3.0 s, (**c**) t = 4.0 s, and (**d**) t = 5.0 s after the gate removal.

4. Discussion

The findings of this study provide important insights into the complex hydraulic behavior of unsteady flows triggered by sudden dam-break events in the presence of downstream geometric constrictions in a channel site. Water–structure interactions during a dam-break in a channel experiencing contraction was also investigated in a couple of previous studies [19,27,30,46], but in the present work, a reasonable geometric range of guide banks along an open channel was chosen for study, which adds a practical aspect to the present research. The results underscore the critical role of channel geometry, specifically the width and shape of contractions in shaping wave reflection patterns, hydraulic jump development, and upstream wave propagation. The considered constrictions included four arrangements of trapezoidal/crescent transitions at the middle of/across the channel. The tested contractions had widths of 66%, 50%, and 33% of the channel width and had lengths of twice, three times, and four times the channel width. The dam-break wave, caused by a sudden emptying of the reservoir, was set to collide with the transitions at a distance of six times the initial reservoir level ($6h_0$). The results show that in all investigated channel contractions, the supercritical unsteady inflow could not cross the transitions without flow chocking and accumulation of water in the transition section. In all cases, formation of an initial hydraulic jump was observed that followed a reflected wave that propagated upstream by time; however, hydraulic jump (caused by flow choking) occurred at different times for different cases of contraction. Moreover, the water rise caused by the flow chocking at the transition section was different between cases. These two facts (time of flow choking and water elevation rise due to flow choking) were decisive in the progress of the hydraulic jump upstream.

Among the examined parameters, contraction width emerged as the most influential in generating hydraulic response. Higher contraction ratios led to stronger wave reflections and more significant upstream effects. The observation of flow profiles and water elevation contours showed that for three contractions occupying 66%, 50%, and 33% of the channel width (namely, Tr 0.9–0.1, Tr 0.9–0.15, and Tr 0.9–0.2, as described in Figure 2), dam-break flow crossing the transition choked at about t = 2.0 s, t = 3.0 s, and t = 4.0 s after the gate (dam) removal, respectively; meanwhile, the water elevations rose up to 20 cm, 15 cm, and 11 cm at the contraction site, respectively. By the formation of hydraulic jump and the reflected wave, the water elevation upstream of the contractions reached 20 cm, 18 cm, and 16 cm, respectively, for these cases at t = 5.0 s. These values for water elevations were extremely noticeable compared with the initial reservoir level (h_0 = 25 cm). Additionally, observations on velocity contours in all the investigated cases indicated that in the area affected by hydraulic jump, flow bifurcation was evident where the upper flow layer had negative velocity (up to -1 m/s) and the bottom layer had positive velocity that was reduced due to being covered by hydraulic jump. Between the investigated cases, there was no significant difference between the velocity interval of the main flow or the reflected wave's velocity. The main difference was the location of the upstream wave's front at various times. At t = 5.0 s, the reflected wave fronts were positioned at x = -0.1 m, x = 0.5 m, and x = 1.1 m for Tr 0.9–0.1, Tr 0.9–0.15, and Tr 0.9–0.2, respectively (where x was measured from the initial gate location downstream). This proves that at a certain time, hydraulic response of the interaction between the dam-break wave and the higher-ratio contraction affected more of the area upstream. These quantitative comparisons are represented in Figure 21, providing a better conclusion about the hydraulic response of a dam-break wave interacting with guide-banks with various compaction ratios. In the diagrams of Figure 21, the water elevation (h) as well as downstream distance from the gate (x) parameters became dimensionless by being divided into the initial reservoir level (h_0 = 25 cm); also, the time (t) parameter was multiplied by $0.01(g/h_0)^{0.5}$ to become dimensionless ($T = 0.01(g/h_0)^{0.5} \times t$, where g is gravitational acceleration; for instance, T = 0.31 indicates the actual time of t = 5.0 s).

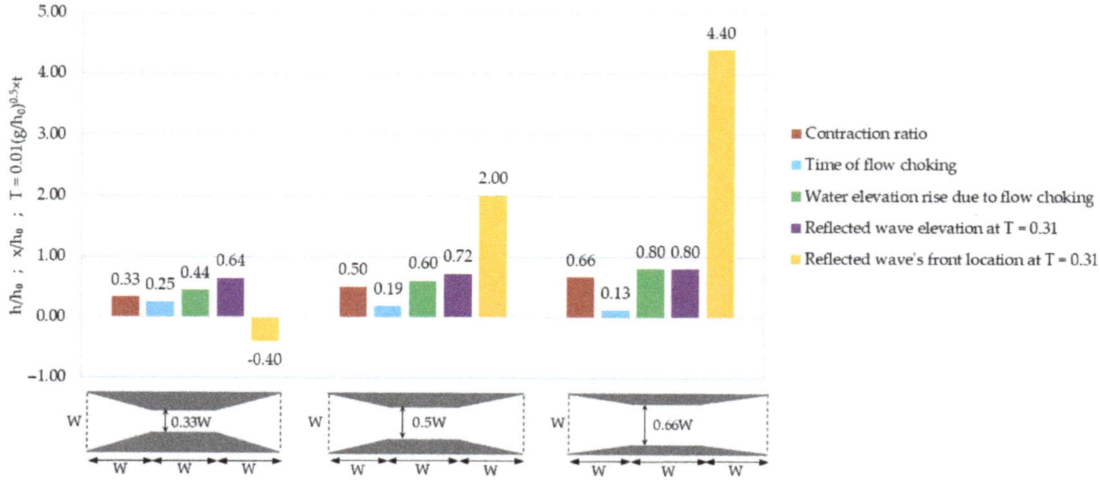

Figure 21. Dam-break wave response interacting with the investigated guide-banks with various contraction ratios; transitions were located $6h_0$ downstream of initial gate location; $W/h_0 = 1.2$.

The type of contraction shape (trapezoidal versus crescent) influenced the flow behavior differently. Trapezoidal geometries generated a more intense convergence of streamlines at the entrance, resulting in a slightly more progressive hydraulic jump, whereas crescent-shaped contractions led to smoother flow transitions. Comparing two cases, Tr 0.9–0.1 and Cr 0.9–0.1, the water elevation rise differed by 1 cm (0.04 times the h_0) when it came to flow choking. Also, at t = 5.0 s, the reflected wave front in cases Tr 0.9–0.1 and Cr 0.9–0.1 were located at x = −0.2 m and x = 0, respectively; this fact further indicates the flow-line convergence effect on the upstream wave propagation. Following that, the contraction length did not directly affect the flow profiles, but the comparison between three crescent transitions with various lengths (namely, Cr 0.6–0.15, Cr 0.9–0.15, and Cr 1.2–0.15) revealed that the shorter transition led to more convergence of the approaching flow lines and therefore created a little more progressive hydraulic response. Moreover, the placement of the contraction (in the middle of/cross the channel) did not significantly cause different hydraulic responses. The comparison of water elevation variations between the Tr 0.9–0.1 and Tr-C 0.9–0.1 cases and also between Cr 0.9–0.1 and Cr-C 0.9–0.1 showed no significant difference in the maximum water rise by flow choking, the time of flow choking, and the propagation pattern of the hydraulic jump. In fact, the convergence of flow lines did not differ neither for Tr 0.9–0.1 and Tr-C 0.9–0.1 cases, nor between Cr 0.9–0.1 and Cr-C 0.9–0.1 cases.

This suggests that in the broader context of flood risk management and hydraulic infrastructure design, careful consideration of channel narrowing is essential to prevent undesirable hydraulic responses and potential damage upstream. These findings have practical implications for the design of energy dissipators, flow control structures, and flood conveyance systems where understanding the nature of reflected waves can inform safer and more efficient designs.

In a broader hydraulic and environmental engineering context, this research highlights the value of using calibrated numerical models to investigate transient flow phenomena with high spatial and temporal resolution. While experimental setups offer controlled observations, numerical tools allow for more extensive parametric explorations under variable and extreme conditions, which are often encountered in real dam-break scenarios.

Future research may expand on these findings by addressing several limitations identified in the current study. For instance, while sediment transport and erosion processes were not incorporated into the simulations, their inclusion is essential for capturing the long-term evolution of channel morphology and the feedback mechanisms that occur during real dam-break events. Additionally, this study only explored a limited range of geometric variations in channel constrictions, namely, specific trapezoidal and crescent- shaped configurations. Investigating a broader spectrum of structural forms, along with varying discharge conditions, would provide a more comprehensive understanding of dam-break hydraulics. Moreover, future work should examine the influence of channel roughness and wall material properties on wave reflection behavior, integrate three-dimensional flow simulations to capture shear stresses and turbulent structures more accurately, and assess the role of suspended materials typically mobilized during floods. Incorporating remote sensing or satellite-based data would also enable model validation across larger spatial domains with natural topographic complexity.

5. Conclusions

In the present study, the hydraulics of unsteady flow resulting from a sudden dam-break in a channel were investigated, with a focus on the interaction between the flow and various guide-bank transitions downstream. A numerical model was applied to estimate water elevations with acceptable accuracy based on experimental data from the literature. Incorporation of Reynolds-Averaged Navier–Stokes (RANS) equations for solving three-dimensional unsteady flow along with the Volume of Fluid (VOF) method for tracking the free surface reproduced the flow profiles and time-history water elevations with Mean Absolute Percentage Errors (MAPEs) of 5.27% and 5.59%, respectively. The model was used to determine flow profiles, analyze water elevation contours, and observe the flow velocity field for numerical cases containing modified guide-banks. The modifications included the shape of contraction (trapezoidal versus crescent), placement of guide-banks (laterally sides of channel versus in the middle of channel), widths of guide-banks, and lengths of guide-banks. The results show that for all cases, the incoming dam-break wave, which faced a contraction at a distance of $6h_0$ from the initial dam location (h_0 = 0.25 m was the initial reservoir level), experienced flow choking with a hydraulic jump following a reflected wave upstream. However, the quantitative aspect of hydraulic response differed for the investigated cases. The determining factor was identified to be the convergence degree of flow lines across the transitions. Therefore, the width of the guide-bank was the most influential item. In the interaction between the dam-break wave with a more compact guide-bank, the time of flow choking was lesser; water elevation rise due to choking was higher; reflected wave elevation was larger; and at a certain time, it propagated more area upstream. Additionally, trapezoidal guide-banks led to more convergence of flow lines compared with crescent ones, therefore generating a noticeable larger response. The guide-bank placement and the length of it did not affect the hydraulic response as much as the convergence of flow lines remaining constant. Finally, quantitative hydraulic response parameters were extracted in a dimensionless form and presented in Figure 21 for three guide-banks with different width ratios. The results of this study have practical implications for predicting the dynamics of dam-break floods around downstream infrastructures, contributing to risk management and hazard analysis efforts. The findings can help improve our understanding of how various contraction geometries affect the propagation and intensity of reflected waves, ultimately aiding in better flood prediction and infrastructure design.

Author Contributions: Conceptualization, A.G. and H.S.; methodology, A.G., H.M. and H.S.; software, H.S., H.H. and J.H.P.; validation, A.G., H.S., H.M. and J.H.P.; investigation, A.G., H.M., H.M., H.H. and J.H.P.; resources, A.G.; data curation, A.G. and H.S.; writing—original draft preparation, A.G., H.M., H.M., H.H. and J.H.P.; writing—review and editing, A.G., H.M., H.M., H.H. and J.H.P.; supervision, A.G.; project administration, A.G.; funding acquisition, A.G. All authors have read and agreed to the published version of the manuscript.

Funding: This research received no external funding.

Data Availability Statement: The data are available to the corresponding authors upon reasonable request.

Conflicts of Interest: The authors declare no conflicts of interest.

Abbreviations

The following abbreviations are used in this manuscript:

CFD	Computational fluid dynamics
RANS	Reynolds-averaged Navier–Stokes
SWE	Shallow-water equations
LES	Large eddy simulation
FVM	Finite volume method
VOF	Volume of fluid
FAVOR	Fractional Area/Volume Obstacle Representation
MAPE	Mean absolute percentage error
ELE	Elevation of water
Tr	Parallel lateral contraction with trapezoidal shape
Cr	Parallel lateral contraction with crescent shape
Tr-C	Parallel mid-width contraction with trapezoidal shape
Cr-C	Parallel mid-width contraction with crescent shape

References

1. Hamidifar, H.; Nones, M. Spatiotemporal variations of riverine flood fatalities: 70 years global to regional perspective. *River* **2023**, *2*, 222–238. [CrossRef]
2. Merz, B.; Blöschl, G.; Vorogushyn, S.; Dottori, F.; Aerts, J.C.; Bates, P.; Bertola, M.; Kemter, M.; Kreibich, H.; Lall, U.; et al. Causes, impacts and patterns of disastrous river floods. *Nat. Rev. Earth Environ.* **2021**, *2*, 592–609. [CrossRef]
3. Hamidifar, H.; Yaghoubi, F.; Rowinski, P.M. Using multi-criteria decision-making methods in prioritizing structural flood control solutions: A case study from Iran. *J. Flood Risk Manag.* **2024**, *17*, e12991. [CrossRef]
4. Murtaza, N.; Pasha, G.A.; Hamidifar, H.; Ghani, U.; Ahmed, A. Enhancing flood resilience: Comparative analysis of single and hybrid defense systems for vulnerable buildings. *Int. J. Disaster Risk Reduct.* **2025**, *116*, 105078. [CrossRef]
5. Wang, Y.; Fu, Z.; Cheng, Z.; Xiang, Y.; Chen, J.; Zhang, P.; Yang, X. Uncertainty analysis of dam-break flood risk consequences under the influence of non-structural measures. *Int. J. Disaster Risk Reduct.* **2024**, *102*, 104265. [CrossRef]
6. Anjaneyulu, R.; Swain, R.; Behera, M.D. Future projections of worst floods and dam break analysis in Mahanadi River Basin under CMIP6 climate change scenarios. *Environ. Monit. Assess.* **2023**, *195*, 1173. [CrossRef]
7. Yerramilli, S. Potential impact of climate changes on the inundation risk levels in a dam break scenario. *ISPRS Int. J. Geo-Inf.* **2013**, *2*, 110–134. [CrossRef]
8. Hooshyaripor, F.; Tahershamsi, A.; Razi, S. Dam break flood wave under different reservoir's capacities and lengths. *Sādhanā* **2017**, *42*, 1557–1569. [CrossRef]
9. Najar, M.; Gül, A. Investigating the influence of dam-breach parameters on dam-break connected flood hydrograph. *Tek. Dergi* **2022**, *33*, 12501–12524. [CrossRef]
10. Wang, B.; Zhang, J.; Chen, Y.; Peng, Y.; Liu, X.; Liu, W. Comparison of measured dam-break flood waves in triangular and rectangular channels. *J. Hydrol.* **2019**, *575*, 690–703. [CrossRef]
11. Yu, M.-H.; Deng, Y.-L.; Qin, L.-C.; Wang, D.-W.; Chen, Y.-L. Numerical simulation of levee breach flows under complex boundary conditions. *J. Hydrodyn. Ser. B* **2009**, *21*, 633–639. [CrossRef]
12. MacDonald, T.C.; Langridge-Monopolis, J. Breaching charateristics of dam failures. *J. Hydraul. Eng.* **1984**, *110*, 567–586. [CrossRef]
13. Cao, Z.; Pender, G.; Wallis, S.; Carling, P. Computational dam-break hydraulics over erodible sediment bed. *J. Hydraul. Eng.* **2004**, *130*, 689–703. [CrossRef]

14. Hogg, A.J.; Pritchard, D. The effects of hydraulic resistance on dam-break and other shallow inertial flows. *J. Fluid Mech.* **2004**, *501*, 179–212. [CrossRef]
15. Oguzhan, S.; Aksoy, A.O. Experimental investigation of the effect of vegetation on dam break flood waves. *J. Hydrol. Hydromech.* **2020**, *68*, 231–241. [CrossRef]
16. Feizi, A. Hydrodynamic study of the flows caused by dam break around downstream obstacles. *Open Civ. Eng. J.* **2018**, *12*, 225–238. [CrossRef]
17. Ismail, H.; Ann Larocque, L.; Bastianon, E.; Hanif Chaudhry, M.; Imran, J. Propagation of tributary dam-break flows through a channel junction. *J. Hydraul. Res.* **2021**, *59*, 214–223. [CrossRef]
18. Lauber, G.; Hager, W. Experiments to dambreak wave: Sloping channel. *J. Hydraul. Res.* **1998**, *36*, 761–773. [CrossRef]
19. Kocaman, S.; Güzel, H.; Evangelista, S.; Ozmen-Cagatay, H.; Viccione, G. Experimental and numerical analysis of a dam-break flow through different contraction geometries of the channel. *Water* **2020**, *12*, 1124. [CrossRef]
20. Chen, Y.H.; Simons, D.B. An experimental study of hydraulic and geomorphic changes in an alluvial channel induced by failure of a dam. *Water Resour. Res.* **1979**, *15*, 1183–1188. [CrossRef]
21. Goutiere, L.; Soares-Frazão, S.; Zech, Y. Dam-break flow on mobile bed in abruptly widening channel: Experimental data. *J. Hydraul. Res.* **2011**, *49*, 367–371. [CrossRef]
22. Kocaman, S.; Ozmen-Cagatay, H. The effect of lateral channel contraction on dam break flows: Laboratory experiment. *J. Hydrol.* **2012**, *432*, 145–153. [CrossRef]
23. Akgun, C.; Nas, S.S.; Uslu, A. 2D and 3D Numerical Simulation of Dam-Break Flooding: A Case Study of the Tuzluca Dam, Turkey. *Water* **2023**, *15*, 3622. [CrossRef]
24. Biscarini, C.; Di Francesco, S.; Ridolfi, E.; Manciola, P. On the simulation of floods in a narrow bending valley: The malpasset dam break case study. *Water* **2016**, *8*, 545. [CrossRef]
25. Haile, T.; Goitom, H.; Degu, A.M.; Grum, B.; Abebe, B.A. Simulation of urban environment flood inundation from potential dam break: Case of Midimar Embankment Dam, Tigray, Northern Ethiopia. *Sustain. Water Resour. Manag.* **2024**, *10*, 46. [CrossRef]
26. Haltas, I.; Tayfur, G.; Elci, S. Two-dimensional numerical modeling of flood wave propagation in an urban area due to Ürkmez dam-break, İzmir, Turkey. *Nat. Hazards* **2016**, *81*, 2103–2119. [CrossRef]
27. Xu, J.; Zhang, Y.; Ma, Q.; Zhang, J.; Hu, Q.; Zhan, Y. Dam-Break Hazard Assessment with CFD Computational Fluid Dynamics Modeling: The Tianchi Dam Case Study. *Water* **2025**, *17*, 108. [CrossRef]
28. Hosseinzadeh-Tabrizi, S.A.; Ghaeini-Hessaroeyeh, M.; Ziaadini-Dashtekhaki, M. Numerical simulation of dam-breach flood waves. *Appl. Water Sci.* **2022**, *12*, 100. [CrossRef]
29. Khoshkonesh, A.; Asim, T.; Mishra, R.; Dehrashid, F.A.; Heidarian, P.; Nsom, B. Study the effect of obstacle arrangements on the dam-break flow. *Int. J. Comadem* **2022**, *25*, 41–50.
30. Kocaman, S.; Evangelista, S.; Guzel, H.; Dal, K.; Yilmaz, A.; Viccione, G. Experimental and numerical investigation of 3d dam-break wave propagation in an enclosed domain with dry and wet bottom. *Appl. Sci.* **2021**, *11*, 5638. [CrossRef]
31. Di Cristo, C.; Greco, M.; Iervolino, M.; Vacca, A. Impact force of a geomorphic dam-break wave against an obstacle: Effects of sediment inertia. *Water* **2021**, *13*, 232. [CrossRef]
32. Maghsoodi, R.; Khademalrasoul, A.; Sarkardeh, H. 3D numerical simulation of dam-break flow over different obstacles in a dry bed. *Water Supply* **2022**, *22*, 4015–4029. [CrossRef]
33. Le, T.T.H.; Nguyen, V.C. Numerical study of partial dam–break flow with arbitrary dam gate location using VOF method. *Appl. Sci.* **2022**, *12*, 3884. [CrossRef]
34. Oodi, S.; Gohari, S.; Di Francesco, S.; Nazari, R.; Nikoo, M.R.; Heidarian, P.; Eidi, A.; Khoshkonesh, A. Wave–Structure Interaction Modeling of Transient Flow Around Channel Obstacles and Contractions. *Water* **2025**, *17*, 424. [CrossRef]
35. Beteille, E.; Larrarte, F.; Boyaval, S.; Demay, E.; Le, M.H. Dam-break flow over various obstacles configurations. *J. Hydraul. Res.* **2025**, *63*, 156–170. [CrossRef]
36. Transportation Association of Canada. *Guide to Bridge Hydraulics*; Thomas Telford: Londen, UK, 2004.
37. Chitale, S.V. Length and Shape of Guide Bunds. *Water Energy Int.* **1980**, *37*, 289–294.
38. Ahmadi, M.; Ghaderi, A.; MohammadNezhad, H.; Kuriqi, A.; Di Francesco, S. Numerical investigation of hydraulics in a vertical slot fishway with upgraded configurations. *Water* **2021**, *13*, 2711. [CrossRef]
39. Bayon, A.; Valero, D.; García-Bartual, R.; López-Jiménez, P.A. Performance assessment of OpenFOAM and FLOW-3D in the numerical modeling of a low Reynolds number hydraulic jump. *Environ. Model. Softw.* **2016**, *80*, 322–335. [CrossRef]
40. Kocaman, S.; Ozmen-Cagatay, H. Investigation of dam-break induced shock waves impact on a vertical wall. *J. Hydrol.* **2015**, *525*, 1–12. [CrossRef]
41. Mirkhorli, P.; Ghaderi, A.; Alizadeh Sanami, F.; Mohammadi, M.; Kuriqi, A.; Kisi, O. An investigation on hydraulic aspects of rectangular labyrinth pool and weir fishway using FLOW-3D. *Arab. J. Sci. Eng.* **2024**, *49*, 6061–6087. [CrossRef]
42. Ozmen-Cagatay, H.; Kocaman, S.; Guzel, H. Investigation of dam-break flood waves in a dry channel with a hump. *J. Hydro-Environ. Res.* **2014**, *8*, 304–315. [CrossRef]

43. Erduran, K.S.; Ünal, U.; Dokuz, A.Ş. Experimental and numerical investigation of partial dam-break waves. *Ocean Eng.* **2024**, *308*, 118346. [CrossRef]
44. Launder, B.; Spalding, D.B. Turbulence modelling. *Com. Mech. Appl. Mech. Eng.* **1974**, *3*, 269. [CrossRef]
45. Yakhot, V.; Orszag, S.A. Renormalization group analysis of turbulence. I. Basic theory. *J. Sci. Comput.* **1986**, *1*, 3–51. [CrossRef]
46. Wilcox, D.C. *Turbulence Modeling for CFD*; DCW Industries: La Canada, CA, USA, 1998; Volume 2, pp. 103–217.
47. Yang, S.; Yang, W.; Qin, S.; Li, Q.; Yang, B. Numerical study on characteristics of dam-break wave. *Ocean Eng.* **2018**, *159*, 358–371. [CrossRef]
48. Issakhov, A.; Borsikbayeva, A. The impact of a multilevel protection column on the propagation of a water wave and pressure distribution during a dam break: Numerical simulation. *J. Hydrol.* **2021**, *598*, 126212. [CrossRef]
49. Hirt, C.W.; Nichols, B.D. Volume of fluid (VOF) method for the dynamics of free boundaries. *J. Comput. Phys.* **1981**, *39*, 201–225. [CrossRef]
50. Hirt, C.; Sicilian, J. A porosity technique for the definition of obstacles in rectangular cell meshes. In Proceedings of the 4th International Conference on Numerical Ship Hydrodynamics, Washington, DC, USA, 24–27 September 1985.
51. Azimi, H.; Heydari, M.; Shabanlou, S. Numerical simulation of the effects of downstream obstacles on malpasset dam break pattern. *J. Appl. Res. Water Wastewater* **2018**, *5*, 441–446.
52. Esmaeeli Mohsenabadi, S. Numerical Modeling of The Initial Stages of Dam-Break Problems. Ph.D. Thesis, Université d'Ottawa/University of Ottawa, Ottawa, ON, Canada, 2021.
53. Ghaderi, A.; Abbasi, S.; Di Francesco, S. Numerical study on the hydraulic properties of flow over different pooled stepped spillways. *Water* **2021**, *13*, 710. [CrossRef]
54. Ghaderi, A.; Dasineh, M.; Aristodemo, F.; Ghahramanzadeh, A. Characteristics of free and submerged hydraulic jumps over different macroroughnesses. *J. Hydroinform.* **2020**, *22*, 1554–1572. [CrossRef]
55. Hien, L.T.T.; Van Chien, N. Investigate impact force of dam-break flow against structures by both 2d and 3d numerical simulations. *Water* **2021**, *13*, 344. [CrossRef]
56. Rong, Y.; Zhang, T.; Peng, L.; Feng, P. Three-dimensional numerical simulation of dam discharge and flood routing in Wudu reservoir. *Water* **2019**, *11*, 2157. [CrossRef]
57. Song, G.; Chen, Y.; Zhao, P.; Yuan, H. Numerical investigation on the evolutionary characteristics of landslide dam-break flow in a wet-bed channel with riparian vegetation. *Front. Mar. Sci.* **2024**, *11*, 1462760. [CrossRef]
58. Te Chow, V. *Open Channel Hydraulics Book*; The Blackburn Press: Caldwell, NJ, USA, 2009; ISBN 07-010776-9.

water

MDPI

Article

Effects of Successive Typhoon Durian and Typhoon Utor on Chlorophyll-*a* Response in South China Sea

Xiaoyang Hou, Zhenxin Ruan, Bo Li * and Yumeng Wang

College of Marine Science and Technology, Zhejiang Ocean University, Zhoushan 316022, China; houxiaoyang@zjou.edu.cn (X.H.); zxruan@zjou.edu.cn (Z.R.); wangyumeng@zjou.edu.cn (Y.W.)
* Correspondence: acelibo@zjou.edu.cn; Tel.: +86-18-3680-91446

Abstract: This paper investigated the effects of the successive Typhoons Durian and Utor on the chlorophyll-*a* concentration in the overlapping regions of the South China Sea in 2006. Satellite observations were employed to analyze the spatial–temporal variability of chlorophyll-*a* concentrations. The results show that the strong vertical mixing and upwelling after the passage of the first Typhoon Durian led to a rapid increase in chlorophyll-*a* concentration, while the effects of the subsequent Typhoon Utor showed regional variability: the chlorophyll-*a* concentration in the area to the right of the path of Typhoon Utor increased significantly, but it did not continue to increase in the area of the overlap with Durian and showed a decreasing trend. Studies have shown that the impacts of successive typhoons on marine ecology are not simply additive but can be modulated by changes in the marine environment caused by the previous typhoon. This study revealed the complexity of the impacts of successive typhoons on marine productivity and provides a new perspective for understanding how typhoons affect marine productivity.

Keywords: remote sensing; spatial data analysis; successive typhoons; chlorophyll-*a* concentration

1. Introduction

The typhoon is the most intense weather system resulting from atmosphere–ocean interactions [1]. The South China Sea, as the largest marginal sea of the western North Pacific in China, is also one of the most active regions for typhoons, where typhoons play a crucial role in driving oceanic variability. Wind stress-induced vertical mixing and upwelling can effectively uplift nutrient-rich subsurface cold waters to the sea surface [2–4]. This process not only leads to a decrease in sea surface temperature (SST) [5–9] but also enhances primary productivity and increases chlorophyll-*a* (chl-*a*) concentration through the input of nutrients [4,10–14]. Wang et al. [15] investigated the changes in chl-*a* concentration before and after typhoon transit and found that chl-*a* concentration increased from 0.74 mg/m^3 to 1.29 mg/m^3. Jung et al. [16] studied the environmental and ecological impacts of summer typhoons entering the East China Sea and found that summer typhoons typically increase surface chl-*a* concentrations by 65.4%. Lin et al. [17] studied Typhoon Kai-tak, which passed through the South China Sea, and revealed a dramatic 30-fold increase in the average sea surface chl-*a* concentration, from 0.1 mg/m^3 before to 3.2 mg/m^3 after the typhoon. Babin et al. [3] conducted a study on 13 tropical cyclones in the North Atlantic and found that these cyclones triggered a significant 5–91% increase in chl-*a* concentrations for 2–3 weeks, which then gradually returned to pre-cyclone levels. Pan et al. [18] investigated the phytoplankton biomass response to typhoons in the South China Sea using a physical–biogeochemical coupled model. Their results demonstrated that among all 79 typhoon events affecting the region between 2000 and 2009, 43 typhoons triggered surface phytoplankton blooms, while 5 cases only induced subsurface blooms without surface blooms. Overall, more than half of the typhoons impacting the South China Sea were found to stimulate phytoplankton blooms and enhance marine primary productivity. Ma et al. [19] examined the phytoplankton response to the successive Typhoons Barijat and Mangkhut in the northern South China Sea during summer 2018. Observations revealed an average 63% increase in chl-*a* concentration following the passage of typhoons.

While the upper ocean's physical and biogeochemical responses to a single typhoon event have been well characterized, the impact of successive typhoons on the ocean remains understudied. Previous studies have primarily focused on SST responses after the passages of typhoons, and the results showed that the SST cooling is significantly modulated when two typhoons consecutively pass through the same region [20,21]. Han et al. [20] conducted a comprehensive study on the SST response to the successive Typhoons Nesat and Nalgae in the South China Sea during summer 2011. Their research revealed that Typhoon Nesat generated a pronounced right-of-track cooling of up to 4.4 °C, whereas the subsequent Typhoon Nalgae produced a comparatively weaker cooling of only 2.2 °C. The analysis further demonstrated that vertical mixing served as the dominant mechanism for SST reduction in areas near the typhoon tracks. Zhang et al. [21] analyzed SST following the passage of the successive Typhoons Sarika and Haima. Their results demonstrated distinct oceanic responses: the first Typhoon Sarika induced approximately 4 °C SST cooling, while the subsequent Typhoon Haima produced relatively weaker effects with a 2 °C SST decrease. Baranowski et al. [22] employed numerical modeling to investigate the successive typhoons Hagupit and Jangmi in September–October 2008. Their simulations revealed that without the preconditioning effect of Typhoon Hagupit, Typhoon Jangmi would have generated more pronounced ocean cooling. Specifically, the cold wake following Jangmi's passage and the left-track temperature anomalies would have been approximately 0.4 °C greater in the absence of prior oceanic disturbance by Hagupit. In addition to SST, the preceding typhoon in a successive event also has an impact on the chl-*a* response processes generated by the subsequent typhoon. Huang et al. [23] analyzed phytoplankton blooms induced by the successive Typhoon Sinlaku and Typhoon Jangmi in the Northwestern Pacific Ocean found that the pre-existing cyclonic eddy provided a relatively unstable thermodynamic structure for Typhoon Sinlaku and facilitated more efficient upwelling of nutrient-rich cold subsurface water to the surface, whereas the deepening of the mixed layer caused by Typhoon Sinlaku prevented deeper nutrient-rich cold water from being carried into the upper ocean mixed layer, resulting in weaker phytoplankton blooms and creating unfavorable preconditions for the chl-*a* response triggered by Typhoon Jangmi. Although the chl-*a* response mechanism of individual typhoons in the South China Sea has been theoretically framed, the mechanism of successive typhoons induced chl-*a* response in the South China Sea is not comprehensive. This paper takes Typhoon Durian and Typhoon Utor in 2006 as examples to analyze the effects of successive typhoons on chl-*a* concentration by multi-source satellite data.

2. Materials and Methods

The typhoon data were derived from the Joint Typhoon Warning Center (JTWC) and included key information such as typhoon location, track, and intensity with a time resolution of 6 h. This study focused on Typhoons Durian and Utor in December 2006, with the investigation area 107° E–118° E and 8° N–19° N in the South China Sea. Typhoon Durian crossed the study area from 2 to 4 December 2006, and Typhoon Utor crossed from 11 to 13 December 2006. The chl-*a* concentration data were acquired using the

Moderate-Resolution Imaging Spectroradiometer (MODIS) Level-3 product with a spatial resolution of 4 km and a daily temporal resolution. The SST data were obtained from the Group for High-Resolution Sea Surface Temperature (GHRSST) Level 4 product with a daily temporal resolution and a spatial resolution of 0.0879°. The 10 m wind field data (U_{10}) were obtained from the Quick Scatterometer (QuikSCAT; NASA/JPL, USA) satellite mission, which makes observations twice a day with a spatial resolution of 0.25°. The current data and temperature profiles were obtained from the Hybrid Coordinate Ocean Model (HYCOM) global reanalysis product with a daily temporal resolution and 0.08° spatial resolution.

Using 16 years of typhoon data from the South China Sea, Wang [24] performed a time series of SST and chl-*a* concentration before and after typhoon passages. The analysis revealed that chl-*a* concentrations fluctuated in two days before the typhoon arrival and remained relatively stable with minimal changes in the 3–8 days before the typhoon. Chl-*a* concentration maintained at a high level for 2–7 days after the typhoon. Combined with the results of this paper, the average chl-*a* concentration of the 3–8 days before the typhoon is defined as the initial condition, the average chl-*a* concentration for 3 to 8 days after the typhoon is defined as the post-typhoon chl-*a* concentration, and the difference between them is the chl-*a* response. According to the results of Wang (2020) [24], SST began to decline after the passage of typhoon, and the maximum decline occurred within 3 days after the pass of the typhoon. Combined with the results of this paper, the SST response was defined as the difference between the average of the SST from t_0 to t_3 after the typhoon and the SST background, and the SST background is defined as the 3–8 days before the typhoon.

The kinetic energy (KE) is calculated using the following formula:

$$\mathrm{KE} = \frac{1}{2}\rho_0\left(u^2 + v^2\right) \tag{1}$$

where u and v are the east–west and north–south current speed, respectively, and ρ_0 is the density of sea water.

The Ekman Pumping Velocity (EPV) is calculated as

$$\mathrm{EPV} = \mathrm{Curl} \times \left(\frac{\tau}{\rho_0 \times f}\right) \tag{2}$$

where f is the Coriolis parameter, and τ is the wind stress, calculated according to the following formula:

$$\tau = \rho_{air} \cdot C_d \cdot |U_{10}| \cdot U_{10} \tag{3}$$

ρ_{air} is the density of air, and C_d is the wind drag coefficient. The chl-*a* response shows regional variability under the influence of successive typhoons. The overlap area 111° E–115° E, 13° N–14.5° N is defined as Region A based on the strong chl-*a* response, and the area to the right of Typhoon Utor, 114° E–116.2° E, 14.3° N–17° N, where the chl-*a* response is strong, is defined as Region B. Regions A and B are labeled with black boxes in Figure 1.

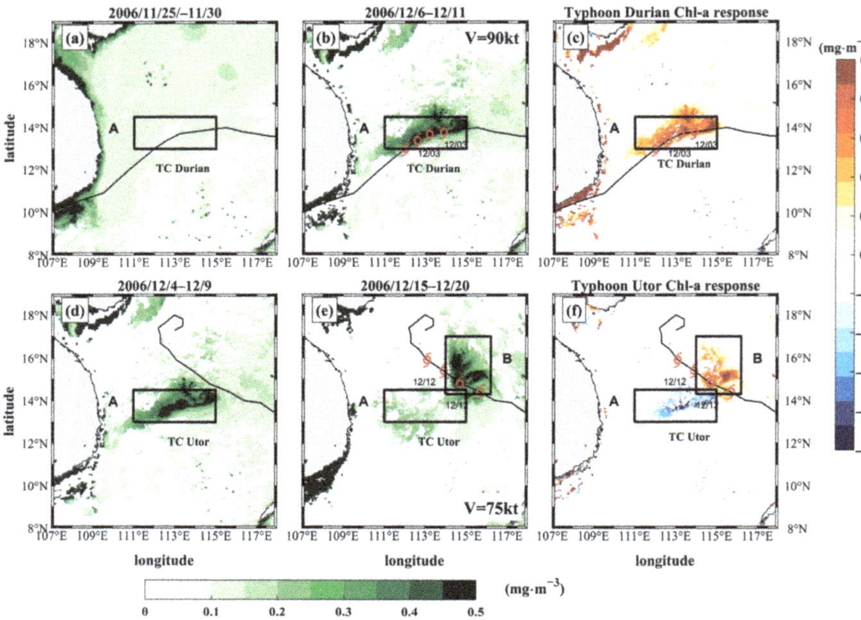

Figure 1. The chl-*a* concentration (**a**) before and (**b**) after the passage of Typhoon Durian, and (**c**) the chl-*a* response induced by Typhoon Durian. (**d–f**) are same as (**a–c**), except for Typhoon Utor. The red typhoon symbol represents the location of the typhoon. The black boxes represent regions A and B.

3. Chlorophyll-a Response Induced by Typhoon Durian and Typhoon Utor

Figure 1 illustrates the distribution of chl-*a* response before, during, and after the passage of Typhoon Durian and Typhoon Utor. The results showed that the chl-*a* concentration in the study area was small before the passage of Typhoon Durian, which was about 0.2 mg/m^3. On December 3, the typhoon intensity was 90 knots and caused a rapid increase in chl-*a* (Figure 1b). The difference in Figure 1c shows that the strongest increase in chl-*a* in Region A was over 0.3 mg/m^3. The strong response was mainly distributed on the right side of the typhoon, which is consistent with previous results that found a rightward bias in the typhoon-induced ocean response [6]. Typhoon Utor affected Region A on December 12, eight days after Typhoon Durian passed, when Typhoon Utor was 75 knots strong and triggered a decrease in chl-*a* concentration of more than 0.2 mg/m^3. Typhoon Utor induced an increase in chl-*a* of 0.3 mg/m^3 in Region B, which was on the right side of the Typhoon Utor path and was not affected by Typhoon Durian.

The time series of chl-*a* in Region A and Region B are presented in Figure 2, and the results show that the chl-*a* concentrations in both Region A ang Region B were low from 3 to 8 days before the passage of Typhoon Durian. The chl-*a* concentration in Region A increased rapidly from after the passage of Typhoon Durian and peaked at about 0.5 mg/m^3 on the 7th day after the passage of the typhoon and then started to decline. Typhoon Utor impacted Region A on December 12 and did not cause an increase in chl-*a* in Region A. At this time, Region A was still in the relaxation stage of Typhoon Durian, which led to the decrease in chl-*a* caused by Typhoon Utor in Region A. In Region B, chl-*a* concentrations remained persistently low before the Typhoon Utor pass and started to increase rapidly after the typhoon, peaking at approximately 0.36 mg/m^3 on the 6th day, which was an increase of 0.2 mg/m^3 in chl-*a* compared with that before the typhoon. This is consistent with the results in Figure 1, where there is a significant difference between the chl-*a* response induced by Typhoon Utor in Region A and Region B.

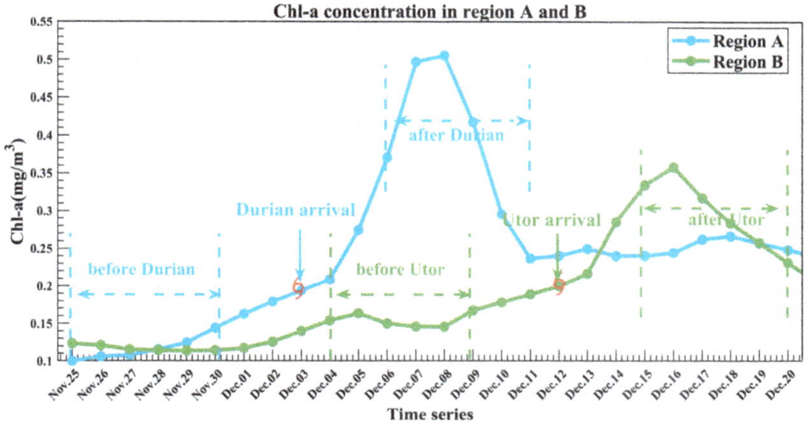

Figure 2. Time series of chl-*a* concentration in Region A (blue solid line) and Region B (green solid line). The red typhoon symbol indicates the time when the typhoon passed.

4. Discussion

To explore how Typhoon Durian affected the chl-*a* response of the succeeding Typhoon Utor, the wind field distributions of the two typhoons as well as the induced EPV responses were analyzed, and the results are shown in Figure 3. The moments that caused a strong chl-*a* response (3 December and 12 December) were taken as the typhoon transit times, and the spatial distribution of the mean EPV superimposed wind field before and after the typhoon passage was mapped.

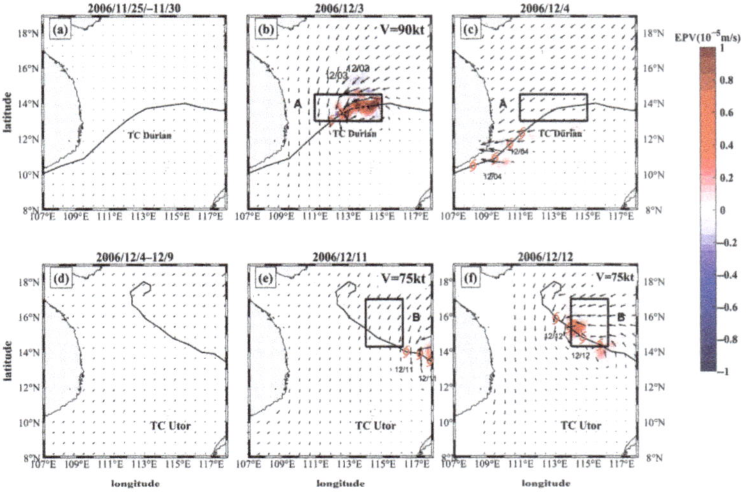

Figure 3. Distribution of wind fields (arrows) and EPVs (filled colors) of Typhoon Durian (**a**) before its transit, (**b**) on 3 December 2006, and (**c**) on 4 December 2006. Distribution of wind fields (arrows) and EPV (filled colors) of Typhoon Utor (**d**) before its transit, (**e**) on 11 December 2006, and (**f**) on 12 December 2006. The red typhoon symbol represents the location of the typhoon. The black boxes represent regions A and B.

Figure 3 presents the spatial distributions of EPV induced by Typhoon Durian and Utor during their respective pre-transit and post-transit periods. The results show low wind speeds and no upwelling areas in the study area prior to the passage of Typhoon Durian. On December 3, the typhoon had an intensity of 90 knots and induced a strong upwelling of about 1×10^{-5} m/s or more in Region A (Figure 3b). Figure 3c shows that by 4 December (one day after the typhoon's passage), the storm had moved out of Region A, resulting in no significant upwelling. Typhoon Utor affected the study area on December 12, following an 8-day interval after Typhoon Durian's transit. In this stage, Typhoon Utor maintained an intensity of 75 knots, generating strong upwelling in Region B (Figure 3f) with EPV exceeding 0.8×10^{-5} m/s. Prior to Typhoon Utor's passage, including on December 11 (one day before transit), no significant upwelling was observed in the study region (Figure 3d,e). Typhoons Durian and Utor generated upwelling through wind vorticity in Region A and B, respectively, transporting nutrient-rich subsurface waters to the sea surface and leading to an increase in chl-*a*.

In addition to Ekman pumping, the strong winds of typhoons can cause strong vertical mixing. Figure 4 presents the SST distributions and SST responses induced by Typhoons Durian and Utor. The results demonstrate that temperatures near the typhoon track averaging approximately 28 °C. After Typhoon Durian's passage, the central region experienced substantial SST cooling exceeding 2 °C (Figure 4c). Notably, the region to the right of the track showed stronger cooling than the left side, with the most pronounced SST reductions spatially coinciding with areas of significant chl-*a* response and elevated EPV. Prior to Typhoon Utor's passage, a low-SST anomaly persisted in the region to the left of the track (Region A) during the relaxation stage of Typhoon Durian and resulted in an SST increase of approximately 0.6 °C after the Typhoon Utor pass. Conversely, in Region B, an SST decrease of approximately 1.5 °C occurred, accompanied by increased chl-*a* concentrations. The SST response patterns following the successive typhoon transits reveal that the ocean response to Typhoon Utor was modulated by the previous Typhoon Durian.

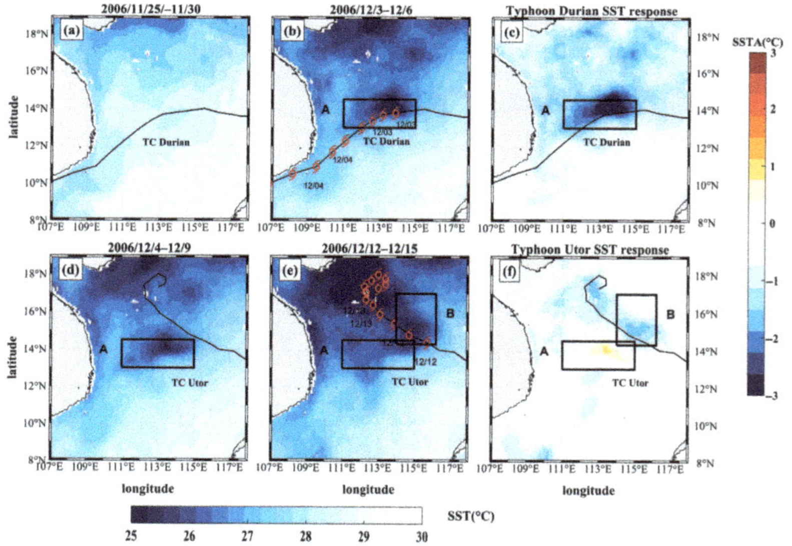

Figure 4. SST fields for Typhoon Durian during (**a**) pre-transit and (**b**) post-transit phases, along with (**c**) the resultant SST response. Panels (**d–f**) present analogous data for Typhoon Utor, following the same sequence. The red typhoon symbol represents the location of the typhoon. The black boxes represent regions A and B.

Figure 5 displays the temporal evolution of depth-averaged temperature in Regions A and B from December 3 to December 18. Following Typhoon Durian's passage, the 14 °C isotherm lifted from 220 m to 200 m, indicating a strong upwelling. The uplift of colder deep waters induced by Durian's wind forcing caused pronounced Ekman pumping on December 5 (Figure 5a). During the Typhoon Utor pass, there was no significant change in the 14 °C isotherm, which means that no intense upwelling was observed in Region A. However, in Region B, Typhoon Utor induced the significant uplift of the 14 °C isotherm from 200 m to 175 m, indicating intense upwelling activity. Overall, Typhoon Durian triggered strong upwelling in Region A, whereas Typhoon Utor failed to generate significant upwelling in Region A but induced intense upwelling in Region B, which also coincided with the chl-*a* responses induced by the two typhoons in Regions A and B.

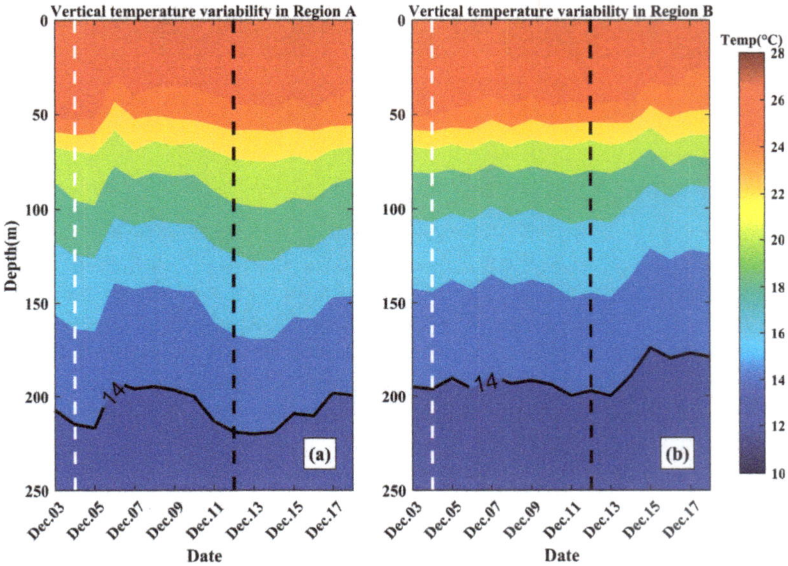

Figure 5. Temporal evolution of depth-averaged temperature in (**a**) Region A and (**b**) Region B before and after typhoon passage. The white and black dashed lines indicate the time of the Typhoon Durian and Typhoon Utor passes, respectively, and the black solid line denotes the 14 °C isotherm.

Figure 6 displays the temporal evolution of depth-averaged KE in Regions A and B from 4 to 18 December 2006. From December 4 to 6, following Typhoon Durian's passage, KE in both Regions A and B reached maximum values exceeding 0.3×10^3 J/m^3. Significant KE increases occurred in the surface layers (0–54 m depth). The strong vertical gradient of KE indicates strong vertical mixing in these layers. After the Typhoon Utor pass on 12 December, KE in the surface of Region A remained nearly unchanged, which means that no vertical mixing was triggered by Typhoon Utor in Region A. There was no mixing and no significant upwelling, which explains why Typhoon Utor did not induce a significant chl-*a* response in Region A. Differently to Region A, the typhoon caused significant KE increases in Region B. Although Typhoon Durian, due to its extensive wind field (the radius of 34 knots was 91.2 nmi) also induced a surface KE response in Region B, it did not trigger significant chl-*a* enhancement in this region because it failed to generate Ekman pumping there.

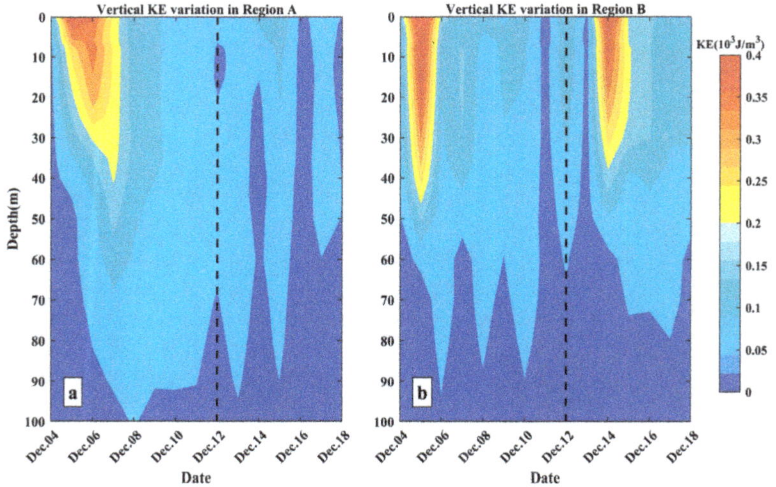

Figure 6. Temporal evolution of depth-averaged KE in (**a**) Region A and (**b**) Region B before and after the typhoon passage. The dashed line indicates the time of the Typhoon Utor pass.

Analysis of EPV, SST, and KE responses to successive typhoons reveal that during Typhoon Durian's impact on Region A, upwelling and vertical mixing drove SST cooling and may have caused an increase in chl-*a* concentrations by bringing nutrients to the surface layer. As Typhoon Durian moved away, when Typhoon Utor affected the same sea area eight days later, the SST and chl-*a* concentration responses induced by Typhoon Durian had not recovered yet. Consequently, Typhoon Utor ultimately caused an increase in SST and a decrease in chl-*a* concentration in Region A. In Region B, which was not affected by Typhoon Durian, a cooling in SST and a rise in chl-*a* concentration were triggered by Typhoon Utor. This was probably due to the change in the ocean environment in the overlapping region caused by the previous typhoon. In a study of successive typhoons, Huang et al. [23] found that the deepening of the mixed layer caused by the previous typhoon prevented deeper nutrient-rich cold water from entering into the upper mixed layer, which created an unfavorable preconditions for the chl-*a* response triggered by next typhoon. Due to the modulation of the oceanic environment by the preceding Typhoon Durian, Typhoon Utor induced significantly different SST and chl-*a* responses on its left and right sides.

5. Conclusions

This study analyzes the differences in chl-*a* responses induced by the two successive typhoons (Durian and Utor) in the South China Sea. The results demonstrate that the preceding Typhoon Durian triggered strong chl-*a* and SST responses in Region A. Analyses of EPV, SST, vertical temperature response, and KE indicate that Typhoon Durian caused strong upwelling and vertical KE gradients in Region A. This led to a decrease in SST on the one hand, and on the other hand, it brought nutrient-enriched seawater from the subsurface to the surface, leading to an increase in chl-*a*. Eight days after Typhoon Durian passing, Typhoon Utor affected Region A. As Region A was still in the relaxation stage of Typhoon Durian and the marine environment was in the recovery phase, there was a weak increase in SST and a weak decrease in chl-*a* concentration in Region A after the passage of Typhoon Utor. In Region B, Typhoon Utor induced Ekman pumping and strong vertical gradients of KE, suggesting a resultant upwelling and vertical mixing, leading to a cooling of the SST and an increase in chl-*a*. The results of this paper illustrate that the preceding typhoon changed the marine environment and modulated the marine response to the passage of the following typhoon.

Author Contributions: X.H. and Z.R. designed this article. X.H. analyzed the data and wrote the manuscript. X.H. and Z.R. improved the figures and manuscript writing. X.H., Z.R., Y.W. and B.L. read and approved the content. All authors have read and agreed to the published version of the manuscript.

Funding: This research received no external funding.

Data Availability Statement: The typhoon data were obtained from the Joint Typhoon Warning Center (JTWC) (https://www.metoc.navy.mil/jtwc/jtwc.html, accessed on 4 April 2023), and the chl-*a* concentration data were derived from the Moderate Resolution Imaging Spectroradiometer (MODIS) aboard NASA's Aqua satellite (https://oceancolor.gsfc.nasa.gov/, accessed on 9 March 2023). The sea surface temperature (SST) data were obtained from the Group for High-Resolution Sea Surface Temperature (GHRSST) (https://www.ghrsst.org/, accessed on 2 April 2024). The 10 m ocean wind vector data were obtained from the Quick Scatterometer (QuikSCAT) mission, available at https://podaac.jpl.nasa.gov/QuikSCAT, accessed on 3 March 2025. The oceanic surface current data and temperature profiles were obtained from the HYbrid Coordinate Ocean Model (HYCOM) (https://www.hycom.org/, accessed on 27 March 2025).

Acknowledgments: We thank the United States Joint Typhoon Warning Center for typhoon data, MODIS Satellite Moderate Resolution Imagery Optical Sensor (MODIS) for chlorophyll data, GHRSST for sea surface temperature data, and QuikScat Fast Scatterometer for wind field data.

Conflicts of Interest: The authors declare that this article was written in the absence of any commercial or financial relationships that could be construed as potential conflicts of interest.

References

1. Emanuel, K.A. An Air-Sea Interaction Theory for Tropical Cyclones. Part I: Steady-State Maintenance. *J. Atmos. Sci.* **1986**, *43*, 585–605. [CrossRef]
2. Geisler, J.E. Linear Theory of the Response of a Two Layer Ocean to a Moving Hurricane. *Geophys. Astrophys. Fluid Dyn.* **1970**, *1*, 249–272. [CrossRef]
3. Babin, S.M.; Carton, J.A.; Dickey, T.D.; Wiggert, J.D. Satellite Evidence of Hurricane-induced Phytoplankton Blooms in an Oceanic Desert. *J. Geophys. Res.* **2004**, *109*, C03043. [CrossRef]
4. Liu, Y.; Tang, D.; Evgeny, M. Chlorophyll Concentration Response to the Typhoon Wind-Pump Induced Upper Ocean Processes Considering Air–Sea Heat Exchange. *Remote Sens.* **2019**, *11*, 1825. [CrossRef]
5. Leipper, D.F. Observed Ocean Conditions and Hurricane Hilda, 1964. *J. Atmos. Sci.* **1967**, *24*, 182–186. [CrossRef]
6. Price, J.F. Upper Ocean Response to a Hurricane. *J. Phys. Oceanogr.* **1981**, *11*, 153–175. [CrossRef]
7. Shay, L.K.; Black, P.G.; Mariano, A.J.; Hawkins, J.D.; Elsberry, R.L. Upper Ocean Response to Hurricane Gilbert. *J. Geophys. Res. Ocean* **1992**, *97*, 20227–20248. [CrossRef]
8. Cione, J.J.; Uhlhorn, E.W. Sea Surface Temperature Variability in Hurricanes: Implications with Respect to Intensity Change. *Mon. Weather Rev.* **2003**, *131*, 1783–1796. [CrossRef]
9. Emanuel, K.; DesAutels, C.; Holloway, C.; Korty, R. Environmental Control of Tropical Cyclone Intensity. *J. Atmos. Sci.* **2004**, *61*, 843–858. [CrossRef]
10. Zheng, G.; Tang, D. Offshore and Nearshore Chlorophyll Increases Induced by Typhoon Winds and Subsequent Terrestrial Rainwater Runoff. *Mar. Ecol. Prog. Ser.* **2007**, *333*, 61–74. [CrossRef]
11. Zhao, H.; Tang, D.; Wang, D. Phytoplankton Blooms near the Pearl River Estuary Induced by Typhoon Nuri. *J. Geophys. Res.* **2009**, *114*, C12027. [CrossRef]
12. Chen, Y.; Tang, D. Eddy-Feature Phytoplankton Bloom Induced by a Tropical Cyclone in the South China Sea. *Int. J. Remote Sens.* **2012**, *33*, 7444–7457. [CrossRef]
13. Ye, H.J.; Sui, Y.; Tang, D.L.; Afanasyev, Y.D. A Subsurface Chlorophyll a Bloom Induced by Typhoon in the South China Sea. *J. Mar. Syst.* **2013**, *128*, 138–145. [CrossRef]
14. Xu, H.; Tang, D.; Sheng, J.; Liu, Y.; Sui, Y. Study of Dissolved Oxygen Responses to Tropical Cyclones in the Bay of Bengal Based on Argo and Satellite Observations. *Sci. Total Environ.* **2019**, *659*, 912–922. [CrossRef]

15. Wang, J.; Guo, B.; Ji, Z.; Che, Y.; Mantravadi, V.S. Effects of Typhoon Chanthu on Marine Chlorophyll a, Temperature and Salinity. *Atmosphere* **2023**, *14*, 1505. [CrossRef]
16. Jung, H.; Ahn, J.; Kang, J.J.; Hwang, J.D.; Youn, S.; Oh, H.; Joo, H.; Kim, C. Typhoon Effects on Surface Phytoplankton Biomass Based on Satellite-Derived Chlorophyll-a in the East Sea during Summer. *J. Mar. Sci. Eng.* **2024**, *12*, 2369. [CrossRef]
17. Lin, I.; Liu, W.T.; Wu, C.; Wong, G.T.F.; Hu, C.; Chen, Z.; Liang, W.; Yang, Y.; Liu, K. New Evidence for Enhanced Ocean Primary Production Triggered by Tropical Cyclone. *Geophys. Res. Lett.* **2003**, *30*, 1718. [CrossRef]
18. Pan, G.; Chai, F.; Tang, D.; Wang, D. Marine Phytoplankton Biomass Responses to Typhoon Events in the South China Sea Based on Physical-Biogeochemical Model. *Ecol. Model.* **2017**, *356*, 38–47. [CrossRef]
19. Ma, Z.; Zhang, Y.; Wu, R.; Na, R. Statistical Characteristics of the Response of Sea Surface Temperatures to Westward Typhoons in the South China Sea. *Remote Sens.* **2021**, *13*, 916. [CrossRef]
20. Han, S.; Wang, M.; Peng, B. Response of Temperature to Successive Typhoons in the South China Sea. *J. Mar. Sci. Eng.* **2022**, *10*, 1157. [CrossRef]
21. Zhang, H.; Liu, X.; Wu, R.; Liu, F.; Yu, L.; Shang, X.; Qi, Y.; Wang, Y.; Song, X.; Xie, X.; et al. Ocean Response to Successive Typhoons Sarika and Haima (2016) Based on Data Acquired via Multiple Satellites and Moored Array. *Remote Sens.* **2019**, *11*, 2360. [CrossRef]
22. Baranowski, D.B.; Flatau, P.J.; Chen, S.; Black, P.G. Upper Ocean Response to the Passage of Two Sequential Typhoons. *Ocean Sci.* **2014**, *10*, 559–570. [CrossRef]
23. Huang, L.; Zhao, H.; Pan, J.; Devlin, A. Remote Sensing Observations of Phytoplankton Increases Triggered by Successive Typhoons. *Front. Earth Sci.* **2017**, *11*, 601–608. [CrossRef]
24. Wang, Y. Composite of Typhoon-Induced Sea Surface Temperature and Chlorophyll—A Responses in the South China Sea. *JGR Ocean* **2020**, *125*, e2020JC016243. [CrossRef]

water

MDPI

Article

Spatial–Temporal Distribution of Offshore Transport Pathways of Coastal Water Masses in the East China Sea Based on GOCI-TSS

Yuanjie Peng and Wenbin Yin *

College of Marine Science and Technology, Zhejiang Ocean University, Zhoushan 316022, China; pengyuanjie@zjou.edu.cn

* Correspondence: yinwenbin@zjou.edu.cn

Abstract: The offshore transport of coastal water masses in the East China Sea is vital for maintaining ecological stability. Understanding its spatial-temporal pathways helps clarify material transport and ecological responses. This study used total suspended sediment (TSS) data from the Korean Geostationary Ocean Color Imager to analyze TSS distribution and anomalies, combined with satellite-derived surface residual currents. Results show significant seasonal variations: coastal water masses expand to the 50 m isobath in winter and contract to the 20 m isobath in summer. Offshore transport pathways vary spatially, extending to the shelf edge north of 28° N but restricted by the Taiwan Warm Current south of 28° N. A persistent transport pathway near 28° N shifts from northeastward to eastward. Other pathways include one south of Hangzhou Bay (spring and autumn) linked to tidal mixing and another north of the Yangtze River estuary (summer) following the Yangtze River Diluted Water. These findings provide crucial observational insights for modeling material cycling in the East China Sea shelf.

Keywords: East China Sea; offshore transport pathways; total suspended sediment anomaly; spatial cumulative frequency

1. Introduction

The East China Sea (ECS) shelf region (Figure 1a) is distinguished by its distinctive ecological environment and is considered to be among China's most prolific fishery resource areas [1]. The maintenance of its high productivity is contingent on terrestrial nutrients carried by the coastal water masses and exogenous nutrients transported by the Kuroshio Current. The cross-shelf transport of these nutrient-rich water masses is of particular significance in ensuring the stability of ecological functions in the ECS shelf region. Research has found that in spring, coastal waters moving south from Zhejiang and Fujian create a large area with a lot of nutrients near the coast of Zhejiang, while the movement and rising of Kuroshio subsurface waters near northeastern Taiwan serve as a key route for bringing in outside nutrients [2]. It is worth noting that the spatial distribution characteristics of marine suspended particulate matter, as an effective carrier and indicator of nutrients, can reflect the transport path of nutrients. Liu's [3] research found that there is a tongue of high suspended matter concentration protruding toward the sea in the waters near 29° N in northern Zhejiang and 27° N in northern Taiwan, which provides direct evidence for the existence of a cross-shelf nutrient transport channel in the East China Sea.

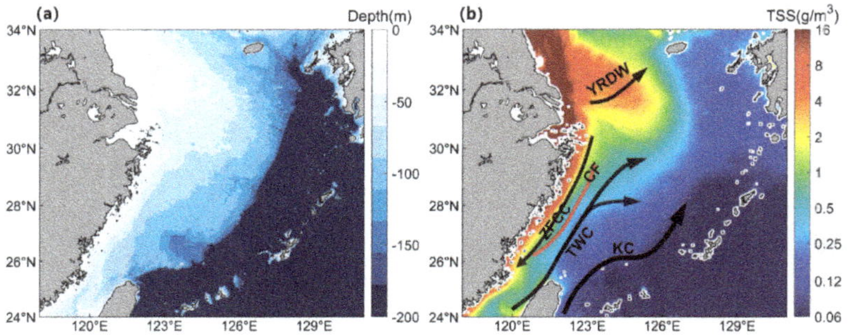

Figure 1. (**a**) Topography and (**b**) schematic diagram of the circulation system in the East China Sea. In (**b**), the background color represents climatological TSS, and the red line indicates the Zhejiang–Fujian coastal turbidity front. **KC**: Kuroshio Current; **TWC**: Taiwan Warm Current; **YRDW**: Yangtze River Diluted Water; **ZFCC**: Zhejiang–Fujian Coastal Current. **CF**: Coastal Front.

The coastal water masses in the ECS are predominantly constituted by the Yangtze River Diluted Water (YRDW) and the Zhejiang−Fujian Coastal Water (ZFCW). The formation of these water masses is a consequence of the mixing of freshwater discharged from coastal rivers and shelf water, exhibiting characteristics such as turbidity, low salinity, and high nutrient concentrations [4]. The dispersal pathway of YRDW exhibits distinct seasonal variations: During summer months, a south−easterly deflection is exhibited, followed by a north−easterly turn [5] with its influence extending as far as Jeju Island or the central South Yellow Sea [6]. Conversely, during winter, the YRDW predominantly flows southward along the coast, eventually merging into the Zhejiang−Fujian Coastal Current [7,8]. Zhu [9] found that, under the influence of wind fields and upwelling, the diluted water of the Yangtze River can form a relatively isolated low−salinity mass in the northern East China Sea. Its separation and outward transport process significantly enhanced the cross-shelf transport efficiency of terrestrial nutrients and organic matter. The ZFCW is predominantly distributed in the western nearshore region of the ECS shelf. Its eastern boundary converges with the northward-flowing Taiwan Warm Current (TWC), forming a distinct Zhejiang–Fujian coastal front due to significant horizontal gradients in hydrographic properties. Conventional studies have suggested that the ZFCW and its associated pollutants, carbon particles, freshwater, and nutrients are unable to undergo large-scale cross-shelf offshore transport due to constraints imposed by frontal barriers and seabed topography. Instead, they were believed to be primarily transported southward to the Nan'ao Island area of Guangdong [10]. However, recent research has revealed the presence of large-scale frontal instability processes (penetrating fronts) in the Zhejiang–Fujian frontal zone. These processes have been shown to facilitate intermittent, large−scale cross-shelf offshore transport of nutrient-rich coastal waters, exerting significant influences on the physical-biogeochemical environment of the ECS shelf [11–14].

Yuan [11] was the first to identify large−scale, horizontally protruding cross−shelf penetrating fronts off the Yangtze Estuary and Zhejiang coast using MODIS SST and CHL imagery. He also confirmed the critical role of these fronts in cross−frontal and cross-shelf water exchange through in situ hydrographic and nutrient data analysis. He [13] analyzed 8−day composite MODIS and SeaWiFS SST/CHL data (1998–2007), revealing 2–11 annual occurrences of such penetrating fronts in this region. The frequency of these penetrating fronts was found to be notably higher during summer and winter months than in spring and autumn, with durations ranging from days to a month and maximum frontal lengths exceeding 300 km. In addition, Wu's [15] examination of MODIS chlorophyll images during the autumn period (2003–2014) led to the identification of a recurrent autumn penetrating front situated along 28.5° N–29.5° N, characterized by a fixed positional pattern. Yin [16] employed GOCI satellite TSS imagery to quantitatively reconstruct the full evolution cycle (∼10 days) of Zhejiang coastal fronts—from submesoscale frontal waves to large-scale penetrating fronts—documenting three developmental phases: growth, maturation, and decay.

Current research on the offshore transport of coastal water masses in the ECS, particularly the ZFCW, remains insufficient and predominantly relies on case studies. There is a notable lack of systematic investigation into the transport locations and pathways. To address this gap, this study utilizes Total Suspended Sediments concentration (TSS) data from the Korean geostationary satellite GOCI (Geostationary Ocean Color Imager). The satellite has significant advantages in monitoring nearshore turbid water bodies. Compared with polar-orbiting satellites such as MODIS/VIIRS, GOCI can more effectively separate the radiation signals of turbid water bodies with its optimized band settings [14], effectively reducing the mean absolute error of suspended sediment concentration (SSC) inversion in high turbidity areas such as the Yangtze River Estuary. At the same time, GOCI's high-frequency observation capability (once per hour, 8 scenes per day) and more than 80% effective data acquisition rate enable it to accurately capture the dynamic changes in tidally driven suspended sediment [17,18], while MODIS/VIIRS's daily coverage rate is less than 30% [19]. Although Sentinel−3A/OLCI is superior to GOCI in the SSC inversion accuracy of Hangzhou Bay [20], its nearly 2−day observation period is difficult to meet the needs of short-term dynamic monitoring. In this paper, by employing positive TSS anomalies (ΔTSS) after removing climatological means as a tracer for nearshore turbid water masses, we systematically analyze their spatiotemporal distribution through the spatial cumulative frequency of ΔTSS ($P_{\Delta TSS}$). This approach aims to reveal the statistical patterns of offshore transport processes in ECS coastal waters.

2. Data and Methodology

2.1. GOCI-TSS

The East China Sea's coastal water masses are characterized by low salinity, rich nutrients and high turbidity, so satellite−observed turbidity data can be used to track the horizontal transport of these water masses. This study utilizes TSS data observed by the GOCI as the tracer dataset. The GOCI has a spatial resolution of 500 m and a temporal resolution of 1 h, covering a time span from 08:16 to 15:45 Beijing Time. In comparison with polar−orbiting satellites, GOCI offers distinct advantages in its ability to monitor variations in ocean color information from a geostationary platform, facilitating the observation of short-term regional marine phenomena. This provides valuable data for studying the intermittent offshore transport processes of coastal waters. Ruddick [21] conducted a comparative analysis of GOCI and MODIS_AQUA's Rrs and TSS products in the Bohai Sea region. The results indicated that the TSS distribution aligned with previous research findings. Although there is room for improvement in GOCI's calibration, atmospheric correction, and inversion algorithms, the existing results demonstrate that GOCI's data products can be effectively used to study dynamic processes in turbid waters. The TSS data from GOCI can be downloaded from the website of the Korea Ocean Satellite Center (https://kosc.kiost.ac.kr) accessed on 1 April 2024.

2.2. Sea Surface Wind

Sea surface winds have been demonstrated to influence the spatial distribution of TSS by driving vertical mixing and horizontal transport [22]. Therefore, this study employs the Cross−Calibrated Multi-Platform (CCMP) gridded sea surface wind data provided by the Remote Sensing System to investigate the impact of wind stress on TSS distribution. This dataset integrates buoy observations, satellite data (QuikSCAT and ASCAT), and

ERA−Interim model outputs using a Variational Analysis Method, with a spatial resolution of 0.25° and a temporal resolution of 6 h. Comparative analyses between CCMP wind data and in situ observations show minimal discrepancies (approximately 0.5 m/s) [23], confirming that CCMP wind data provide reliable meteorological data for studying regional ocean dynamics. The dataset is available for download from the website: https://www.remss.com/measurements/ccmp/ accessed on 1 April 2024.

2.3. Tide Current

Tidal-induced mixing is a crucial dynamic process responsible for the resuspension of bottom sediments, significantly influencing the spatial distribution of TSS [24]. To minimize interference from TSS variations caused by tidal vertical mixing in this study, it is necessary to identify regions where tidal mixing is fully developed. This research employs tidal current data along the Zhejiang–Fujian coast, predicted by the global tidal model TPXO developed by Oregon State University, to calculate the Simpson–Hunter parameter for tidal mixing [25]. Based on thresholds established in previous studies, the boundaries of fully mixed tidal regions are delineated. The TPXO global tidal model is a widely used tidal simulation model developed by assimilating satellite altimetry data (e.g., TOPEX/Poseidon, Jason series) and tide gauge observations. It employs harmonic analysis and data assimilation techniques, optimizing tidal parameters via least-squares methods to enhance simulation accuracy. Recognized for its high precision and reliability, the TPXO model has been extensively applied in global tidal studies, demonstrating strong consistency with observations across various oceanic regions [26]. The model data can be downloaded from the website https://www.tpxo.net/global accessed on 1 April 2024.

2.4. Topographic Data

The bathymetric data used in this study were obtained from the latest global seabed and land elevation model, GEBCO_2023, released by the General Bathymetric Chart of the Oceans (GEBCO). This dataset integrates multi−source data, including shipborne multibeam bathymetry, remote sensing satellite observations, sub-ice topography estimations, and terrestrial elevation data [27,28], providing global elevation coverage at a 15-arc-second grid resolution in meters. Its accuracy has been validated, particularly in shallow waters, where comparisons with ICESat-2 satellite-derived seabed measurements demonstrate excellent performance of GEBCO_2023 [28]. As such, this dataset is widely applied in marine scientific research, ocean engineering, climate modeling, and ecological conservation, serving as a fundamental tool for analyzing oceanic and terrestrial topography [29].

2.5. Calculation of Residual Current Field

In order to analyze the influence of the ECS's circulation system on the offshore transport pathways of coastal water masses, this study employed the Maximum Cross−Correlation (MCC) algorithm [30] to derive the surface current fields along the Zhejiang–Fujian coast from continuous TSS satellite imagery. The MCC method estimates the displacement of surface water features between sequential satellite images. It identifies the shift (in direction and distance) that produces the highest correlation between image pairs, effectively capturing the horizontal movement of water masses. This approach allows for the derivation of surface residual currents without direct velocity measurements. To validate the reliability of the retrieved current fields, the tidal components within them were verified using the ECS tidal current fields simulated by the FVCOM model, as presented by Xuan et al. [31]. The M_2 tidal constituent in Xuan's model data aligns well with cotidal charts obtained from satellite altimeter and coastal tide gauge observations, demonstrating high accuracy in simulating the tidal current fields of the ECS and confirming the model's reliability. Figure 2 compares the M_2 tidal ellipses extracted from the retrieved current fields with those from the model data. The results demonstrate a strong agreement across most of the study area, with the exception of shallow nearshore regions, thereby indicating the high credibility of the retrieved surface current fields. The validated current fields were then processed by subtracting the concurrent tidal currents predicted by TPXO, yielding the residual current fields.

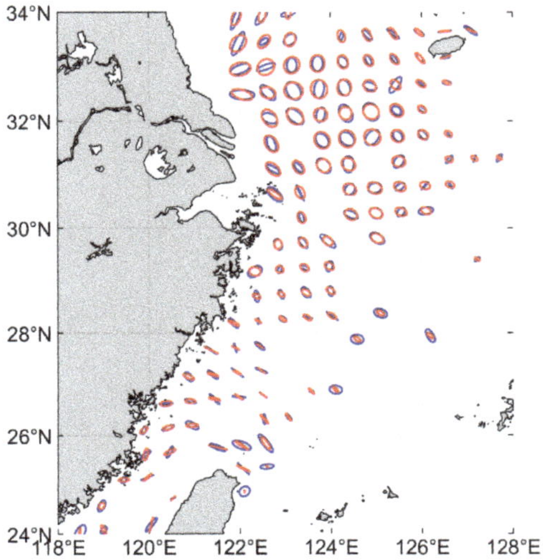

Figure 2. Comparison of the M_2 tidal current ellipse of the retrieved flow field (blue) and the model flow field (red).

2.6. Calculation of Spatial Cumulative Frequency of ΔTSS

In the East China Sea, water turbidity is typically high in coastal areas and low in offshore regions. When turbid coastal water masses are transported offshore, their TSS values are higher than the background TSS values of the surrounding areas. Therefore, positive TSS anomalies (ΔTSS) obtained by subtracting climatological means can serve as tracers for the offshore transport of coastal water masses. The principle behind using ΔTSS lies in anomaly detection. By subtracting the climatological monthly mean from hourly TSS values, we isolate short-term turbidity increases likely linked to offshore transport events. A pixel with ΔTSS > 0 suggests the presence of turbid coastal water moving into clearer offshore areas. By counting the frequency of such events ($P_{\Delta TSS}$), we derive a statistical proxy for the intensity and recurrence of offshore transport over time. In order to statistically analyze the spatio-temporal distribution characteristics of the offshore transport pathways, the spatial cumulative frequency of ΔTSS ($P_{\Delta TSS}$) was calculated in this study using the following methodology:

Data pre-processing: To eliminate noise, TSS values greater than 100 g/m^3 or equal to 0 g/m^3 were set as NaN. The monthly climatological mean TSS ($CSST_{month}$) was then calculated.

Anomaly calculation: Hourly TSS satellite data were subtracted from the corresponding monthly climatological mean TSS to obtain hourly anomalies ($\Delta TSS_{hourly} = TSS_{hourly} - CSST_{month}$) (Figure 3b). If $\Delta TSS_{hourly} > 0$, it indicates that the TSS value at that pixel and hour was higher than the background level, possibly due to offshore transport of highly turbid coastal waters (Figure 3a,b). In this study, sensitivity analysis was performed on the ΔTSS_{hourly} (Figure S1 of the support information), and the results showed that the selection of ΔTSS values did not affect the spatial distribution morphology of the transport path in the study area.

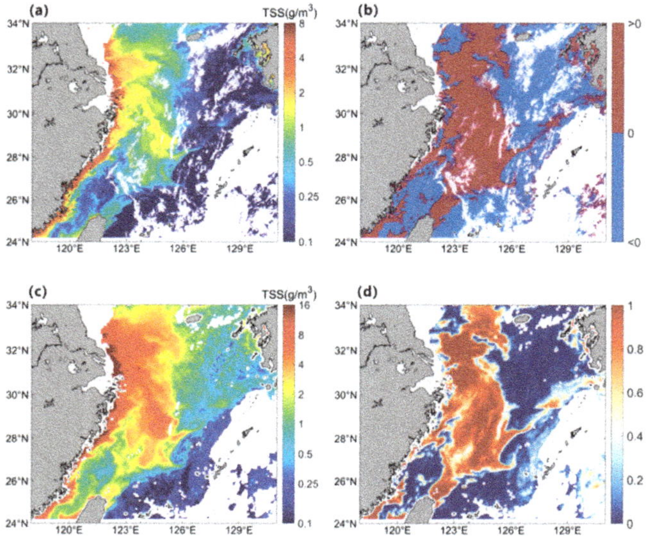

Figure 3. Spatial distribution of (**a**) hourly TSS at 05:00, (**b**) ΔTSS_{hourly} at 05:00, (**c**) daily TSS, and (**d**) $P_{\Delta TSS}$ on 2 August 2015.

Cumulative frequency calculation: For each pixel, the number ($N_{\Delta TSS_{hourly}>0}$) of instances where ΔTSS_{hourly} is greater than 0 was divided by the total number ($N_{cloud-free}$) of cloud−free satellite images, yielding the spatial cumulative frequency of ΔTSS ($P_{\Delta TSS}$) (Figure 3d):

$$P_{\Delta TSS} = \frac{N_{\Delta TSS_{hourly}>0}}{N_{cloud-free}} \tag{1}$$

Using the above methodology, the climatological annual mean ($P_{\Delta TSS}(climatology)$), seasonal mean ($P_{\Delta TSS}(seasonal)$), and monthly mean ($P_{\Delta TSS}(monthly)$) spatial cumulative frequencies of ΔTSS were calculated.

2.7. Simpson–Hunter Tidal Mixing Index

Since this study focuses on the horizontal offshore transport processes of nearshore high−turbidity water masses, it is necessary to identify the extent of tidal mixing−dominated zones to avoid interference from tidal−induced TSS variations in the results. This study employs the Simpson–Hunter stratification parameter K [25] to determine the boundary of the fully tidally mixed zone, where $K = \log_{10}\left(\frac{H}{U^3}\right)$, with H representing water depth and U denoting the characteristic tidal current velocity. This parameter has been widely used to locate tidal fronts (i.e., the boundaries of fully tidally mixed zones). Generally, when K > 3, the water column exhibits strong stratification, whereas when K < 1.5, it is fully mixed [16]. To delineate the maximum influence area of tidal vertical mixing, this study selects the maximum tidal current velocity as the characteristic tidal velocity. Following the approach of Zhao et al. [32] in their investigation of tidal mixing fronts in the Yellow Sea and northern ECS, a critical value of 1.8 is adopted to define the position of the tidal front.

3. Results and Discussion

3.1. Distribution Characteristics of Climatological TSS and Influencing Factors

3.1.1. Distribution Characteristics of Climatological TSS

The climatological monthly mean TSS data from May 2012 to December 2020 (Figure 4) reveal that the coastal waters of the East China Sea exhibit a typical spatial distribution pattern of high TSS concentrations nearshore and low concentrations offshore. Within the 20 m isobath, TSS concentrations are generally high and show an exponential decrease with increasing distance from the coast, nearly dropping to zero in shelf areas where water depth exceeds 80 m.

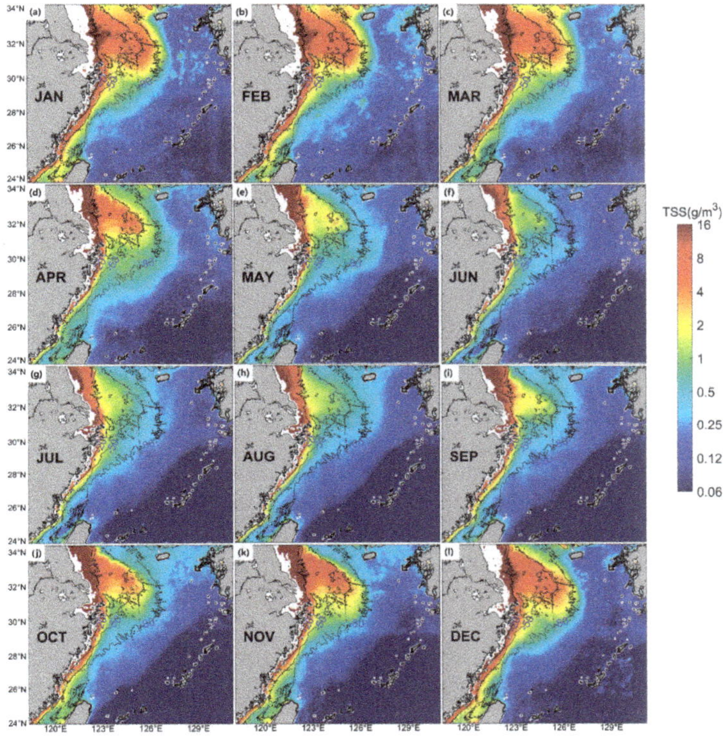

Figure 4. (**a**–**l**) Climatological monthly mean TSS distribution from May 2012 to December 2020 (background color), with black contours indicating the 20 m, 50 m, and 80 m isobaths.

Additionally, TSS concentrations in the coastal East China Sea display pronounced seasonal variations. In the Subei Shoal and adjacent shelf areas north of the Yangtze River Estuary, TSS concentrations exhibit a "summer accumulation and winter transport" temporal pattern [33]. During summer, high TSS values are primarily confined within the 20 m isobath (Figure 4i), whereas in winter, high-TSS water masses expand southeastward to approximately the 80 m isobath on the shelf (Figure 4b). South of the Yangtze River Estuary along the Zhejiang–Fujian coastal waters, TSS concentrations generally follow a distribution pattern parallel to the coastline. High-turbidity coastal water masses are confined within the 50 m isobath, forming a distinct Zhejiang–Fujian coastal turbidity front at their boundary with low-turbidity shelf waters. In summer, the turbidity front is located closer to the shoreline, typically near the 20 m isobath (Figure 4i), while in winter, it extends outward to the vicinity of the 50 m isobath (Figure 4b).

3.1.2. Influencing Factors of TSS Climate State Distribution Characteristics

Previous studies have demonstrated that the distribution of surface TSS in the ECS is primarily influenced by tidal mixing, riverine input, wind-induced stirring, and circulation systems [33,34]. The coastal ECS is globally renowned as a strong tidal region, dominated by semidiurnal tides. The maximum tidal amplitude of the M2 constituent exceeds 1 m, with tidal current velocities reaching 1 m/s. Additionally, the S2 constituent exhibits a maximum amplitude of 0.6 m and a tidal current velocity of 0.5 m/s [35,36]. The vertical mixing caused by the strong interaction between tidal currents and coastal topography resuspends bottom sediments into surface waters, thereby dominating the spatial distribution of high TSS concentrations in coastal areas. In Figure 5, the contour of 1.8 (thick black line in Figure 5), representing the boundary of the tidally well−mixed zone [37], nearly coincides with the boundary of climatological high-TSS regions, further confirming the dominant role of tidal forcing in TSS distribution.

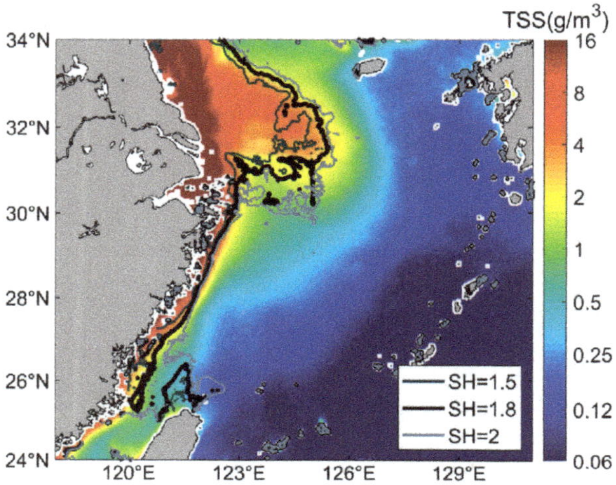

Figure 5. Distribution of the Simpson–Hunter tidal mixing parameter K, with background colors representing the mean TSS from May 2012 to December 2020. Black contour lines denote K values of 1.5, 1.8, and 2, respectively.

Although the high-TSS distribution in the coastal East China Sea exhibits seasonal variations characterized by onshore contraction in summer and offshore expansion in winter [38,39], the underlying driving factors differ between the northern and southern sides of the Yangtze estuary.

North of the estuary on the Subei Shoal, TSS seasonality is primarily governed by vertical mixing intensity [40] and cross−shelf transport driven by the Yellow Sea Coastal Current (YSCC) [41,42]. During summer, southerly monsoon winds weaken the YSCC [14], thereby preventing the northward deflection of turbid coastal waters from reaching the shelf area between the 20–50 m isobaths. Furthermore, enhanced thermal stratification has been shown to reduce vertical mixing [37,43], thereby limiting sediment resuspension and confining high-TSS waters to the shoreward side of the 20 m isobath. Conversely, during winter, the northerly monsoons intensify the southward−flowing YSCC (Figure 6). When this current deflects northward near the Yangtze estuary, it transports substantial turbid coastal water masses onto the shelf [44]. Vigorous wind stirring and surface cooling induce full-depth mixing within the 50 m isobath, entraining bottom sediments into the surface layers [42,45]. This process drives the offshore expansion of high-TSS waters across the shelf region within the 50 m isobath [46].

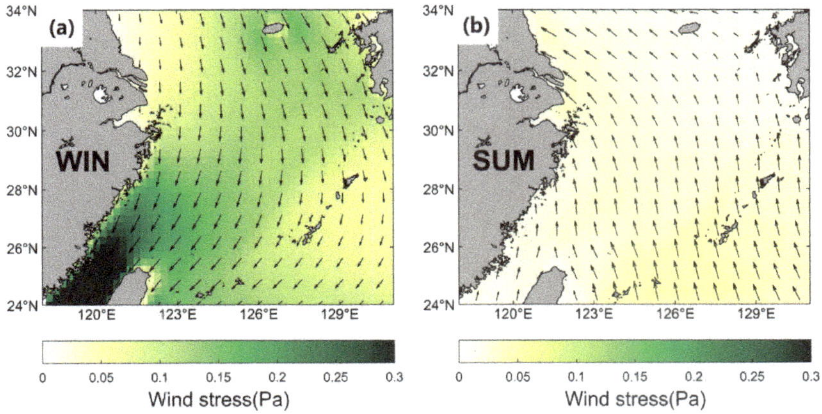

Figure 6. Climatological winter (**a**) and summer (**b**) sea surface wind fields from May 2012 to December 2020. The background color represents wind stress, with arrows indicating wind vectors.

South of the Yangtze River estuary along the Zhejiang–Fujian coast, the seasonal variability of TSS is primarily controlled by the intensity variations in the Zhejiang–Fujian Coastal Current (ZFCC) and the TWC. During summer, the northeastward deflection of YRDW and the weakened southward ZFCC under southerly monsoons [47], coupled with the significantly strengthened northward TWC [48], shift the Zhejiang–Fujian coastal turbidity front closer to shore, typically positioned near the 20 m isobath [49–51]. In winter, the opposite pattern emerges: the YRDW extends southward alongshore and merges with the ZFCC, while northerly monsoons intensify the coastal current [52]. Concurrently, the TWC experiences a weakening under the influence of monsoon forcing [53]. Consequently, the turbidity front migrates offshore, typically located near the 50 m isobath.

3.2. Spatiotemporal Distribution of the $P_{\Delta TSS}$

The climatological distribution of TSS (Figures 4 and 5) demonstrates that coastal water masses in the ECS are primarily confined to nearshore areas within the 50 m isobath throughout the year. However, the image of hourly TSS (Figure 3) reveals intermittent offshore transport processes in this region. These processes have the capacity to transport high-turbidity coastal water masses extensively to the shelf area. Given the higher turbidity of coastal water masses compared to shelf water masses, the offshore transport of coastal water can be tracked by analyzing positive TSS anomalies (ΔTSS). Furthermore, through spatiotemporal analysis of the spatial cumulative frequency of ΔTSS $P_{\Delta TSS}$, we can elucidate the distribution patterns of offshore transport pathways for coastal water masses in the ECS.

3.2.1. The Climatological Spatial Distribution Characteristics of $P_{\Delta TSS}$

As demonstrated in Figure 7, the climatological mean of the spatial cumulative frequency $P_{\Delta TSS}$ is predominantly minimal within the 20 m isobath. Between the 20 and 50 m isobaths, a band of high values extends from north to south, while beyond the 50 m isobath on the continental shelf, high−value zones are mainly distributed north of 28° N, extending as far as the edge of the ECS shelf. Conversely, a low-value zone is evident south of 28° N, extending northwestward from the shelf break northeast of Taiwan to the area between the 50 and 80 m isobaths. Furthermore, regions influenced by the Kuroshio Current, particularly along the shelf break of the ECS, generally exhibit low $P_{\Delta TSS}$ values. This suggests that long−distance cross-shelf offshore transport of high-turbidity coastal water masses primarily occurs in the northern ECS, reaching as far as the shelf edge, while cross-shelf transport in the southern ECS is largely constrained by the 50 m isobath. This phenomenon is further corroborated by the climatological surface chlorophyll concentration distribution in the ECS (Figure 7b). In the southeastern region, high chlorophyll concentrations are mainly confined to nearshore areas, whereas in the northern ECS, high chlorophyll zones extend from the coast to the shelf region.

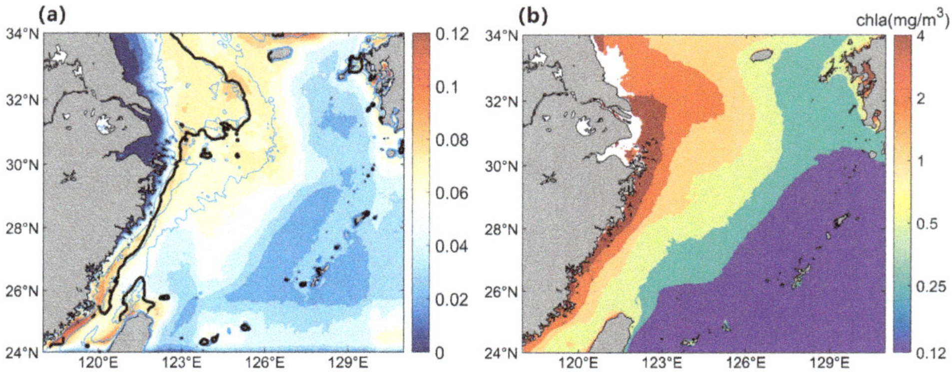

Figure 7. (**a**) Climatological distribution of $P_{\Delta TSS}$ (May 2012–December 2020); (**b**) climatological distribution of Chl-a (June 2002–March 2024). In (**a**), the black contours indicating the K = 1.8 isoline and blue contours showing the 20 m, 50 m, and 80 m isobaths.

3.2.2. Seasonal Variability of $P_{\Delta TSS}$ Distribution

From Figure 8, it can be seen that the distribution of $P_{\Delta TSS}$ exhibits not only significant spatial differences but also pronounced seasonal variations, characterized by higher values in summer and lower values in winter. In winter, $P_{\Delta TSS}$ remains generally low, both in nearshore areas within the 50 m isobath and in the shelf region beyond it (Figure 8a). In spring, $P_{\Delta TSS}$ remains low south of 29° N, while it increases significantly to 0.08 in Hangzhou Bay and the outer Yangtze River estuary north of 29° N, with the high-value zone extending as far as approximately 127.5° E (Figure 8b). During summer, $P_{\Delta TSS}$ is about 0.12−0.14, reaching its annual peak in the ECS, forming a high−value belt between the 20 m and 80 m isobaths from north to south. Around 30° N, the high-value zone extends beyond the 80 m isobath, reaching as far as approximately 129° E in the Kuroshio region (Figure 8c). In autumn, the spatial distribution of $P_{\Delta TSS}$ is similar to that in summer, but its overall values significantly decrease to 0.06–0.1. The high-value belt north of Hangzhou Bay contracts within the 80 m isobath, while the high-value belt south of Hangzhou Bay retreats within the 50 m isobath (Figure 8d). Although this study mainly focuses on the offshore transport pathways and physical mechanisms of nearshore highly turbid waters, these processes may also trigger important responses at the ecosystem level. Taking summer as an example, there are the most transport channels and the widest coverage, indicating that a large amount of nearshore water masses containing nutrients may be transported to the shelf area. After these nitrogen− and phosphorus−rich waters enter the relatively clear surface of the shelf, they may induce local or regional phytoplankton algal bloom events [54], thereby changing the pattern of primary productivity [55]. In addition, the characteristics of particulate organic carbon (POC) attached to suspended sand also make these transport pathways potential carbon export pathways [56], which help promote the cross−shelf carbon flux process and affect the regional carbon cycle. However, excessive suspended matter transported to the open sea will block sunlight, inhibit photosynthesis, and affect primary production [57,58].

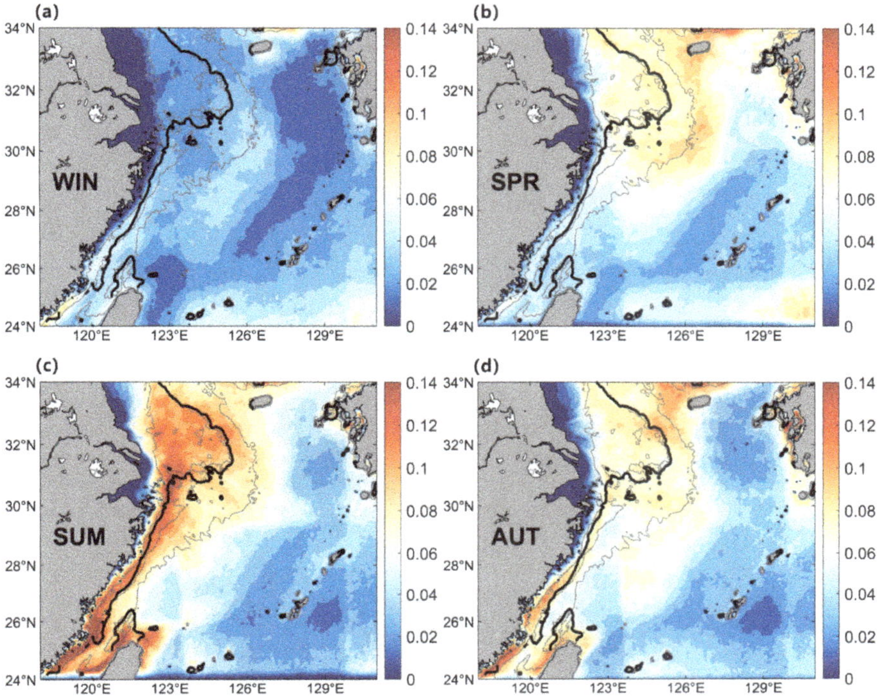

Figure 8. Seasonal distribution of the climatological $P_{\Delta TSS}$ from May 2012 to December 2020. (**a**) Winter. (**b**) Spring. (**c**) Summer. (**d**) Autumn. The black contours indicate the K = 1.8 isoline and gray contours show the 20 m, 50 m, and 80 m isobaths.

3.3. Spatial-Temporal Distribution of Offshore Transport Pathways of Coastal Water Masses

As shown in Figures 7 and 8, the spatial cumulative frequency of suspended sediment concentration anomalies (ΔTSS), denoted as $P_{\Delta TSS}$, exhibits pronounced spatial heterogeneity and seasonal variability. This reflects the highly dynamic nature of offshore transport processes of coastal turbid water masses. To comprehensively examine the spatiotemporal distribution of these offshore transport pathways, this study integrates the monthly mean spatial distribution of $P_{\Delta TSS}$ with satellite-derived sea surface residual current fields (Figure 9). To isolate the influence of horizontal transport, the analysis focuses on regions outside the tidally well−mixed zone, delineated by the Simpson–Hunter parameter K = 1.8 (black line in Figure 9). Persistent high-value $P_{\Delta TSS}$ zones were observed beyond this boundary, indicating frequent offshore movement of coastal high-turbidity waters. Source-tracking analysis was conducted on these high-value zones to identify the origin and pathways of offshore transport, as indicated by the purple arrows in Figure 9.

Between January and June, $P_{\Delta TSS}$ values across the East China Sea (ECS) remained generally below 0.1. In January and February, one to two offshore transport pathways were identified along the Zhejiang and Fujian coasts between 26° N and 30° N, extending either southeastward or northeastward. In March, $P_{\Delta TSS}$ increased significantly in the northern ECS, from 0.02 to 0.08, particularly near the Yangtze River estuary and Subei Shoal, with three offshore pathways emerging. These patterns largely persisted through April, although the high-value zone of about 0.1 near 28° N disappeared in May, and the number of pathways decreased to two. In June, offshore transport activity reached its minimum, with only a single alongshore pathway forming near the tidally mixed region of the Taiwan Strait. Transport activity intensified in July and August, with $P_{\Delta TSS}$ exceeding

0.12 and reaching 0.16 in some areas, reaching its annual peak in the northern part of the ECS. The High−value zones exceeding 0.12 extended to the continental shelf edge, and the number of offshore transport pathways increased to three or four. These originated from the Yangtze River estuary, Subei Shoal, and near 28° N, extending southeastward or northeastward. In September, the number of pathways decreased to two as transport activity declined. In November and December, offshore transport pathways were largely absent, with only a persistent eastward pathway observed near 28° N.

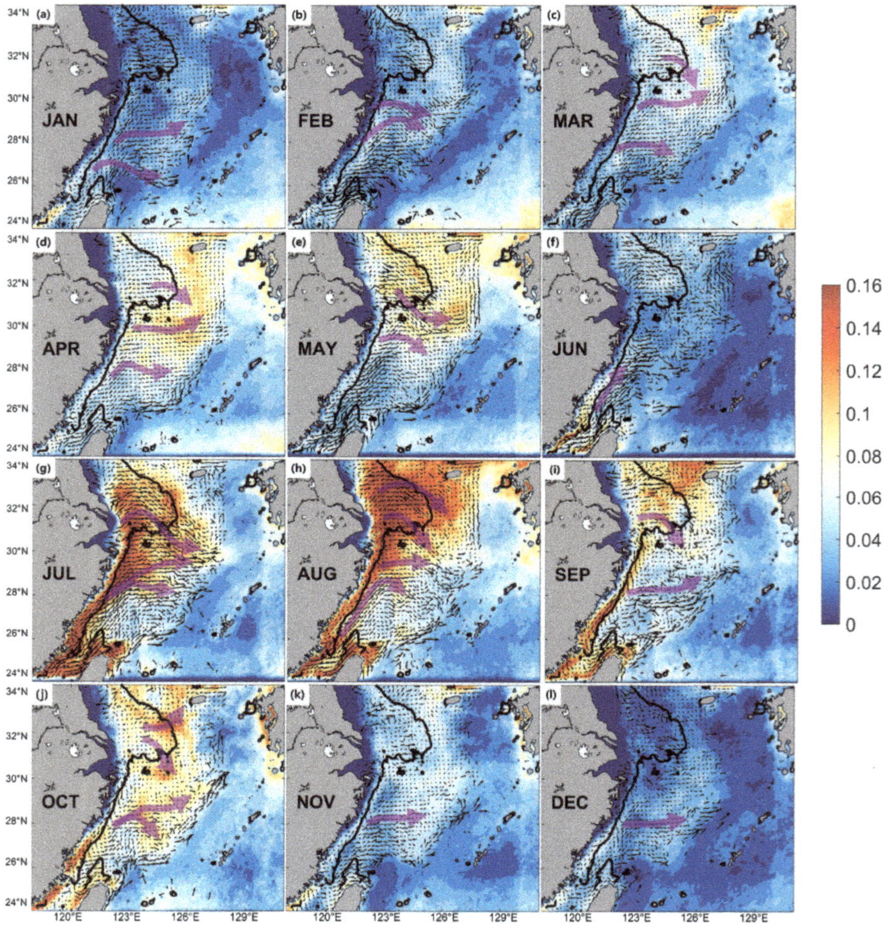

Figure 9. (**a**–**l**) Climatological monthly mean distribution of $P_{\Delta TSS}$ from May 2012 to December 2020. Black arrows represent the residual current field derived from satellite observations, the black line denotes the K = 1.8 tidal mixing parameter contour, and semi-transparent purple arrows indicate potential offshore transport pathways.

The offshore transport pathways of the Zhejiang–Fujian coastal waters exhibit clear spatiotemporal heterogeneity, with the frequency and number of transport events peaking in summer (up to four pathways) and dropping to a minimum in winter (one pathway). Except for January, all identified pathways are located north of 28° N. Among these, the pathway near 28° N is the most persistent, present in all months except May and June. Its formation is closely linked to the dynamics of the Taiwan Warm Current (TWC). Originating from coastal waters south of 28° N, this pathway is driven northeastward by the northward flow of the TWC, then turns eastward under the influence of its bifurcation, eventually reaching the ECS shelf edge. An additional pathway between 29° N and 30° N occurs during February–May and August, extending eastward and reaching waters near 126° E. This aligns well with the position of the penetrating front described by Wu [15], which is typically observed under prevailing northerly monsoon conditions and is attributed to thermocline oscillations driven by spatially uneven tidal mixing. Moreover, a southeastward transport pathway off the north bank of the Yangtze estuary appears in most months except January–February, June, and November–December. This finding deviates from the traditional "summer accumulation, winter transport" paradigm of Subei Shoal sediment dynamics. Instead, it suggests a summer−dominant transport mechanism possibly linked to the seasonal extension of the Yangtze River Diluted Water (YRDW). This interpretation is supported by Lie and Cho [59], who reported that the YRDW flows northeastward then southeastward in summer after exiting the estuary, consistent with the pathway identified in this study.

4. Conclusions

The offshore transport process of coastal water masses in the East China Sea (ECS) plays a crucial role in supporting the regional marine ecosystem. This study employs TSS concentration data from the Korean GOCI satellite. By analyzing the climatological distribution of TSS and the spatial cumulative frequency of positive TSS anomalies ($P_{\Delta TSS}$), in conjunction with satellite−derived surface residual currents, we systematically investigate the spatio-temporal characteristics of offshore transport for coastal water masses in the ECS. The main findings are summarized as follows:

1. **Dominant spatial pattern of nearshore turbidity waters**

Nearshore waters in the ECS consistently exhibit a "high nearshore, low offshore" TSS distribution pattern. High-turbidity water masses are mainly confined within the 20–50 m isobath range, with a clear seasonal shift—extending southeastward to the 50 m isobath in winter and retreating shoreward to the 20 m isobath in summer.

2. **Spatio-temporal heterogeneity of offshore transport patterns**

The offshore transport of high-turbidity water masses shows pronounced spatial and seasonal variability.

Seasonally, summer (June–August) is the peak transport period, featuring up to four active pathways and the broadest spatial coverage. In contrast, winter (December–February) is marked by only one main pathway and reduced transport frequency, while spring and autumn display transitional characteristics.

Spatially, pathways north of 28° N can extend to the shelf edge (~129° E), indicating relatively unobstructed offshore movement. South of 28° N, however, the northward branch of the Taiwan Warm Current (TWC) acts as a physical barrier, confining transport within the 50 m isobath.

3. **Identification of persistent and seasonal transport pathways**

A persistent transport pathway near 28° N is observed year-round (except during May–June), likely driven by the eastward branch of the TWC. This pathway extends northeastward to ~126° E, then turns eastward toward the shelf edge (~129° E).

Two additional seasonal pathways are identified:

- One located south of Hangzhou Bay (~30° N), active during February–May and August, is likely linked to tidal-induced thermocline oscillations. Its location aligns with the penetrating front reported by Wu [15].

- Another located off the north bank of the Yangtze River Estuary appears exclusively in summer and follows a southeastward trajectory consistent with the expansion of the Yangtze River Diluted Water (YRDW). This challenges the conventional "summer accumulation, winter transport" paradigm and highlights the significance of summer offshore export processes in the northern ECS.

These findings provide new observational insights into the offshore transport dynamics of coastal water masses in the ECS. They enhance our understanding of the spatial structure and seasonal variability of transport pathways and carry important implications for material exchange, nutrient cycling, and the ecological stability of the continental shelf system. Future research should integrate numerical modeling with field observations to better quantify the roles of various physical drivers and to improve the predictive capability of material transport processes under evolving environmental conditions.

Supplementary Materials: The following supporting information can be downloaded at: https://www.mdpi.com/article/10.3390/w17091370/s1, Figure S1: Sensitivity test diagram of ΔTSS, the figure represents $P_{\Delta TSS}$ on 2 August 2015. (a) takes ΔTSS > 0, (b) takes ΔTSS > 0.05, (c) takes ΔTSS > 0.1.

Author Contributions: Methodology, W.Y.; software, Y.P.; validation, Y.P.; formal analysis, W.Y.; data curation, Y.P., writing—original draft preparation, Y.P. and W.Y.; review and editing, W.Y., visualization, Y.P.; supervision, W.Y.; funding acquisition, W.Y. All authors have read and agreed to the published version of the manuscript.

Funding: Yuanjie Peng and WenBin Yin was funded by the Science and Technology Project of Zhoushan (Grant No. 2023C41020) and the National Natural Science Foundation of China (Grant No. 41906025).

Data Availability Statement: The TSS data from GOCI can be downloaded from the website of the Korea Ocean Satellite Center (https://kosc.kiost.ac.kr) (accessed on 1 April 2025) The CCMP dataset is available to download from the website http://www.remss.com/measurements/ccmp/ (accessed on 1 April 2025). The TPXO model data can be downloaded from the website https://www.tpxo.net/global (accessed on 2 April 2025). The FVCOM model data were obtained from Xuan [31] in 2016. The $P_{\Delta TSS}$ data can be downloaded from the website https://figshare.com/account/home (accessed on 20 April 2025).

Conflicts of Interest: The authors declare no conflicts of interest.

References

1. Liu, S.; Han, S.; Wei, Y. Analysis of water mass in the northwest of the East China Sea and its relationship with fishing grounds. *J. Fish. China* **1984**, *8*, 125–133.
2. Wu, L.; Wei, Q.; Xin, M. Spatial distribution patterns of nutrients and controlling mechanisms in the East China Sea during spring. *Adv. Mar. Sci.* **2023**, *41*, 622–636.
3. Liu, D.; Qiao, L.; Li, G. Suspended matter transport, flux and seasonal variation in the inner shelf of the East China Sea. *Oceanol. Et Limnol. Sinica* **2022**, *49*, 24–39.
4. Guan, B. Patterns and Structures of the Currents in Bohai, Huanghai and East China Seas. In *Institute of Oceanology*; Springer: Berlin/Heidelberg, Germany, 1994.
5. Guo, Y.; Rong, Z.; Chi, Y.; Li, X.; Na, R. Numerical Study on the Interannual Variation of the Changjiang Diluted Water in Summer. *Ocean. Limnol. Bull.* **2020**, *4*, 30–41.
6. Mao, H.; Gan, Z.; Lan, S. Preliminary study on the Changjiang diluted water and its mixing. *J. Oceanol. Limnol.* **1963**, *5*, 183–206.
7. Beardsley, R.; Limeburner, R.; Yu, H.; Cannon, G. Discharge of the Changjiang (Yangtze River) into the East China sea. *Cont. Shelf Res.* **1985**, *4*, 57–76. [CrossRef]
8. Yuan, Y.; Su, J.; Zhao, J. A single-layer model of continental shelf circulation in the East China Sea. *Acta Oceanol. Sin.* **1982**, *4*, 1–11.
9. Zhu, B.; Yang, W.; Jiang, C.; Wang, T.; Wei, H. Observations of turbulent mixing and vertical diffusive salt flux in the Changjiang Diluted Water. *J. Oceanol. Limnol.* **2022**, *40*, 1349–1360. [CrossRef]
10. Wang, C.; Guo, X.; Fang, J.; Li, Q. Seasonal characteristics of the extension range of Fujian-Zhejiang Coastal Current and its influence on typical bays. *J. Appl. Oceanogr.* **2018**, *37*, 1–8.

11. Yuan, D.; Qiao, F.; Su, J. Cross-shelf penetrating fronts off the southeast coast of China observed by MODIS. *Geophys. Res. Lett.* **2005**, *32*, L19603. [CrossRef]
12. Yuan, D.; Li, Y.; He, L.; Zhou, H.; Li, R.; Wang, F.; Lei, H.; Hu, D. An observation of the three-dimensional structure of a cross-shelf penetrating front off the Changjiang mouth. *Deep. Sea Res. Part II Top. Stud. Oceanogr.* **2010**, *57*, 1827–1834. [CrossRef]
13. He, L.; Li, Y.; Zhou, H.; Yuan, D. Variability of cross-shelf penetrating fronts in the East China Sea. *Deep. Sea Res. Part II Top. Stud. Oceanogr.* **2010**, *57*, 1820–1826. [CrossRef]
14. Li, Y. Summertime Circulation Characteristics and Its Dynamic Mechanism in the Coastal Waters of Eastern China. Ph.D. Thesis, Institute of Oceanography, Chinese Academy of Sciences, Qingdao, China, 2010.
15. Wu, H. Cross-shelf penetrating fronts: A response of buoyant coastal water to ambient pycnocline undulation. *J. Geophys. Res. Ocean.* **2015**, *120*, 5101–5119. [CrossRef]
16. Yin, W.; Huang, D. Evolution of submesoscale coastal frontal waves in the East China Sea based on geostationary ocean color imager observational data. *Geophys. Res. Lett.* **2016**, *43*, 9801–9809. [CrossRef]
17. Peng, X.; Shen, F. Comparative Analysis of Suspended Particulate Matter Concentration in Yangtze Estuary Derived by Several Satellite Sensors. *Infrared* **2014**, *35*, 31.
18. Yeom, J.-M.; Kim, H.-O. Comparison of NDVIs from GOCI and MODIS data towards improved assessment of crop temporal dynamics in the case of paddy rice. *Remote Sens.* **2015**, *7*, 11326–11343. [CrossRef]
19. Liu, X.; Yang, Q.; Liu, Q. Adaptability analysis of various versions of GDPS in GOCI data processing in the Yellow Sea based on QA Score. *Spectrosc. Spectr. Anal.* **2021**, *41*, 2233–2239.
20. Yu, Z.; Wang, J.; Li, Y. Remote sensing of suspended sediment in high turbid estuary from sentinel-3A/OLCI: A case study of Hangzhou Bay. *Front. Mar. Sci.* **2022**, *9*, 1008070. [CrossRef]
21. Ruddick, K.; Vanhellemont, Q.; Yan, J.; Neukermans, G.; Wei, G.; Shang, S. Variability of suspended particulate matter in the Bohai Sea from the geostationary Ocean Color Imager (GOCI). *Ocean. Sci. J.* **2012**, *47*, 331–345. [CrossRef]
22. Yang, Z.; Lei, K.; Guo, Z.; Wang, H. Effect of a winter storm on sediment transport and resuspension in the distal mud area, the East China Sea. *J. Coast. Res.* **2007**, *23*, 310–318. [CrossRef]
23. Zheng, C. Based on CCMP wind field, the characteristics of sea surface wind field in China seas in the past 22 years are analyzed. *Res. Meteorol. Disaster Reduct.* **2011**, *34*, 41–46.
24. Zhang, W.; Zhu, S.; Li, X.; Ruan, K.; Guan, W.; Peng, J. The effects of tidal residual current and tidal mixing on the low-salinity water mass in the northeastern sea area outside the Yangtze Estuary. *Acta Oceanol. Sinica* **2014**, *36*, 9–18.
25. Simpson, J.H.; Hunter, J. Fronts in the Irish sea. *Nature* **1974**, *250*, 404–406. [CrossRef]
26. Tian, Z.; Wang, C.; Yu, Z.; Liu, H.; Lin, P.; Li, Z. Tide simulation in a global eddy-resolving ocean model. *Acta Oceanol. Sin.* **2024**, *43*, 1–10. [CrossRef]
27. Morlighem, M. *Measures Bedmachine Antarctica, Version 2*; National Snow and Ice Data Center: Boulder, CO, USA, 2020.
28. Giribabu, D.; Hari, R.; Sharma, J.; Ghosh, K.; Padiyar, N.; Sharma, A.; Bera, A.K.; Srivastav, S.K. Performance assessment of GEBCO_2023 gridded bathymetric data in selected shallow waters of Indian ocean using the seafloor from ICESat-2 photons. *Mar. Geophys. Res.* **2024**, *45*, 1. [CrossRef]
29. Smith, W.H.; Sandwell, D.T. Global sea floor topography from satellite altimetry and ship depth soundings. *Science* **1997**, *277*, 1956–1962. [CrossRef]
30. Ma, Y.; Yin, W.; Guo, Z.; Xuan, J. The ocean surface current in the East China Sea computed by the Geostationary Ocean Color Imager satellite. *Remote Sens.* **2023**, *15*, 2210. [CrossRef]
31. Xuan, J.; Huang, D.; Pohlmann, T.; Su, J.; Mayer, B.; Ding, R.; Zhou, F. Synoptic fluctuation of the Taiwan Warm Current in winter on the East China Sea shelf. *Ocean. Sci.* **2017**, *13*, 105–122. [CrossRef]
32. Zhao, B. Distribution of tidal shelf fronts in the Yellow Sea. *Bohai Seas* **1987**, *5*, 16–23.
33. Qiao, L. Suspended Sediment Transport, Flux, and Seasonal Variation on the East China Sea Inner Shelf. *Ocean. Lakes* **2018**, *49*, 16.
34. Zhou, Y.; Xuan, J.; Huang, D. Tidal variation of total suspended solids over the Yangtze Bank based on the geostationary ocean color imager. *Sci. China Earth Sci.* **2020**, *63*, 1381–1389. [CrossRef]
35. Zhao, B.; Fang, G.; Cao, D. Numerical simulation of tidal current in Bohai Sea, Yellow Sea and East China Sea. *J. Oceanogr. (Chin. Version)* **1994**, *16*, 10.
36. Lin, Z.; Zhu, X.; Bao, X.; Liu, Q. Numerical simulation of three-dimensional tide and tidal current in Quanzhou Bay based on FVCOM. *J. Oceanogr. (Chin. Version)* **2013**, *35*, 15–24.
37. Zhao, B. Yellow Sea Cold Water Mass front mixed with tide. *Ocean. Lakes* **1985**, *16*, 451–460.
38. Chang, P.H.; Isobe, A. A numerical study on the Changjiang diluted water in the Yellow and East China Seas. *J. Geophys. Res. Ocean.* **2003**, *108*, 3299. [CrossRef]
39. Chang, Y.; Lee, M.; Shimada, T.; Sakaida, F.; Kawamura, H.; Chan, J.; Lu, H. Wintertime high-resolution features of sea surface temperature and chlorophyll—A fields associated with oceanic fronts in the southern East China Sea. *Int. J. Remote Sens.* **2008**, *29*, 6249–6261. [CrossRef]

40. Cheng, X.; Sun, Q.; Wang, Y.; Yang, Y. Seasonal Variation and Structural Analysis of the Tidal Front Outside the Jiangsu Bank Radial Sand Ridges. *Mar. Sci.* **2017**, *41*, 1–8.
41. Guo, Z.; Yang, Z.; Zhang, D.; Fan, D.; Lei, K. Winter, summer distribution of suspended matter in the northern East China Sea and the blocking effect of ocean currents on suspended matter transport. *J. Oceanogr.* **2002**, *24*, 71–80.
42. Yang, Z.; Guo, Z.; Wang, Z.; Xu, J.; Gao, W. The macroscopic pattern of the transport of suspended matter from the continental shelf of the Yellow Sea and East China Sea to its eastern deep sea area. *Acta Oceanol. Sin.* **1992**, *14*, 81–90.
43. Tang, Y.; Zou, E.; Li, X.; Li, Z. Some characteristics of the circulation in the South Yellow Sea. *Acta Oceanol. Sin.* **2000**, *22*, 1–16.
44. Bian, C. Sediment Transport in the Nearshore Area of China in the Bohai Sea, Yellow Sea and East China Sea. Ph.D. Dissertation, Ocean University of China, Qingdao, China, 2012.
45. Bao, X.; Li, Z.; Wang, Y.; Li, N. Winter and Summer Distribution Characteristics of Suspended Matter in the Northern Yellow Sea. *Sediment Res.* **2010**, *2*, 48–56.
46. Chen, Y. Numerical Simulation of Seasonal Continuous Variation of the Yellow Sea-East China Sea Circulation and the Changjiang Diluted Water. Master's Thesis, East China Normal University, Shanghai, China, 2007.
47. Qi, J. Characteristics of the East Sea Water Mass and a Study of the Kuroshio Current's Exchange with East China Sea Shelf Water. Ph.D. Thesis, School of the Chinese Academy of Sciences (Institute of Oceanology), Qingdao, China, 2014.
48. Lin, X.; Hou, L.; Liu, M.; Li, X.; Yin, G.; Zheng, Y.; Deng, F. Gross nitrogen mineralization in surface sediments of the Yangtze Estuary. *PLoS ONE* **2016**, *11*, e0151930. [CrossRef] [PubMed]
49. Shi, X.; Li, H.; Wang, H.; Wang, L.; Zhang, C. Preliminary Study on the Hydrochemical Characteristics of the Taiwan Warm Current in Summer and Its Impact on the Frequent Occurrence Area of Red Tide in the East China Sea. *Ocean. Lakes* **2013**, *44*, 1208–1215.
50. Luo, Y.; Yu, G. Numerical calculation of upwelling along the East China Sea caused by wind and Taiwan warm current. *J. Qingdao Ocean. Univ. (Nat. Sci. Ed.)* **1998**, *28*, 536–542.
51. Bao, M.; Guan, W.; Cao, Z.; Chen, Q.; Yang, Y. Marine Ecological Disasters and Their Physical Controlling Mechanisms in Jiangsu Coastal Area. In *Coastal Environment, Disaster, and Infrastructure—A Case Study of China's Coastline*; IntechOpen: London, UK, 2018.
52. Wang, J.; Si, G.; Yu, F. Research Progress on Variation Characteristics and Mechanism of Taiwan Warm Current. *Mar. Sci.* **2020**, *44*, 141–148.
53. Sun, X.; Fang, M.; Huang, W. The temporal and spatial variation of suspended matter transport in the Yellow Sea and East China Sea continental shelf area. *Ocean. Lakes* **2000**, *6*, 581–587.
54. Yuan, Y. Study on the Distribution and Key Processes of Nutrients in Typical Sea Areas Under Different Backgrounds of Human Activities and Natural Driving. Ph.D. Thesis, School of the Chinese Academy of Sciences (Institute of Oceanology), Qingdao, China, 2016.
55. Liu, K.; Chao, S.; Lee, H.; Gong, G.; Teng, Y. Seasonal variation of primary productivity in the East China Sea: A numerical study based on coupled physical-biogeochemical model. *Deep. Sea Res. Part II Top. Stud. Oceanogr.* **2010**, *57*, 1762–1782. [CrossRef]
56. Zhao, B.; Yao, P.; Bianchi, T.; Yu, Z. Controls on organic carbon burial in the Eastern China marginal seas: A regional synthesis. *Glob. Biogeochem. Cycles* **2021**, *35*, e2020GB006608. [CrossRef]
57. Wang, Y.; Wu, H.; Lin, J.; Zhu, J.; Zhang, W.; Li, C. Phytoplankton blooms off a high turbidity estuary: A case study in the Changjiang River Estuary. *J. Geophys. Res. Ocean.* **2019**, *124*, 8036–8059. [CrossRef]
58. Llames, M.E.; Lagomarsino, L.; Diovisalvi, N.; Fermani, P.; Torremorell, A.M.; Pérez, G. The effects of light availability in shallow, turbid waters: A mesocosm study. *J. Plankton Res.* **2009**, *31*, 1517–1529. [CrossRef]
59. Lie, H.J.; Cho, C.H.; Lee, J.H.; Lee, S. Structure and eastward extension of the Changjiang River plume in the East China Sea. *J. Geophys. Res. Ocean.* **2003**, *108*, 3077. [CrossRef]

water

MDPI

Article

Selection of a Turbulence Model for Wave Evolution on a New Ecological Hollow Cube

Haitao Zhao [1,2], Junwei Ye [1,2,*], Kaifang Wang [3], Yian Zhou [4], Zhen Zeng [3], Qiang Li [1] and Xizeng Zhao [5]

1 School of Hydraulic Engineering, Zhejiang University of Water Resources and Electric Power, Hangzhou 310018, China; zhaoht@zjweu.edu.cn (H.Z.); liq@zjweu.edu.cn (Q.L.)
2 School of Marine Engineering Equipment, Zhejiang Ocean University, Zhoushan 316022, China
3 Zhejiang Design Institute of Water Conservancy and Hydroelectric Power, Hangzhou 310002, China; wkf8707@mail.zdwp.net (K.W.); zz8003@mail.zdwp.net (Z.Z.)
4 Hangzhou Forestry and Water Conservancy Bureau, Hangzhou 310014, China; 13706515118@163.com
5 Ocean College, Zhejiang University, Zhoushan 316021, China; xizengzhao@zju.edu.cn
* Correspondence: junweiyegg@gmail.com; Tel.: +86-17398593260

Abstract: A suitable turbulence model is needed for numerical simulations to accurately simulate the wave evolution and hydrodynamic performance of the new ecological hollow cube. The new ecological hollow cube is an improvement upon traditional designs, as it can grow plants to dissipate wave energy. In this study, the open-source computational fluid dynamics (CFD) software OpenFOAM v2206 is used as the computational platform to analyze and evaluate the numerical results of four turbulence models, i.e., the standard k-ε, steady k-ω shear stress transfer (SST), buoyancy-corrected k-ω SST, and large eddy simulation (LES) models, by using three mesh systems (with grid counts of 0.89, 2.92, and 8.91 million grids, respectively). Comparison of the numerical results from the four turbulence models reveals that the stabilized k-ω SST turbulence model provides better results for simulating the complex wave evolution process on the cube and effectively captures the wave free surface. In contrast, the other models exhibit a greater grid dependency. The stabilized k-ω SST model more accurately captures the wave run-up and reflection coefficient better than other turbulence models do. Therefore, the stabilized k-ω SST model is selected as the most suitable turbulence model for hydrodynamic modeling of the new ecological hollow cube.

Keywords: numerical modeling; turbulence models; hollow cube; wave run-up; wave reflection; computational fluid dynamics

1. Introduction

Offshore infrastructure in coastal cities is facing increasing threats from both rapid urbanization and a rise in severe weather events, including the destructive forces of typhoons and storm surges. It is a major engineering challenge to protect coastal structures from being eroded by waves [1]. In coastal engineering, armor blocks are frequently utilized as a form of protective armor unit. Dams, embankments, seawalls, and breakwaters are frequently shielded from wave erosion through the use of armor blocks. In particular, based on the traditional hollow cube, the improved new ecological hollow cube has been proven to have excellent wave dissipation [2]. It is important to protect critical coastal infrastructure from wave erosion and impact.

Due to the complexities of wave behavior, physical model tests and numerical simulations are standard techniques for analyzing how waves evolve around armor blocks and how effectively these blocks perform under wave action [3]. However, performing physical model tests for armor blocks in wave flumes is expensive, and it takes time to fabricate the blocks. Numerical simulation is important in researching armor blocks as an effective and complementary tool. The complex hydrodynamics associated with wave interaction over novel ecological hollow cube structures necessitates a comprehensive evaluation of the resulting wave turbulence. To guarantee the reliability of numerical simulations that depict wave interaction with innovative ecological hollow cubes, it is crucial to utilize an appropriate model for turbulence during the calculation process.

Two major categories of turbulence models, RANS and LES, have undergone substantial development in recent decades. Within the RANS category, several specific model types exist, including those based on zero-equation, one-equation, two-equation, and Reynolds stress approaches. However, this model is seldom used in practical engineering because turbulent pulsation is neglected, and the mixing length can be easily obtained only for simple flows. The Spalart–Allmaras model [4] is a notable instance of a one-equation turbulence model, prized for its utility in simulating flow phenomena in aerospace and mechanical engineering applications. Due to their versatility and relative simplicity, two-equation turbulence models, with the k-ε and k-ω models being prominent examples, are not only applied extensively across diverse disciplines, but also continue to be the focus of ongoing research and enhancement efforts. In practical applications, the standard k-ε model [5], the RNG k-ε model [6], the standard k-ω model [7], and the k-ω SST model [8] are frequently selected as examples of two-equation turbulence models, in addition to several other available alternatives. The classical Reynolds stress models include the k-ε model [9] and the Wilcox k-ω model [10]. However, the Reynolds stress model is computationally intensive and difficult to converge, so it is not widely used. In the LES turbulence model, the larger, more energetic turbulent eddies are directly computed using a subgrid-scale approach, while the effects of smaller, subgrid-scale eddies are represented through appropriate turbulence models. Within the realm of LES turbulence modeling techniques, the Smagorinsky model [11] is a particularly celebrated and frequently employed approach. The LES approach to turbulence modeling often proves impractical for many engineering applications due to its intense computational requirements and difficulties in resolving flow near solid walls. In contrast, RANS models offer a computationally affordable and versatile alternative, making them a popular choice for a broad spectrum of engineering problems. While the RANS model offers advantages in terms of computational efficiency, it also exhibits recognized shortcomings in wave hydrodynamics simulations. Notably, many researchers have documented a tendency for this model to predict excessive wave decay and over-predict the levels of turbulent kinetic energy, resulting in results that deviate substantially from experimental measurements [12]. Seeking to remedy the aforementioned issues, Ref. [13] introduced a particular modification to the standard k-ω model. Recognizing the tendency for the k-ω shear stress transfer (SST) model to generate artificially high levels of turbulent kinetic energy in regions near free surfaces, Ref. [14] proposed an improvement in the form of a buoyancy correction term [15] designed to mitigate this issue. Recognizing that the k-ω SST model tended to overestimate turbulent kinetic energy, Ref. [16] proposed a refinement to the model by correcting the formulation of its vortex viscosity, a change that effectively suppressed the excessive generation of turbulence.

Obtaining trustworthy and meaningful insights from computational fluid dynamics (CFD) simulations hinges on the careful selection and implementation of a turbulence model that is appropriate for the specific physical phenomena under investigation. Aiming to elucidate the hydrodynamic behavior of a sloped surf zone, Ref. [17] conducted a series of simulations using various RANS turbulence models, and the accuracy of these simulations was assessed by comparing the predicted results with experimental data obtained by [18]. Based on their analysis, the standard k-ε model demonstrated a greater capacity to accurately simulate turbulent phenomena within the surf zone compared to the performance of the RNG k-ε model. In their investigation, Ref. [19] compared numerical simulations with experimental data obtained from a wave flume and found that the k-ω SST model more accurately represented the free surface compared to the k-ε model. Ref. [20] conducted a study to characterize the forces, velocities, and other hydrodynamic properties of waves as they propagate past a cylindrical structure, utilizing a selection of turbulence models to simulate the flow. The hydrodynamic characteristics of the buoyancy-modified k-ω SST model and the stabilized k-ω SST model effectively reduce the turbulent kinetic energy of the waves before they pass through the cylinder. However, the obtained wave impact force is smaller than that in the experimental data. Ref. [21] studied the wave evolution and hydrodynamic characteristics of a coral reef berm breakwater. A comprehensive comparison of the numerical output generated by three distinct turbulence models indicated that the stabilized k-ω SST model exhibited a superior capability in producing consistent and accurate simulations of the complex wave propagation and breaking processes that occur on a reef beach.

Prior investigations have employed initial physical model testing and numerical simulations to explore the hydrodynamic characteristics of armor blocks and to gain a better understanding of the intricate processes involved in the evolution of complex waves. Typical approaches in the study of armour blocks, typical methods include physical modeling tests [22] and numerical simulations [23]. In the existing literature, numerical simulations have often been limited to the application of a single turbulence model, such as the standard k-ε, the standard k-ω, or the stabilized k-ω SST formulation. Furthermore, these studies have not typically addressed the important issue of grid sensitivity, meaning they have not evaluated whether the results are influenced by the size and spacing of the computational cells used in the simulation. To the authors' knowledge, there is no relevant work on the comparative hydrodynamic characteristics of a new ecological hollow cube using different turbulence models, which is used as the research condition in this study, and its hydrodynamic characteristics are comparatively investigated by four different turbulence models.

This paper is organized as follows: Section 2 presents the numerical model setup. Section 3 describes the new ecological hollow cube, numerical model meshing, and data postprocessing of wave run-up and the reflection coefficient. Section 4 evaluates the effects of grid and turbulence modeling on various parameters of the hydrodynamic model, such as the relative wave run-up and reflection coefficient. Section 5 presents the summary and conclusions of this research.

2. Numerical Model

2.1. Governing Equations

The olaFlow, a two-phase solver, conducted the numerical model of the new ecological hollow cube. Ref. [24] developed the computational module IHFOAM as a platform, which was later developed into olaFlow [24]. As one of the extended solvers for OpenFOAM v2206, olaFlow uses the finite volume method (FVM) for the numerical discretization of the Reynolds-averaged Navier–Stokes (RANS) equations and relies on the volume of fluid (VOF) approach to accurately represent the interface between two distinct fluid phases. It can effectively handle structures with complex shapes and various wave interactions. Furthermore, it incorporates sophisticated wave generation and active absorption boundary algorithms [25], contributing to both a decrease in computational expense and an increase in the precision of the simulations. Considering the complexity of the new ecological hollow cube, olaFlow is chosen as the computational platform in this study. In order to reduce the computational cost of the simulations, a variable time-stepping scheme was employed. Furthermore, to maintain numerical stability and accuracy, the Courant number, a measure of numerical diffusion, was kept below a value of 0.25.

2.2. Turbulence Models

To determine the applicability of different turbulence models in simulating the hydrodynamic characteristics of new ecological hollow cubes, the standard *k*-*ε* model, the stabilized *k*-*ω* SST model, the buoyancy-modified *k*-*ω* SST model, and the LES model were initially used in this study. Ref. [2] conducted a series of physical modeling experiments for new ecological hollow cubes, and Ref. [26] conducted numerical simulations adopting the RNG *k*-*ε* turbulent model. Despite the satisfactory agreement between numerical and experimental results, the RNG *k*-*ε* turbulence model was not employed in this investigation due to its slow convergence and prohibitive computational expense in large-scale simulations. Ultimately, this study utilizes only four turbulent models: the standard *k*-*ε* model, the stabilized *k*-*ω* SST model, the buoyancy-modified *k*-*ω* SST model, and the LES model.

2.2.1. Standard *k*-*ε* Model

The standard *k*-*ε* turbulence model assumes that the flow is fully turbulent and that the eddy viscosities are isotropic, so the model applies to a flow field that fully develops into a turbulent state. The two equations for k and ε are as follows:

$$\frac{\partial(\rho k)}{\partial t}+\frac{\partial(\rho k u_i)}{\partial x_i}=\frac{\partial}{\partial x_j}\left[\frac{\mu_t}{\sigma_k}\frac{\partial k}{\partial x_j}\right]+2\mu_t S_{ij}S_{ij}-\rho\varepsilon \tag{1}$$

$$\frac{\partial(\rho\varepsilon)}{\partial t}+\frac{\partial(\rho\varepsilon u_i)}{\partial x_i}=\frac{\partial}{\partial x_j}\left[\frac{\mu_t}{\sigma_\varepsilon}\frac{\partial \varepsilon}{\partial x_j}\right]+C_{1\varepsilon}\frac{\varepsilon}{k}2\mu_t S_{ij}S_{ij}-C_{2\varepsilon}\frac{\varepsilon^2}{k} \tag{2}$$

where S_{ij} is the tensor of the average strain rate, $S_{ij}=\frac{1}{2}\left(\frac{\partial u_i}{\partial x_j}+\frac{\partial u_j}{\partial x_i}\right)$, and the dynamic eddy viscosity μ_t is determined by the following:

$$\mu_t=\rho C_\mu\frac{k^2}{\varepsilon} \tag{3}$$

The constants $C_{1\varepsilon}$, $C_{2\varepsilon}$, C_μ, ∂_k, and ∂_ε involved in this turbulent model were calibrated by fitting a large amount of turbulence test data, as listed in Table 1.

Table 1. Reference values of the coefficients in the *k*-*ε* turbulence model.

Coefficient	C_μ	σ_K	σ_ϵ	$C_{1\epsilon}$	$C_{2\epsilon}$
Value	0.09	1	1.30	1.44	1.92

2.2.2. Stabilized *k*-*ω* SST Model

To mitigate the excessive increase in turbulent energy and eddy viscosity within the fluid domain, [16] explored the implementation of both buoyancy and eddy viscosity corrections within the framework of the *k*-*ω* SST model [8]. The subsequent equations of the stabilized *k*-*ω* SST model are presented below:

$$\frac{\partial(\rho k)}{\partial t}+\frac{\partial(\rho k u_i)}{\partial x_i}=\frac{\partial}{\partial x_j}\left[(\mu+\sigma_K\mu_t)\frac{\partial k}{\partial x_j}\right]+P_k+G_b-\beta^*\rho\omega k \tag{4}$$

$$\frac{\partial(\rho\omega)}{\partial t}+\frac{\partial(\rho\omega u_i)}{\partial x_i}=\frac{\partial}{\partial x_i}\left[(\mu+\sigma_\omega\mu_t)\frac{\partial\omega}{\partial x_i}\right]+\alpha\rho S^2-\beta\rho\omega^2+2(1-F_1)\frac{\rho\sigma_{\omega 2}}{\omega}\frac{\partial k}{\partial x_i}\frac{\partial\omega}{\partial x_i} \tag{5}$$

Within this particular turbulence model, the correction factor $\alpha_{\beta s}$, used in the buoyancy term G_b, is assigned a value of 1.36. Subsequently, the calculation of turbulent viscosity proceeds as follows:

$$v_t = a_1 k / max(a_1\omega, F_2\sqrt{P_0}, a_1\lambda_2 \frac{\beta}{\beta^* a}\frac{P_0}{P_\Omega}\omega) \quad (6)$$

where the stress limitation factor λ_2 is 0.05, $P_0 = 2S_{ij}S_{ij}$,$P_\Omega = 2\Omega_{ij}\Omega_{ij}$, and the average rotation rate tensor $\Omega_{ij} = (\partial u_i/\partial x_j - \partial u_j/\partial x_i)/2$.

2.2.3. Buoyancy-Modified *k*-ω SST Model

Building upon the foundation of the *k*-ω SST model [8,14] engineered a variant that accounts for buoyancy, resulting in a buoyancy-modified *k*-ω SST model, which avoids excessive turbulent kinetic energy at the gas–liquid interface by introducing a buoyancy correction term. The *k* and ω equations are as follows:

$$\frac{\partial(\rho k)}{\partial t} + \frac{\partial(\rho k u_i)}{\partial x_i} = \frac{\partial}{\partial x_j}\left[(\mu + \sigma_K \mu_t)\frac{\partial k}{\partial x_j}\right] + P_k + G_b - \beta^* \rho\omega k \quad (7)$$

$$\frac{\partial(\rho\omega)}{\partial t} + \frac{\partial(\rho\omega u_i)}{\partial x_i} = \frac{\partial}{\partial x_i}\left[(\mu + \sigma_\omega \mu_t)\frac{\partial \omega}{\partial x_i}\right] + \frac{\alpha}{v_t}G - \beta\rho\omega^2 + 2(1 - F_1)\frac{\rho\sigma_{\omega 2}}{\omega}\frac{\partial k}{\partial x_i}\frac{\partial \omega}{\partial x_i} \quad (8)$$

F_1 is defined as:

$$F_1 = \tanh\left\{\left\{min\left[max\left(\frac{\sqrt{k}}{\beta^*\omega y}, \frac{500v}{y^2\omega}\right), \frac{4\rho\sigma_{\omega 2}k}{CD_{k\omega}y^2}\right]\right\}^4\right\} \quad (9)$$

$$CD_{k\omega} = max\left(2\frac{\rho\sigma_{\omega 2}}{\omega}\frac{\partial k}{\partial x_i}\frac{\partial \omega}{\partial x_i}, 10^{-10}\right) \quad (10)$$

where y represents the distance to the closest solid boundary. The dynamic eddy viscosity, denoted as μ_t, is then calculated by the following:

$$\mu_t = \frac{\rho\alpha_1 k}{\max(\alpha_1\omega, SF_2)} \quad (11)$$

$$F_2 = tanh\left[\left[\max\left(\frac{2\sqrt{k}}{\beta^*\omega y}, \frac{500v}{y^2\omega}\right)\right]^2\right] \quad (12)$$

where term $\alpha_{\beta s}$ represents the buoyancy correction factor, and its standard value is set to 1.176.

2.2.4. LES Model

The LES (large eddy simulation) technique is employed to model turbulence, while the volume of fluid (VOF) method is used to track the boundary between the liquid and gaseous phases. The continuity equations of this model can be written as follows:

$$\frac{\partial \overline{u_i}}{\partial t} = 0 \quad (13)$$

Momentum equation:

$$\frac{\partial \overline{u_i}}{\partial t} + \frac{\partial}{\partial x_j}\left(\overline{u_i u_j}\right) = -\frac{1}{\rho}\frac{\partial \overline{p}}{\partial x_i} + g_i + 2(v + v_t)\frac{\partial \overline{S}_{ij}}{\partial x_j} \quad (14)$$

where ρ represents the mass density; ν represents the kinematic viscosity; x_i represents the coordinate component; u_i represents the filtered velocity; g_i represents the gravitational acceleration (i = 1, 2, 3); p represents the filtered pressure; and the tensor of the fluid strain rate S_{ij} is denoted as (i, j = 1, 2, 3):

$$\overline{S_{ij}} = \frac{1}{2}\left(\frac{\partial \overline{u_i}}{\partial x_j} - \frac{\partial \overline{u_j}}{\partial x_i}\right) \quad (15)$$

Eddy viscosity, denoted as v_t, was calculated using the one-equation model developed by [27].

$$v_t = C_t \Delta \sqrt{k_s} \quad (16)$$

where Δ represents the characteristic size of the computational grid, and the subgrid kinetic energy, denoted as k_s, is then calculated as follows:

$$\frac{\partial k_s}{\partial t} + \overline{u_j}\frac{\partial k_s}{\partial x_j} = \tau_{ij}\frac{\partial \overline{u_i}}{\partial x_j} - \frac{\partial}{\partial x_j}(\tau_{ij}\overline{u_i}) + \frac{\partial}{\partial x_i}\left(\frac{v}{P_r}\frac{\partial k_s}{\partial x_i}\right) + C_\varepsilon \frac{k_s{}^{3/2}}{\Delta} \quad (17)$$

with coefficients C_k = 0.094, C_ε = 0.916, and P_r = 0.9.

3. Geometric Model, Mesh, and Post Processing

3.1. The New Ecological Hollow Cube

The new ecological hollow cube is transformed from the traditional hollow cube. On the basis of traditional hollow cubes with a raised frame and appropriate digging down to form a planting groove, nine cavities are arranged in the planting groove for wave dissipation. The geometry and dimensions of the block are visually presented in Figure 1. Its wave dissipation properties stem from the generation of turbulent vortices as waves enter the interconnected chambers within the block structure during their run-up on the slope. These vortices interact with the chamber walls, resulting in significant energy losses and effective wave attenuation. This new cube design provides a trifecta of advantages: high porosity, strong wave dissipation performance, and excellent structural stability, making it a promising solution for coastal protection.

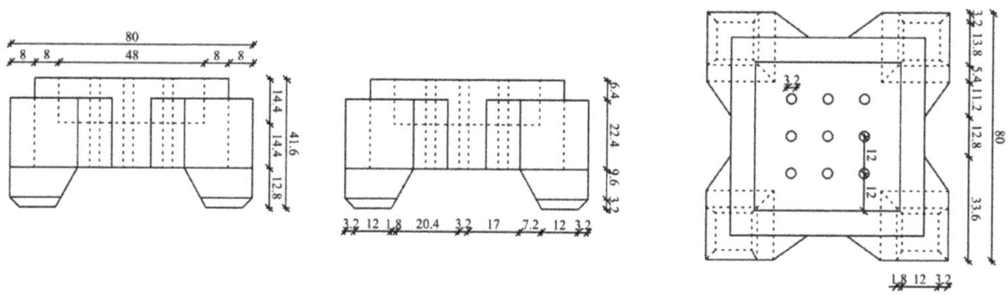

(**a**) Front View (**b**) Side View (**c**) Top View

Figure 1. Schematic diagram of the new ecological hollow cube (unit: mm).

3.2. Mesh

A three-dimensional numerical wave flume was designed, with the goal of accurately reproducing the experimental setup presented in Figure 2. To achieve this, the computational domain was configured with a length of 25.9 m, a height of 1 m, and a width of 0.08 m. The breakwater was positioned 24.7 m from the inlet boundary, mirroring the breakwater-wavemaker spacing in the experimental setup described by [2]. Three wave gauges (WG1, WG2, and WG3) used for incident wave analysis were defined at the exact locations used in the laboratory experiments. In addition, three wave gauges (G1, G2, and G3) were used in the numerical model to measure the wave surface changes. Nonslip boundary conditions were applied to all modeled impermeable structures, including the breakwater, cubes, and the flume bottom. At the atmospheric boundary, a fixed pressure was maintained, permitting bidirectional airflow while restricting water flow to an outflow condition only. The computational mesh was generated using blockMesh and snappy-HexMesh, specifically for the breakwater and the cube structures. Within the OpenFOAM framework, blockMesh and snappyHexMesh are used as mesh generation tools. blockMesh was initially used to create the base mesh, considering the overall structure. Subsequently, snappyHexMesh was used to refine the mesh resolution in the vicinity of the free surface and around the cube structures.

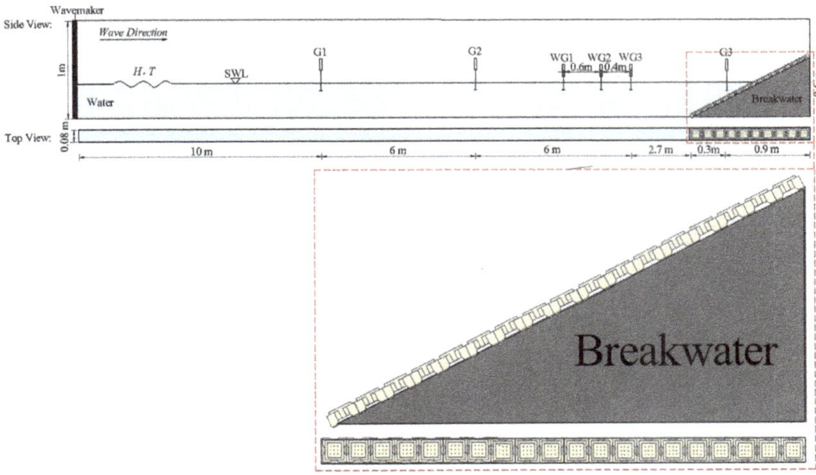

Figure 2. Layout of the numerical wave flume.

A nonuniform mesh is used in the mesh generation process of the numerical model to simulate the wave propagation process effectively and the interaction between the wave and the structure. In Figure 3, the mesh near the water-air interface and around the structure was relatively dense, whereas the mesh in other areas was relatively loose. The number of meshes would be large (nearly 89 million grids would be needed to simulate the new ecological hollow cube in the experiment entirely) if the mesh around the armor block were generated on the basis of the configuration used in the physical modeling tests (flume width 0.8 m). Therefore, to simplify the calculations and improve computational efficiency, only one row of armored blocks was installed on the sloping breakwater, which is a numerical model that was fully verified in the study of [26]. Notably, Ref. [28] also used the one-row block method for their study and achieved good accuracy. The number of grids used in this study ranged from 891,844 (Mesh I), 2,922,119 (Mesh II), to 8,909,544 (Mesh III), from coarse to dense, respectively. Table 2 provides an overview of the mesh characteristics, including the total cell count and the smallest grid spacing in each spatial dimension (x, y, and z) for the three different grid configurations used in this study. Table 3 shows the wave conditions used for the numerical studies.

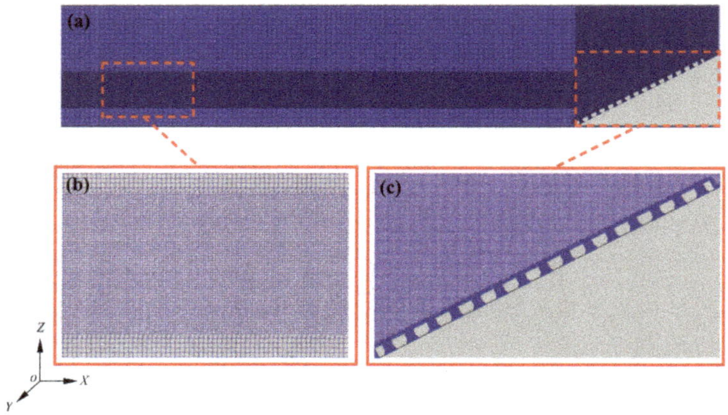

Figure 3. Mesh of the numerical wave flume: (**a**) side view, (**b**) free surface, (**c**) cube and breakwater.

Table 2. Grid number and grid size in the computational domain.

Grid Information	Cell Number (Million)	Minimum Size in the x, y, and z Direction (m)
Mesh-I	891,844	0.005 × 0.005 × 0.005
Mesh-II	2,922,119	0.0025 × 0.0025 × 0.0025
Mesh-III	8,909,544	0.00125 × 0.00125 × 0.00125

Table 3. Wave conditions in the numerical simulation.

Test No.	Water Depth (m)	Period (s)	Wave Height (m)
Wave-I		2.46	0.08
Wave-II	0.3	1.79	0.06
Wave-III		1.12	0.04

3.3. Boundary Conditions

Inlet and outlet boundaries apply OlaFlow active generation and absorption boundaries. Non-slip velocity conditions are included for the flume bottom, blocks, and breakwater. The upper boundary, i.e., the atmosphere, is treated as an open boundary where water and air can leave the domain but only air may enter the domain again. The side boundaries are specified as periodic boundaries.

3.4. Post Processing of the Wave Run-Up and Reflection Coefficient

In Figure 2, three wave gauges, WG1, WG2, and WG3, were placed in front of a new ecological hollow cube to obtain reflection coefficients. In accordance with [29], the incident and reflected waves were separated. The proposed relationships are shown in Equations (18) and (19).

$$X_{12} = \frac{L_0}{10} \tag{18}$$

$$\frac{L_0}{6} < X_{13} < \frac{L_0}{3} \text{and } X_{13} \neq \frac{L_0}{5} \text{and } X_{13} \neq \frac{3L_0}{10} \tag{19}$$

where X_{12} is the distance between wave gauges WG1 and WG2, X_{13} is the distance between wave gauges WG1 and WG3, and L_0 is the shallow wavelength. According to the water depth at the breakwater, the wavelengths corresponding to different wave periods at different water levels are calculated via Equation (20).

$$L_0 = \frac{gT^2}{2\pi}\tanh(2\pi h/L) \tag{20}$$

In this study, wave reflection is quantified by the reflection coefficient C_r:

$$C_r = \frac{H_r}{H_i} \tag{21}$$

where H_i [m] and H_r [m] are the incident and reflected wave heights, respectively.

In Figure 4, the wave run-up height was obtained by placing a plurality of probes near the surface of a new ecological hollow cube. Each probe is associated with a single computational cell, positioned at the cell's centroid. The alpha water of the cell was recorded with each probe (alpha water indicates the percentage of water contained in the cell). For regular waves, the height of the run-up is usually expressed as $R_{u2\%}$, defined as the height of the run-up exceeded by 2% of the incoming wave [30]. After analysis, the value of alpha water = 0.1 was used as the run-up height $R_{u2\%}$. The probes with alpha water = 0.1 at all moments are counted, and the maximum value of the vertical coordinate of all the probes that satisfy the condition of the wave run-up height $R_{u2\%}$ is taken. For ease of understanding, $R_{u2\%}$ is denoted by R below.

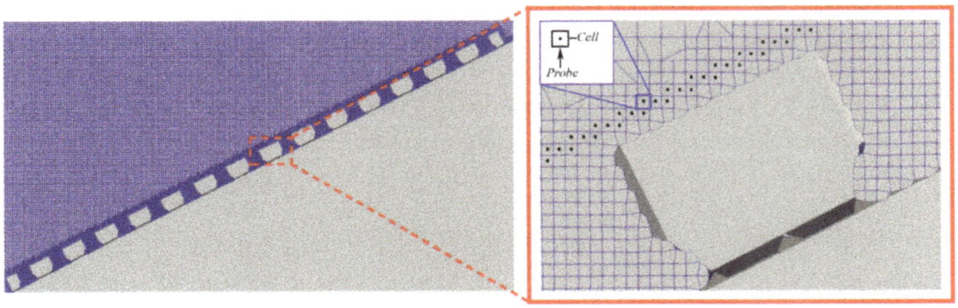

Figure 4. Probe to capture the wave run-up height.

4. The Results and Discussion

4.1. Wave Height and Period

In Figure 5, to validate the accuracy of numerical simulations, the numerically simulated wave features for all selected conditions were carefully compared with the corresponding target wave features. Figure 5b presents the results of the comparison between the numerically simulated and target wave periods, and it can be observed that there is a high degree of consistency between the two, which shows excellent agreement, and the quantitative metrics show that the mean absolute percentage error (MAPE) is only 0.68%. These findings indicate that the numerical model used in this study is highly reliable for wave height prediction. However, there are some deviations between the numerically simulated wave periods and the experimental measurements in Figure 5a. Specifically, the numerical model underestimates the wave heights with an average absolute error (MAPE) of 6.9%. The MAPE value is defined as follows:

$$\text{MAPE} = \frac{1}{n_{tests}}\left(\frac{H_{tar} - H_{num}}{H_{tar}}\right)*100\% \tag{22}$$

where n_{test} is the number of tests on which the root mean square error (RMSE) is based. H_{tar} is the target value, and H_{num} represents the numerical value measured by the numerical model.

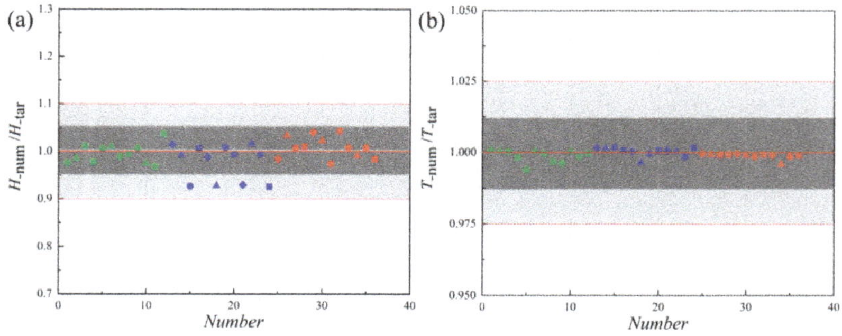

Figure 5. Comparisons of the numerical model and target incident height and period. (**a**) Incident wave height. The light gray areas depict the 10% error, and the dark gray areas depict the 5% error. (**b**) Incident wave period. The light gray areas depict the 2.5% error, and the dark gray areas depict the 1.25% error. The diamond, triangle, circle, and square markers represent the standard k-ε, stabilized k-ω SST, buoyancy-corrected k-ω SST, and LES, respectively. Green, blue, and red represent Mesh I, Mesh II, and Mesh III, respectively.

This error may be a run-up from the following aspects. First, the numerical model may fail to adequately capture the complex nonlinear effects or dissipation mechanisms in the wave propagation process, resulting in the wave energy not being correctly attenuated in the simulation. Second, the limitations of the grid resolution and the choice of the time step may also impact the wave period's simulation accuracy. Therefore, although the numerical model exhibits satisfactory accuracy in wave period prediction, the prediction of wave height still needs to be further improved, and more refined numerical methods and parameter settings need to be explored in future studies to improve the overall simulation accuracy.

In Figure 5a, most data points fall in the dark gray region (within 5% error), indicating that all turbulence models better predict the incident wave heights. Overall, the performances of all the models in terms of incident wave heights are relatively close and show that the grid resolution has a small effect on the wave heights. In Figure 5b, most data points fall in the dark gray region (within a 1.25% error), indicating that all turbulence models predict the incident wave period more accurately, with little or no bias. The differences between different turbulence models (shapes) are negligible, and the differences between different grids (colors) are small, suggesting that the incident wave period is less sensitive to grid and model selection.

4.2. Validation of Numerical Model

The numerical model's accuracy was assessed by comparing its predictions of wave run-up height and reflection coefficients against experimental data, focusing on a block structure under various incident wave conditions. Figure 6 presents the comparison, specifically detailing the results obtained using two turbulence models: the stabilized k-ω SST model and the RNG k-ε model. The simulations were conducted using Mesh II. Quantitative comparison reveals that the stabilized k-ω SST model yields more accurate results.

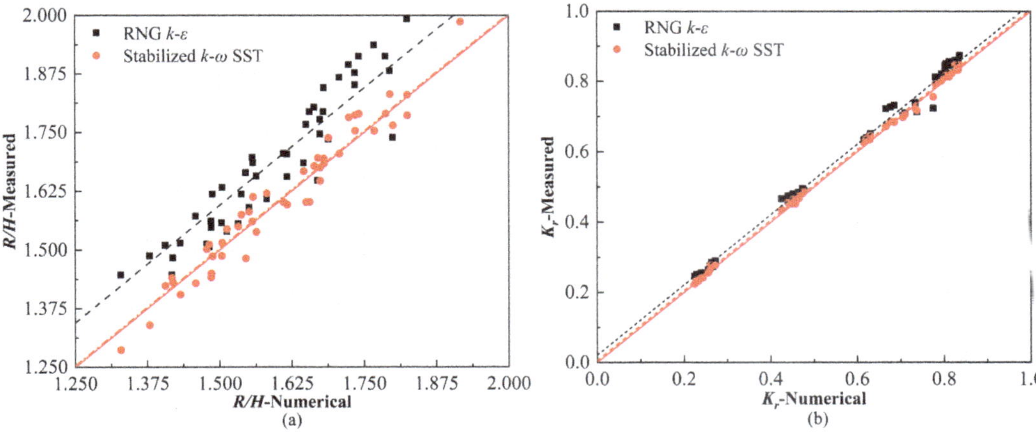

Figure 6. Comparison of experimental and numerical wave run-up and reflections (stabilized *k*-*ω* SST and RNG *k*-*ε*). (**a**) Relative wave run-up; (**b**) reflection coefficients.

Figure 6a presents the "Wang et al. [26] (RNG *k*-*ε*)" data points (black squares) which are scattered around the diagonal line, but generally tend to overestimate the measured values (they lie above the diagonal line). The "Stabilized *k*-*ω* SST" data points (red circles) are generally below the diagonal, which means it is underestimating the R/H ratio. The root mean square error (RMSE) for wave run-up was 0.10914 for the stabilized *k*-*ω* SST model and 0.11114 for the RNG *k*-*ε* model. Figure 6b presents the both data sets ("Wang et al. [26] (RNG *k*-*ε*)" and "Stabilized *k*-*ω* SST") which appear to be clustered much closer to the diagonal line, indicating better agreement between the numerical simulations and the measured data compared to Figure 6a. The data points are less scattered and more concentrated around the diagonal, suggesting higher accuracy in the numerical predictions for K_r. Similarly, the RMSE for wave reflection coefficients was 0.02109 for the stabilized *k*-*ω* SST model and 0.02799 for the RNG *k*-*ε* model. These values demonstrate that the numerical model, when implemented with the stabilized *k*-*ω* SST turbulence model, exhibits a better correlation with the experimental data, particularly in accurately predicting both wave run-up. Both models (RNG *k*-*ε* and stabilized *k*-*ω* SST) show very good agreement between the numerical and measured K_r values (Figure 6b). The choice of turbulence model (RNG *k*-*ε* and stabilized *k*-*ω* SST) has a greater impact on the accuracy of R/H ratio predictions than on K_r prediction.

4.3. Wave Profile Evolution

To obtain reliable numerical results, a comparative numerical modeling result under three different grid systems (Mesh I, Mesh II, and Mesh III) is performed, and three typical wave conditions (Wave-I, Wave-II, and Wave-III) are used. First, the time histories of the wave profiles at G1, G2, and G3 are comparatively analyzed under four different turbulence models (standard *k*-*ε*, stabilized *k*-*ω* SST, buoyancy-modified *k*-*ω* SST, and LES). G1 is located in the open sea in the fluid domain; G2 is in front of the breakwater; and G3 is in the front part of the breakwater. Notably, the wavefronts of the different turbulence models for both Wave-I and Wave-II conditions are very close to each other, so only the wavefronts of Wave-III are analyzed in this section (the RMSEs for Wave-I and Wave-II were 2.2154×10^{-5} and 6.16985×10^{-5}, respectively). The numerical results using the standard *k*-*ε*, stabilized *k*-*ω* SST, buoyancy-modified *k*-*ω* SST, and LES turbulence models shown in Figure 7 indicate that there is significant nonlinearity after 30 s in the wave profile results at G1, G2, and G3 obtained via different grids. Overall, Figure 7 shows that the waves exhibit regular periodic fluctuations, and the results simulated with different grid densities are relatively close in most cases (the RMSE values for Mesh I and Mesh II, Mesh II and Mesh III are 6.13131×10^{-5}, 6.56766×10^{-5}, respectively), indicating good grid convergence. The wave profiles simulated with the standard k-ε, buoyancy-modified k-ω SST, and LES models show a slight decrease in wave height on three different grid systems (the RMSE values were 1.35112×10^{-5}, 2.12065×10^{-5}, 2.00269×10^{-5}). In contrast, the wave profiles obtained with the stabilized k-ω SST model maintain good consistency (the RMSE value is 0.17832×10^{-5}), with almost no wave height drop observed, suggesting better stability and grid independence under this condition.

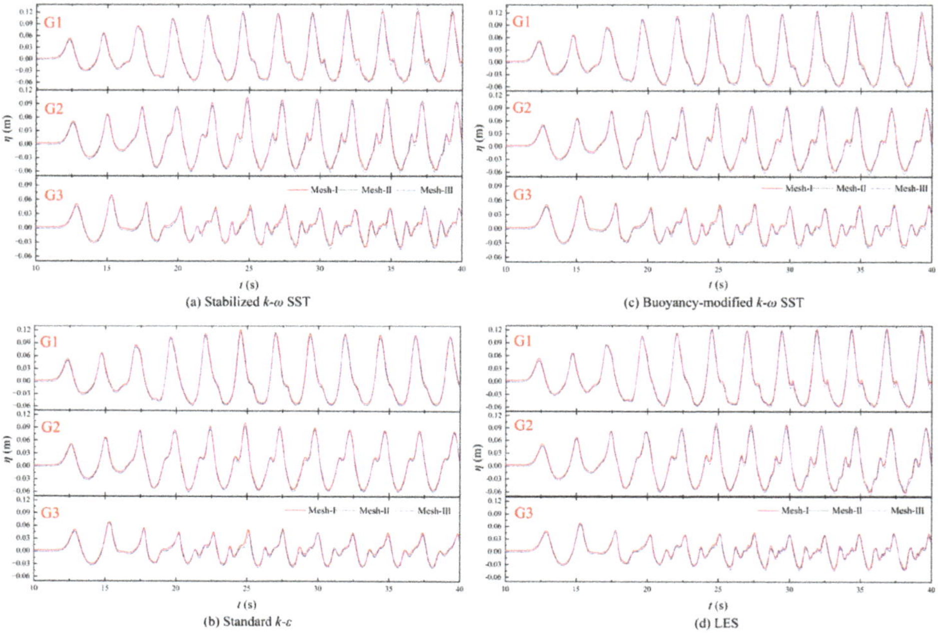

Figure 7. Comparison of wave profiles at locations G1, G2, and G3 from numerical model test results under different turbulence models at Wave-III (third wave condition, Table 3).

To observe the numerical results under different turbulence models and grids more clearly, we select the time histories of the wave heights of G1, G2, and G3 in one period for comparison, focusing on evaluating their performance in capturing the free surface. In Figure 8, for Wave-I, the results of the standard k-ε model, the buoyancy-modified k-ω SST model, and the LES model are significantly correlated with the grid (the RMSE values for Mesh I and Mesh II are 6.13131×10^{-5}, 6.56766×10^{-5}, respectively). In contrast, the correlation of the stabilized k-ω model is not significant (the RMSE value for Mesh I and Mesh II is 3.91784×10^{-5}). For both Wave-II and Wave-III, the grid dependence is insignificant for the stabilized k-ω model, standard k-ε model, buoyancy-modified k-ω SST model, and LES model. The stabilized k-ω model outperforms the standard k-ε model, the buoyancy-modified k-ω SST model, and the LES model in reliably capturing the free surface, which shows higher accuracy and reliability in simulating the wave free-surface flow.

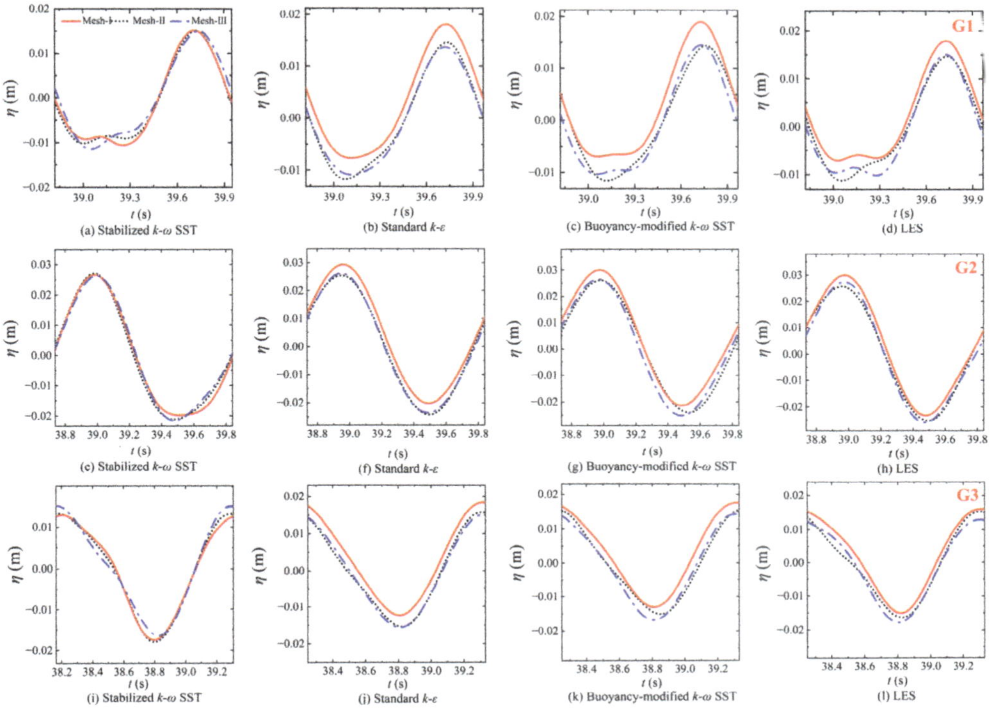

Figure 8. Comparison of wave profiles in one period when different turbulence models are adopted for Wave-II (second wave condition, Table 3).

4.4. Wave Run-Up

In practical engineering applications, wave run-up heights help in understanding the interaction between waves and structures. In addition, wave run-up heights are valuable for the placement and maintenance of critical infrastructure and for ensuring the safety of people in coastal areas. In Table 4, considering the influence of wave conditions on wave run-up heights, three groups of probes were designed in this study: P1, P2, and P3 (Figure 9). Therefore, a more accurate numerical model evaluation is achieved by selecting sounding points at different locations to ensure that sufficient variability data are available for different wave conditions.

Table 4. Position of the probes.

Label	Coordinates (x, z, y) (cm)
P1	(x = 0.74525, z = 0.004, y = 0.40512)
P2	(x = 0.7515, z = 0.004, y = 0.40825)
P3	(x = 0.7565, z = 0.004, y = 0.41075)

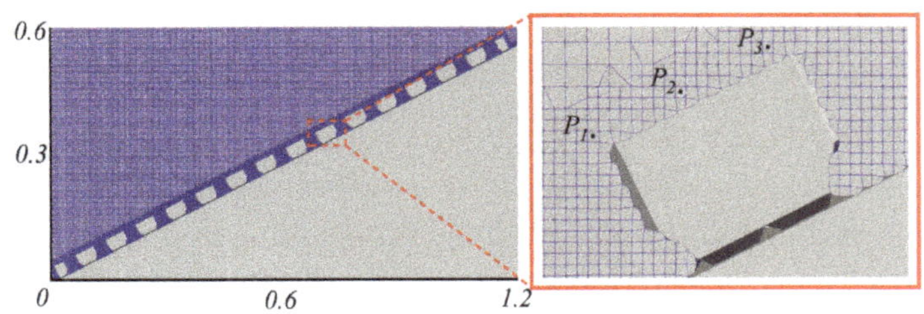

Figure 9. P1, P2, and P3 point maps.

This study examines the simulation results of different grids and turbulence models (standard k-ε, steady k-ω SST, buoyancy-modified k-ω SST, and LES) for wave run-up heights (alpha water) at each point from P1 to P3 under different wave conditions (Wave-I, Wave-II, and Wave-III) by comparing and analyzing the results. Under Wave-I and Wave-II conditions, the alpha water values of the grids are in good agreement overall (the RMSEs for Wave-I and Wave-II are 7.3621×10^{-5} and 8.7849×10^{-5}, respectively). The results show that in Figure 10, under Wave-III conditions, the alpha water is more sensitive to the grids, and the models are less consistent among different grids, but all of them can capture the maximum value of alpha water better, among which the stable k-ω SST model can capture more alpha water values and shows a prediction advantage (the RMSE value of stable k-ω SST under Mesh I and Mesh II is 0.000362). However, there is an overestimation of the alpha water values in Mesh I at 25–35 s. Under Wave III conditions, the simulated alpha water values of Mesh I are generally higher than those of the other grids, especially the standard k-ε and LES models (the RMSE values of standard k-ε and LES models are 0.00166, 0.00125 for Mesh I and Mesh II, respectively), which are more significant. These findings emphasize the influence of grid resolution and turbulence model selection on the simulation results of wave run-up heights, particularly the dominance of the stabilized k-ω SST model in predicting alpha water.

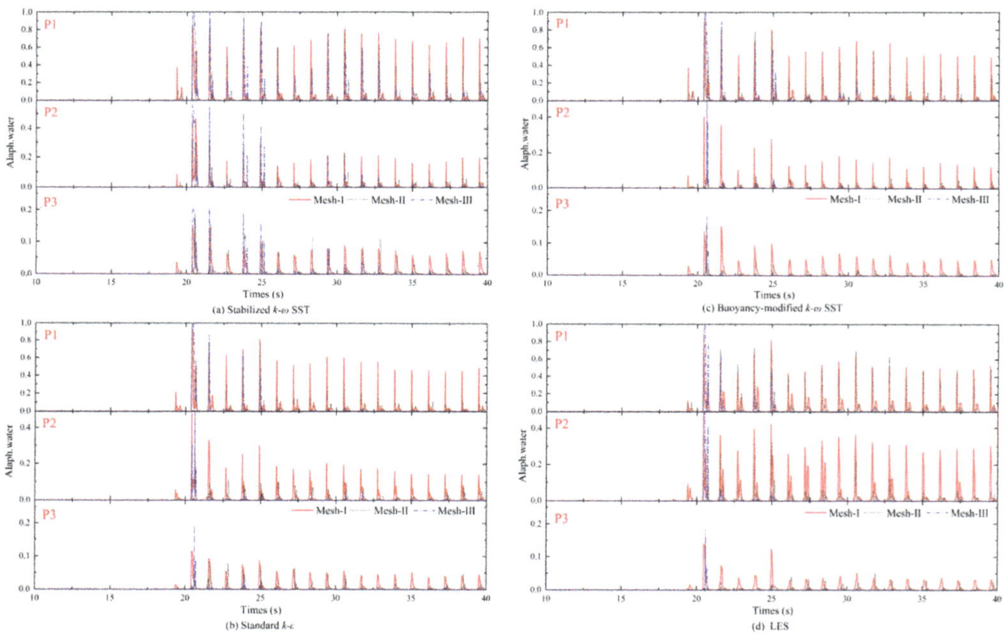

Figure 10. Comparison of alpha water on new ecological hollow cubes with different turbulent models under different grids in Wave-II (second wave condition, Table 3).

To analyze the effects of different turbulence models, wave conditions, and grid resolutions on the wave run-up height in detail, the time histories of one cycle at each measurement point from P1 to P3 are selected for comparison in this study. In Figure 11, all four turbulence models exhibit significant grid dependence under different wave conditions, and this dependence is more significant the farther the measurement point P is from the structure. However, the consistency of the alpha water content over time is poor under different models and conditions, especially when the measurement point P is farther from the structure (e.g., P3). Overall, the alpha water parameter is more sensitive to the grid in the wave run-up calculation. The difference in the results between different turbulence models is not apparent. However, the results are related to the location of the measurement point: the closer the measurement point is to the structure, the smaller the dependence of the results on the grid, and vice versa, the dependence increases, which indicates that the effects of the turbulence model, the grid resolution, and the location of the measurement point need to be considered comprehensively in simulating the height of the wave run-up.

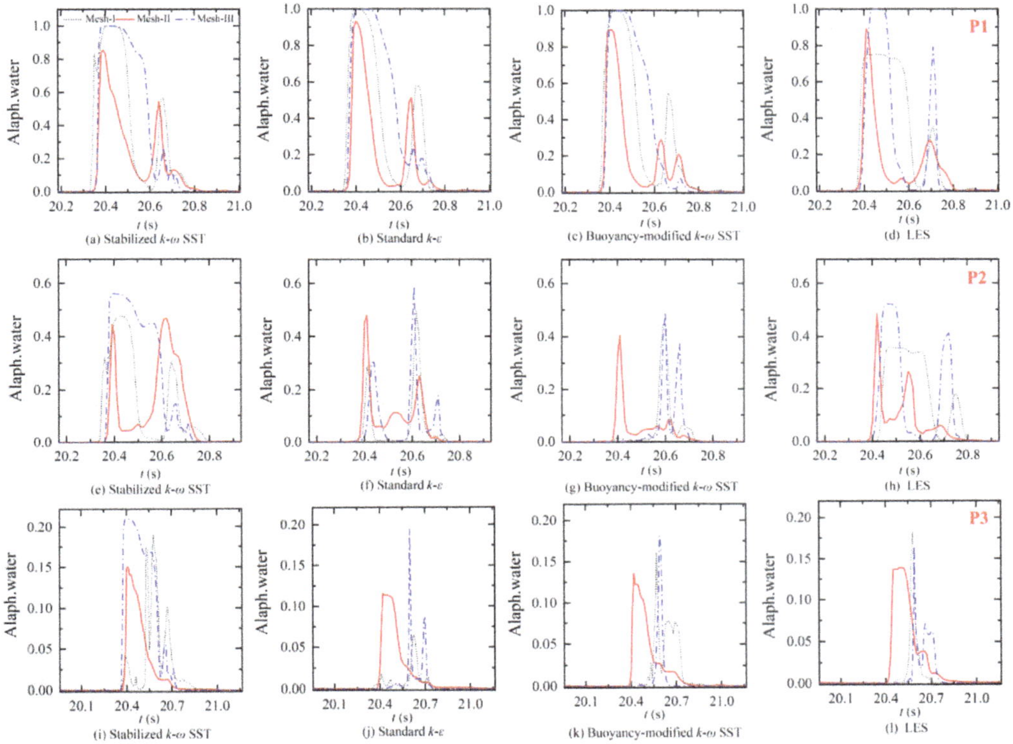

Figure 11. Comparison of alpha water in one period with different turbulence models at P1, P2, and P3 during Wave-III (third wave condition, Table 3).

4.5. Wave Velocity Distribution and Dynamic Pressure Fields

Figure 12 shows the velocity distributions and pressure fields of different turbulence models (stabilized k-ω SST, standard k-ε, buoyancy-modified k-ω SST, and LES) on the new ecological hollow cube under Mesh III. Since the velocity distribution and pressure field of steady k-ω SST and standard k-ε are very close to each other, only the velocity distribution and pressure field plots of steady k-ω SST are given in Figure 12. The results show that the pressure fields of all four models tend to increase in pressure near the cube, and the overall pressures of the stabilized k-ω SST and buoyancy-modified k-ω SST models are slightly greater than those of the other models in terms of the velocity distribution. The stable k-ω SST and standard k-ε models behave similarly. Tiny vortices are present on the surface of the wave block and the front of the wave concentrate velocities. The k-ω SST model shows smaller eddies and velocities in the vicinity of the eddies. The LES model exhibits more pronounced wave surface undulations, reflecting the excessive generation of turbulent kinetic energy. This latter reflection leads to wave energy attenuation affecting in turn the wave surface and wave motion. The stabilized k-ω SST, standard k-ε, and buoyancy-modified k-ω SST models can better solve the problem of turbulent kinetic energy generation and produce a smoother free surface.

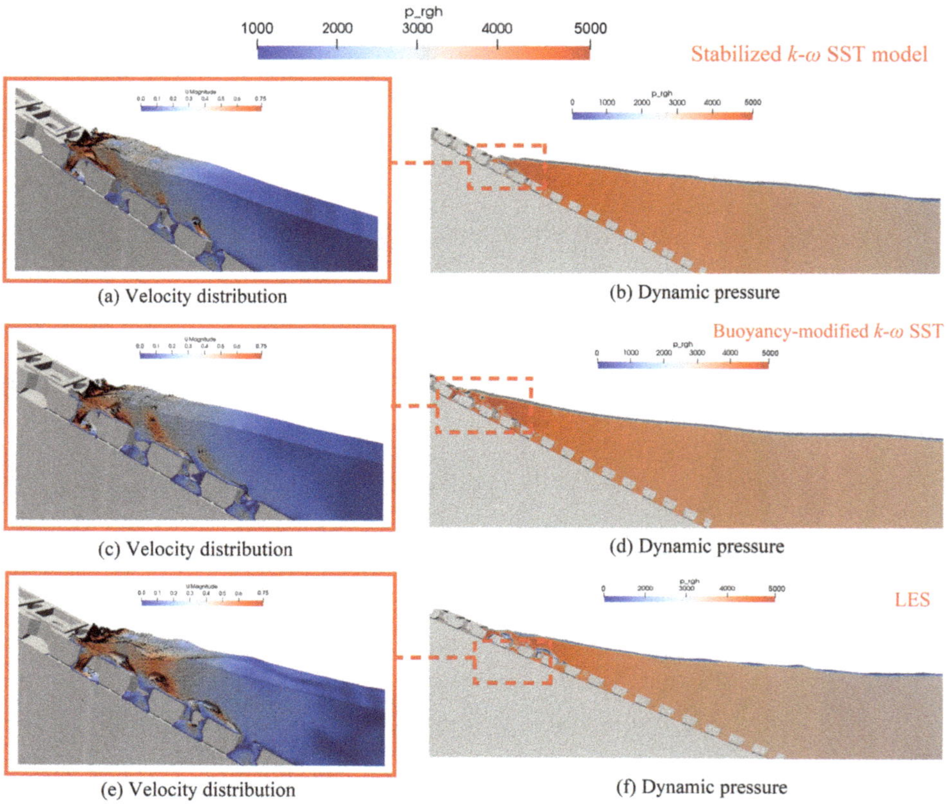

Figure 12. Velocity distributions (left column, including (**a**,**c**,**e**)) and pressure fields (right column, including (**b**,**d**,**f**)) near the blocks (Wave-III, Mesh III).

The velocity distributions and pressure fields of different turbulence models (stabilized k-ω SST, standard k-ε, buoyancy-modified k-ω SST, and LES) on the new ecological hollow cube under Mesh III are shown in Figure 13. The results show that the velocities of the stabilized k-ω SST and standard k-ε models are mainly concentrated on the surface of the new ecological hollow cube during wave run-up. In the wave run-down phase, the four models have similar characteristics. However, the free-surface velocity distributions of the stabilized k-ω SST and standard k-ε models are larger. In comparison, the velocity distributions of the buoyancy-modified k-ω SST model are smaller and smoother. In contrast, the LES model has a larger free-surface velocity distribution and more dramatic surface undulations. Overall, the free-surface velocity distributions of the stabilized k-ω SST model are very close to those of the standard k-ε model in the wave run-up and run-down phases. The buoyancy-modified k-ω SST model has the most minor and smoothest free-surface velocity distributions. The suggest that the different turbulence models present different flow field characteristics when simulating free-surface flows.

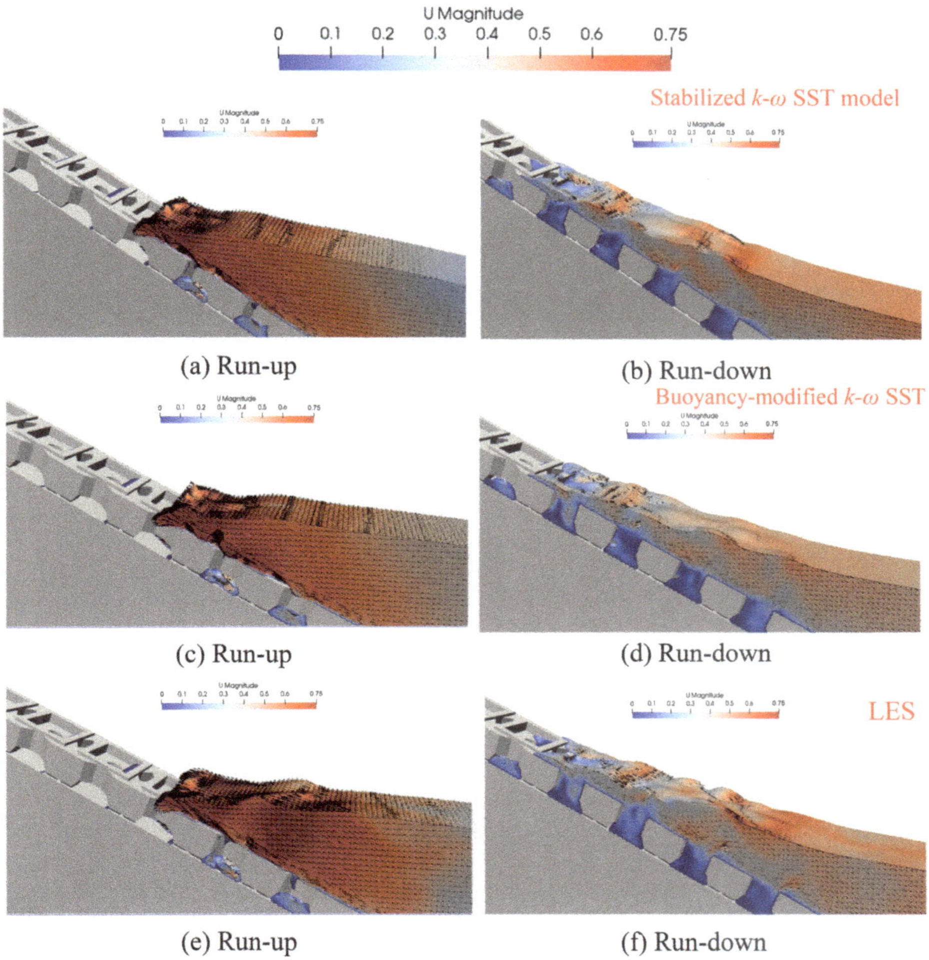

Figure 13. Run-up (left column, including (**a**,**c**,**e**)) and run-down (right column, including (**b**,**d**,**f**)) near the blocks (Wave-III, Mesh III).

4.6. Sensitivity Analysis of the Wave Run-Up and Reflection Coefficient

In order to assess the influence of grid resolution on the numerical solutions, the extrapolated relative error (ERE) and grid convergence index (GCI) were computed to quantify the discretization error [31]. In this context, 's' denotes the characteristic grid size for the numerical model:

$$s = \left[\frac{1}{N}\sum_{i=1}^{N}(\Delta V_i)\right]^{0.5} \tag{23}$$

where ΔV_i is the volume of the ith cell and N is the total number of cells.

For the grid independence study, three different grid resolutions were implemented at the free surface: s_1 = 0.005 m, s_2 = 0.0025 m, and s_3 = 0.00125 m. Here, 's' denotes the grid size, and the subscripts 1, 2, and 3 correspond to grid partition schemes Mesh I, Mesh II, and Mesh III, respectively.

The GCI is defined as follows:

$$GCI_{21} = \frac{1.25e_{21}}{r_{21}^{n} - 1} \tag{24}$$

where

$$e_{21} = \left| \frac{\Psi_1 - \Psi_2}{\Psi_1} \right| \tag{25}$$

$$n = \frac{1}{lnr_{21}} |\ln(\delta_{32}/\delta_{21}) + q(n)| \tag{26}$$

$$q(n) = \ln(\frac{r_{21}^n - p'}{r_{32}^n - p'}) \tag{27}$$

$$p' = Sign(\delta_{32}/\delta_{21}) \tag{28}$$

Here, $\delta_{32} = \Psi_3 - \Psi_1$, $\delta_{21} = \Psi_2 - \Psi_1$, $r_{21} = s_2/s_1$, and $r_{32} = s_3/s_2$. The variable Ψ represents the numerical solution obtained on a given grid. The function "Sign" refers to the sign function. Based on these definitions, the solution is expected to demonstrate either monotone convergence or divergence when $p' = 1$, and oscillatory convergence when $p' = -1$.

The extrapolated value is

$$\Psi_{ext}^{21} = (r_{21}^n \Psi_1 - \Psi_2)/(r_{21}^n - 1) \tag{29}$$

Moreover, the ERE is

$$ERE_{21} = \left| \frac{\Psi_{ext}^{21} - \Psi_1}{\Psi_{ext}^{21}} \right| \tag{30}$$

Table 5 shows in detail the discretization errors of the relative wave run-up and reflection coefficients of the new ecological hollow cube for three different wave conditions (Wave-I, Wave-II, and Wave-III) as well as for four turbulence models (e.g., standard k-ε, stabilized k-ω SST, buoyancy-modified k-ω SST, and LES). The results show that the discretization errors for the four turbulence models exhibit a high degree of consistency across the three wave conditions: the highest error values are generally found in the Wave-I condition, whereas the lowest values are found in the Wave-III condition. This trend suggests that wave conditions significantly impact the accuracy of numerical simulations, with Wave-I representing the most complex wave dynamics scenario, leading to large uncertainties in model predictions. The stabilized k-ω SST turbulence model exhibits relatively better performance in terms of discretization error, and its predictions are in better agreement with the experimental data. These results suggest that the stabilized k-ω SST model may be a more robust and efficient choice for wave interaction simulations for this type of ecological hollow cube. However, other turbulence models may also exhibit acceptable accuracy in specific cases. To evaluate the performance of each turbulence model more comprehensively, it still needs to be considered in conjunction with other factors (e.g., computational cost and model robustness) and further explored in subsequent studies.

To better understand wave run-up, wave reflection coefficients are important because wave reflection affects the sea state and can impede safe navigation [32]. In addition, increased scour due to a high reflection coefficient can affect the stability of coastal structures [33]. Figure 14 shows the relative wave run-up and reflection coefficients when different turbulence models are used. Under the same wave conditions, the relative wave run-up heights derived from the four different turbulence models increase with decreasing grid size, and the reflection coefficient decreases with decreasing grid size. The differences in the wave rise heights and reflection coefficients for different grids are shown in Figure 15, which reveals that the wave run-up height and reflection coefficient of the new ecological hollow cube have some grid dependence. The RMSE value is defined as follows:

$$\mathrm{RMSE} = \sqrt{\frac{\sum_{i=1}^{n_{tests}} (K_{measured} - K_{numerical})^2}{n_{tests}}} \quad (31)$$

where n_{test} represents the number of experimental or simulation runs used to calculate the root mean square error (RMSE). $K_{measured}$ is the experimentally measured value, and $K_{numerical}$ represents the numerical value measured by the numerical model.

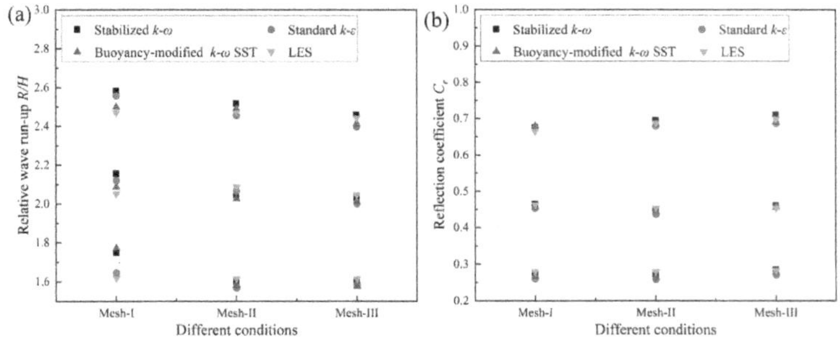

Figure 14. Comparison between the different mesh and turbulence models (**a**) R/H and (**b**) C_r.

Table 5. Analysis of the discretization error.

Type	Partition Scheme	Values	Wave-I	Wave-II	Wave-III
Standard k-ε	Mesh I	R/H	1.75000	1.60125	1.57000
		C_r	0.27773	0.26991	0.28575
	Mesh II	R/H	2.15667	2.04667	2.03125
		C_r	0.46507	0.44991	0.461
	Mesh III	R/H	2.58219	2.51781	2.45969
		C_r	0.67653	0.69569	0.71023
	Ψ_{ext}^{21}	R/H	−7.02350	−6.11255	−4.45310
		C_r	−1.17734	−0.22264	−0.12940
	ERE_{21}	R/H	1.25%	1.26%	1.35%
		C_r	1.24%	2.21%	3.20%
	GCI_{21}	R/H	6.27%	6.02%	4.80%
		C_r	6.55%	2.28%	1.82%
Stabilized k-ω SST	Mesh I	R/H	1.64688	1.56875	1.58438
		C_r	0.26136	0.25905	0.27081
	Mesh II	R/H	2.12125	2.06667	1.95933
		C_r	0.45292	0.43634	0.431241
	Mesh III	R/H	2.5575	2.45625	2.39813
		C_r	0.67056	0.6798	0.68706
	Ψ_{ext}^{21}	R/H	−3.78187	−0.22172	−0.61746
		C_r	−1.14567	−0.21597	0.05407
	ERE_{21}	R/H	1.44%	8.08%	3.57%
		C_r	1.23%	2.20%	4.16%
	GCI_{21}	R/H	4.12%	1.43%	1.74%
		C_r	6.73%	2.29%	1.01%

Table 5. *Cont.*

Type	Partition Scheme	Values	Wave-I	Wave-II	Wave-III
Buoyancy-modified k-ω SST	Mesh I	R/H	1.58438	1.57688	1.77188
		C_r	0.26432	0.28003	0.26852
	Mesh II	R/H	2.02667	2.01458	2.08667
		C_r	0.44362	0.45617	0.4583
	Mesh III	R/H	2.48969	2.41406	2.49975
		C_r	0.68529	0.69106	0.68056
	Ψ_{ext}^{21}	R/H	−0.84036	−0.25113	−0.24806
		C_r	−7.85221	−2.99801	0.76371
	ERE_{21}	R/H	1.32%	2.05%	2.13%
		C_r	1.20%	1.53%	1.32%
	GCI_{21}	R/H	5.16%	2.44%	2.36%
		C_r	7.45%	3.63%	0.71%
LES	Mesh I	R/H	1.61875	1.61563	1.61563
		C_r	0.27988	0.28047	0.28308
	Mesh II	R/H	2.05125	2.08708	2.04583
		C_r	0.46338	0.45438	0.45369
	Mesh III	R/H	2.46969	2.47188	2.44406
		C_r	0.665	0.6878	0.6991
	Ψ_{ext}^{21}	R/H	−11.25289	−3.74310	−0.47801
		C_r	−1.57841	−0.22776	−0.10606
	ERE_{21}	R/H	1.14%	1.43%	4.38%
		C_r	1.18%	2.23%	3.67%
	GCI_{21}	R/H	9.94%	4.15%	1.62%
		C_r	8.30%	2.27%	1.72%

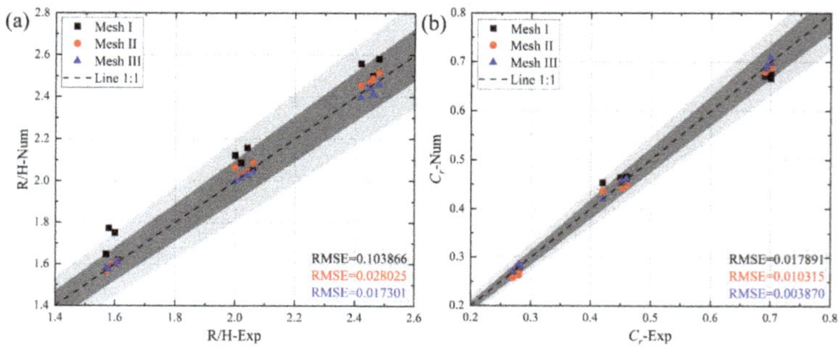

Figure 15. Comparison between the numerical and experimental (**a**) R/H and (**b**) C_r. The light gray areas depict the 30% error, and the dark gray areas depict the 15% error.

Wave run-up and reflection have been extensively studied through both experimental and numerical approaches [34]. However, few works have discussed the grid sensitivity of wave run-up and reflection coefficients. Therefore, the comparative analysis of the wave run-up height and reflection coefficients of the new ecological hollow cubes with different turbulence models for different grid systems derived in this section is informative.

Moreover, turbulence models that are not available in this study or more precise grid sizes can be used for further research to obtain more credible results.

5. Conclusions

This study presents a comparative analysis of hydrodynamic modeling of the new ecological hollow cube, utilizing the open-source solver olaFlow as the computational platform, with three mesh systems and four RANS turbulence models. The characteristics of wave evolution on the new ecological hollow cube, including wave run-up, velocity distribution, pressure fields, and reflection coefficients, were comprehensively compared. The key findings of this study are summarized as follows:

(1) The stabilized k-ω model demonstrates superior performance in capturing wave evolution on the new ecological hollow cube. The wave profiles simulated with the standard k-ε, buoyancy-modified k-ω SST, and LES models show a slight decrease in wave height on three different grid systems (the RMSE values were 1.35112×10^{-5}, 2.12065×10^{-5}, 2.00269×10^{-5}). In contrast, the wave profiles obtained with the stabilized k-ω SST model maintain good consistency (the RMSE value is 0.17832×10^{-5}), with almost no wave height drop observed, suggesting better stability and grid independence under this condition.

(2) The parameter "alpha water" is more sensitive to the grids, and the models are less consistent among different grids, but all of them can capture the maximum value of alpha water better, among which the stable k-ω SST model can capture more alpha water values and shows a prediction advantage (the RMSE value of stable k-ω SST under Mesh I and Mesh II is 0.000362).

(3) The pressure fields predicted by all four turbulence models are generally similar, with higher pressures observed near the new ecological hollow cube. The stabilized k-ω and buoyancy-modified k-ω SST models predict slightly higher overall pressures compared to the standard k-ε and LES models. Regarding velocity distributions, the stabilized k-ω and standard k-ε models show very similar patterns. The LES model, however, exhibits excessive turbulent kinetic energy generation. The stabilized k-ω, standard k-ε, and buoyancy-modified k-ω SST models are better suited for mitigating this issue of excessive turbulence kinetic energy production, resulting in smoother free-surface representations.

(4) The numerical results for the new ecological hollow cube exhibit some grid dependence. The analysis of discretization errors across the four turbulence models reveals a high degree of consistency across the three wave conditions. The highest error values are consistently observed under the Wave-I condition (the average GCI_{21} values for wave run-up and reflection were 6.3725 and 7.2575, respectively), while the lowest error values are found under the Wave-III conditions (the average GCI_{21} values for wave run-up and reflection were 2.63 and 1.315, respectively).

This study focused on numerical model calculations and corresponding physical model tests conducted under regular wave conditions. Future research should investigate the influence of irregular waves on cube performance. Additionally, this study employed a simplified slope breakwater and did not account for the complexities of breakwaters in real-world engineering scenarios. This aspect warrants further investigation in future work.

Author Contributions: Conceptualization, H.Z., J.Y., K.W., Q.L. and X.Z.; Methodology, H.Z., J.Y., K.W. and Z.Z.; Software, Y.Z. and J.Y.; Validation, H.Z.; Investigation, H.Z., J.Y., K.W., Z.Z. and Q.L.; Formal Analysis, H.Z., J.Y., K.W., Z.Z., Q.L. and X.Z.; Resources, H.Z., K.W. and Z.Z.; Data Curation, H.Z., J.Y. and K.W.; Writing—Original Draft, H.Z. and J.Y.; Writing—Review and Editing, H.Z. and

J.Y.; Visualization, J.Y.; Supervision, H.Z., J.Y. K.W., Z.Z., Q.L. and X.Z.; Funding Acquisition, H.Z. and Z.Z. All authors have read and agreed to the published version of the manuscript.

Funding: This research was funded by the Natural Science Foundation of Zhejiang province (No. ZJWZ23E090009), and the Science and Technology Project of the Department of Water Resources of Zhejiang Province in 2020 (No. RA2012).

Data Availability Statement: The data presented in this study are available upon request from the corresponding author.

Acknowledgments: The authors acknowledge the editors and reviewers for their valuable comments and suggestions.

Conflicts of Interest: Kaifang Wang and Zhen Zeng were employed by Zhejiang Design Institute of Water Conservancy and Hydroelectric Power. The remaining authors declare that the research was conducted in the absence of any commercial or financial relationships that could be construed as a potential conflict of interest.

Abbreviations

C_r	Reflection coefficient [-]
RMSE	Root mean square error [-]
R	Wave run-up height related to regular waves [m]
$R_{u,2\%}$	Wave run-up height exceeded by 2% of incident waves [m]
$R_{u,2\%}/H$	Relative run-up [-]
T	Wave period
$T_{m-1,0}$	Spectral wave period [s]
L	Wavelength in deep water based on T ($=gT^2/2\pi$) [m]
$L_{m-1,0}$	Spectral wavelength in deep water ($=gT_{m-1,0}{}^2/2\pi$) [m]
g	Acceleration due to gravity [m/s^2]
H	Wave height for regular waves [m]
H_{m0}	Incident spectral significant wave height at the toe of the structure [m]
h	Water depth at toe of the structure [m]
ξ	Breaker parameter (regular waves) ($=\tan\alpha/(H/L)^{0.5}$) [-]
α	Angle between structure slope and horizontal [∘]
ERE	Extrapolated relative error
GCI	Grid convergence index

References

1. Bruce, T.; Van Der Meer, J.; Franco, L.; Pearson, J.M. Overtopping performance of different armour units for rubble mound breakwaters. *Coast. Eng.* **2009**, *56*, 166–179. [CrossRef]
2. Zhao, H.; Ding, F.; Ye, J.; Jiang, H.; Chen, W.; Gu, W.; Yu, G.; Li, Q. Physical Experimental Study on the Wave Reflection and Run-Up of a New Ecological Hollow Cube. *J. Mar. Sci. Eng.* **2024**, *12*, 664. [CrossRef]
3. Campos, Á.; Molina-Sanchez, R.; Castillo, C. Damage in rubble mound breakwaters. Part II: Review of the definition, parameterization, and measurement of damage. *J. Mar. Sci. Eng.* **2020**, *8*, 306. [CrossRef]
4. Spalart, P.; Allmaras, S. A one-equation turbulence model for aerodynamic flows. In Proceedings of the 30th Aerospace Sciences Meeting and Exhibit, Reno, NV, USA, 6–9 January 1992; p. 439.
5. Launder, B.E.; Spalding, D.B. The numerical computation of turbulent flows. In *Numerical Prediction of Flow, Heat Transfer, Turbulence Combustion*; Elsevier: Amsterdam, The Netherlands, 1983; pp. 96–116.
6. Yakhot, V.; Orszag, S.A.; Thangam, S.; Gatski, T.; Speziale, C. Development of turbulence models for shear flows by a double expansion technique. *Phys. Fluids A Fluid Dyn.* **1992**, *4*, 1510–1520. [CrossRef]
7. Wilcox, D.C. Reassessment of the scale-determining equation for advanced turbulence models. *AIAA J.* **1988**, *26*, 1299–1310. [CrossRef]
8. Menter, F.R. Two-equation eddy-viscosity turbulence models for engineering applications. *AIAA J.* **1994**, *32*, 1598–1605. [CrossRef]
9. Launder, B.E.; Reece, G.J.; Rodi, W. Progress in the development of a Reynolds-stress turbulence closure. *J. Fluid Mech.* **1975**, *68*, 537–566. [CrossRef]

10. Wilcox, D.C. *Turbulence Modeling for CFD (Third Edition) (Hardcover)*; DCW Industries: La Canada, CA USA, 2006.
11. Smagorinsky, J. General circulation experiments with the primitive equations: I. The basic experiment. *Mon. Weather. Rev.* **1963**, *91*, 99–164. [CrossRef]
12. Brown, S.A.; Greaves, D.M.; Magar, V.; Conley, D.C. Evaluation of turbulence closure models under spilling and plunging breakers in the surf zone. *Coast. Eng.* **2016**, *114*, 177–193. [CrossRef]
13. Mayer, S.; Madsen, P.A. Simulation of breaking waves in the surf zone using a Navier-Stokes solver. *Coast. Eng.* **2001**, *2000*, 928–941.
14. Devolder, B.; Troch, P.; Rauwoens, P. Performance of a buoyancy-modified k-ω and k-ω SST turbulence model for simulating wave breaking under regular waves using OpenFOAM®. *Coast. Eng.* **2018**, *138*, 49–65. [CrossRef]
15. Devolder, B.; Rauwoens, P.; Troch, P. Application of a buoyancy-modified k-ω SST turbulence model to simulate wave run-up around a monopile subjected to regular waves using OpenFOAM®. *Coast. Eng.* **2017**, *125*, 81–94. [CrossRef]
16. Larsen, B.E.; Fuhrman, D.R.; Roenby, J. Performance of interFoam on the simulation of progressive waves. *Coast. Eng. J.* **2019**, *61*, 380–400. [CrossRef]
17. Bradford, S.F. Numerical simulation of surf zone dynamics. *J. Waterw. Port Coast. Ocean. Eng.* **2000**, *126*, 1–13. [CrossRef]
18. Ting, F.C.; Kirby, J.T. Observation of undertow and turbulence in a laboratory surf zone. *Coast. Eng.* **1994**, *24*, 51–80. [CrossRef]
19. Galera-Calero, L.; Blanco, J.M.; Izquierdo, U.; Esteban, G.A. Performance assessment of three turbulence models validated through an experimental wave flume under different scenarios of wave generation. *J. Mar. Sci. Eng.* **2020**, *8*, 881. [CrossRef]
20. Qu, S.; Liu, S.; Ong, M.C. An evaluation of different RANS turbulence models for simulating breaking waves past a vertical cylinder. *Ocean. Eng.* **2021**, *234*, 109195. [CrossRef]
21. Zhou, H.; Ye, J. Numerical study on the hydrodynamic performance of a revetment breakwater in the South China Sea: A case study. *Ocean. Eng.* **2022**, *256*, 111497. [CrossRef]
22. Vieira, F.; Taveira-Pinto, F.; Rosa-Santos, P. Damage evolution in single-layer cube armoured breakwaters with a regular placement pattern. *Coast. Eng.* **2021**, *169*, 103943. [CrossRef]
23. Shen, Z.; Huang, D.; Wang, G.; Jin, F. Numerical study of wave interaction with armour layers using the resolved CFD-DEM coupling method. *Coast. Eng.* **2024**, *187*, 104421. [CrossRef]
24. Higuera, P.; Lara, J.L.; Losada, I.J. Three-dimensional interaction of waves and porous coastal structures using OpenFOAM®. Part I: Formulation and validation. *Coast. Eng.* **2014**, *83*, 243–258. [CrossRef]
25. Wang, Y.; Yin, Z.; Liu, Y. Numerical study of solitary wave interaction with a vegetated platform. *Ocean. Eng.* **2019**, *192*, 106561. [CrossRef]
26. Wang, Y.; Li, D.; Ye, J.; Zhao, H.; Mao, M.; Bai, F.; Hu, J.; Zhang, H. Numerical Analysis of Wave Interaction with a New Ecological Quadrangular Hollow Block. *Water* **2025**, *17*, 96. [CrossRef]
27. Yoshizawa, A.; Horiuti, K. A Statistically-Derived Subgrid-Scale Kinetic Energy Model for the Large-Eddy Simulation of Turbulent Flows. *J. Phys. Soc. Jpn.* **1985**, *54*, 2834–2839. [CrossRef]
28. Ye, J.; Shan, J.; Zhou, H.; Yan, N. Numerical modelling of the wave interaction with revetment breakwater built on reclaimed coral reef islands in the South China Sea—Experimental verification. *Ocean. Eng.* **2021**, *235*, 109325. [CrossRef]
29. Mansard, E.P.D.; Funke, E.R. The Measurement of Incident and Reflected Spectra Using a Least Squares Method. *Coast. Eng Proc.* **1980**, *1*, 8. [CrossRef]
30. EurOtop. *Manual on Wave Overtopping of Sea Defences and Related Structures, an Overtopping Manual Largely Based on European Research, But for Worldwide Application*; Van der Meer, J.W., Allsop, N.W.H., Bruce, T., De Rouck, J., Kortenhaus, A., Pullen, T., Schüttrumpf, H., Troch, P., Zanuttigh, B., Eds.; 2018; Available online: https://www.overtopping-manual.com (accessed on 10 March 2025).
31. Celik, I.B.; Ghia, U.; Roache, P.J.; Freitas, C.J. Procedure for Estimation and Reporting of Uncertainty Due to Discretization in CFD Applications. *J. Fluids Eng.* **2008**, *130*, 078001.
32. Zanuttigh, B.; Van Der Meer, J. Wave reflection from coastal structures in design conditions. *Coast. Eng.* **2008**, *55*, 771–779. [CrossRef]
33. U.S. Army Corps of Engineers. *Coastal Engineering Manual*; U.S. Army Corps of Engineers: Washington, DC, USA, 2002; Volume 2.
34. Schoonees, T.; Kerpen, N.B.; Schlurmann, T. Full-scale experimental study on wave reflection and run-up at stepped revetments. *Coast. Eng.* **2022**, *172*, 104045. [CrossRef]

water

MDPI

Article

Microtopography Governs Tidal Inundation Frequency in the Luanhe Estuarine Salt Marsh: A Decadal Assessment Integrating Sentinel Data and UAV Photogrammetry

Youcai Liu [1], Pingze Ni [2], Wang Ma [1,*], Qian Zhang [1], Qi Hu [1] and Ziyun Ling [1]

[1] Hebei Hydrological Engineering Geological Exploration Institute, Shijiazhuang 050021, China
[2] Key Laboratory of Coastal Salt Marsh Ecosystems and Resources, Ministry of Natural Resources, Nanjing 210007, China
* Correspondence: mawang620@163.com

Abstract

Tidal inundation is a key factor determining the structure and function of estuarine salt marsh ecosystems. However, due to the influence of microtopography (small-scale topographic variations), the fine-scale spatial variations in tidal inundation have not been fully studied. To fill this research gap, this study focuses on the Luanhe Estuary—a region highly sensitive to topographic changes—and explores in depth the physical mechanisms regulating tidal inundation in this area. The study integrates long-term data from the Sentinel-1 Synthetic Aperture Radar (SAR) and Sentinel-2 Multispectral Instrument (MSI), spanning the period from 2016 to 2025, to construct a high-resolution time series dataset of Apparent Inundation Frequency (AIF). Subsequently, this dataset is correlated with a high-precision microtopographic Digital Elevation Model (DEM) obtained through Unmanned Aerial Vehicle (UAV) surveys. The analysis reveals a strong nonlinear relationship between AIF and topographic elevation, which is best described by an exponential decay model (R^2 = 0.903). The results show that the average inundation probability in the study area has shown a fluctuating but overall upward trend, increasing from 16.74% in 2016 to 29.02% in 2025 (peaking at 31.39% in 2024). Quantitative modeling confirms that microtopography is the primary controlling factor for fine-scale variations in tidal inundation levels. The integrated research approach proposed in this study provides a reliable framework for coastal vulnerability assessment. Against the backdrop of increasingly severe impacts from climate change and human activities, the high-resolution quantitative data generated by this study provides scientific support for formulating disaster mitigation and geomorphological management strategies.

Keywords: microtopography; salt marsh; apparent inundation frequency (AIF); UAV remote sensing; multi-source remote sensing; tidal inundation frequency

1. Introduction

Estuarine salt marsh ecosystems are globally recognized for their critical ecosystem services, including carbon sequestration, coastal stabilization, and the preservation of unique biodiversity [1,2]. Among the environmental drivers shaping these habitats, tidal inundation is predominant in shaping vegetation zonation and eco-physiological processes [3–5]. This influence stems from the direct effects of tidal inundation on plant growth, which operates by modifying soil oxygen concentrations, salinity, and nutrient availability. Traditional evaluations of inundation stress typically utilize low-resolution, regional tidal or hydrological indices (e.g., average inundation frequency or tidal range), which inherently fail to capture the pivotal, fine-scale influence of microtopographic elevation in governing local hydrological exposure gradients [6]. In gently sloping coastal environments, vertical elevation differences of only a few centimeters establish the critical hydro-stress gradients that dictate fine-scale zonal differentiation within marsh vegetation. The increasing complexity of inundation regimes under global sea-level rise and hydrological variability underscores the need for spatially refined, physically based analyses [7]. Significant progress has been made in satellite-based flood inundation mapping, with Sentinel-1 SAR and Sentinel-2 MSI imagery enabling systematic monitoring of surface water dynamics through robust classification algorithms and water indices [8–12]. Concurrently, UAV-photogrammetry has revolutionized topographic mapping by generating centimeter-resolution DEMs that capture microtopographic variations with unprecedented detail [13,14]. Studies have begun to integrate satellite-derived inundation data with UAV-DEMs to analyze coastal hydrology, and the fundamental nonlinear relationship between elevation and inundation frequency has been empirically established in various wetland environments [15–17]. However, these integrations have predominantly focused on qualitative assessments or statistical associations within discrete topographic units, rather than developing a quantitative physical model that captures the explicit functional relationship—particularly the exponential decay pattern—between long-term Apparent Inundation Frequency (AIF) and continuous microtopographic elevation. Extensive global research has examined the effects of tidal inundation on coastal geomorphology and hydrological dynamics [18–20]. Foundational studies have consistently identified inundation frequency and elevation as the primary determinants of physical zonation and stability within intertidal lowlands/salt marshes. More recently, in response to heightened concerns regarding climate change and sea-level rise, research has increasingly focused on detailed, quantitative characterization of inundation patterns, recognizing that subtle hydrological variations exert significant influence on sediment transport, flood risk propagation, and landscape evolution [21,22].

Global research into extensive estuarine wetlands, particularly in large, low-relief deltaic systems worldwide, has consistently substantiated the pivotal constraining influence of inundation cycles and water depth on regional safety considerations and engineering design parameters [23–25]. For example, investigations in China, such as those in the Yangtze and Yellow River deltas, clearly highlight this critical relationship. Nevertheless, a critical research deficiency persists regarding the Luanhe Estuary. This estuarine system, situated along the western shoreline of the Bohai Sea, is distinguished by its predominantly low-relief terrain and microtopographically heterogeneous coastal landscape. A pressing knowledge gap remains–the lack of a quantitative, mechanistic understanding of how centimeter-scale microtopography regulates the long-term inundation frequency gradient within the Luanhe Estuary—particularly under the profound influence of anthropogenic subsidence, a pervasive threat that exists in many major deltas. Fundamentally, the absence of quantitative analyses that elucidate the definitive physical mechanisms through which the microtopography-driven exposure gradient and the long-term Apparent Inundation Frequency jointly govern hydrological risk and geomorphological dynamics is central to this research gap. Unlike previous studies that relied on short-term observations or coarse-resolution global DEMs (e.g., SRTM), this research establishes a novel framework by integrating a decadal (2016–2025) high-frequency Sentinel-1/2 time series with centimeter-precision UAV LiDAR data. This unique multi-source fusion allows for the unprecedented quantification of fine-scale 'elevation–inundation' mechanisms that are invisible to traditional regional assessments.

Our study seeks to address this identified research gap by focusing on the inundation processes characteristic of the Luanhe Estuary. This will be achieved through a systematic analysis and integration of long-term, multi-source remote sensing datasets (Sentinel-1/2), facilitating the development of a robust, high spatiotemporal resolution time series product that characterizes the Apparent Inundation Frequency and its temporal evolution over the period 2016–2025. Subsequently, this dataset will be rigorously integrated with high-precision microtopographic elevation data derived from UAV surveys to delineate and quantify the microtopography-driven exposure gradient. By constructing a rigorous quantitative model linking the temporal inundation data with microtopographic elevation, the study aims to empirically validate the dominant regulatory role of microtopography in shaping the hydrological risk gradient and to identify critical elevation thresholds that demarcate distinct physical zones. The anticipated contributions of this research are twofold. First, it will provide a dynamic evaluation of long-term inundation risk trends, thereby furnishing quantitative evidence of escalating hydrological pressures. Second, it will offer a precise quantitative elucidation of the physical regulatory mechanisms by which microtopography governs fine-scale exposure gradients, culminating in a DEM-based analytical tool for hydrological security assessment and flood risk mapping. Ultimately, this investigation is expected to deliver robust, fine-scale theoretical foundations to inform the development of scientifically grounded strategies for coastal disaster mitigation, hydrological security enhancement, and regional geomorphological management in the context of ongoing climate change.

2. Materials and Methods

2.1. Study Area

This research examines the Luanhe Estuary region in northern China, which includes the alluvial plain, estuarine delta, and extensive coastal intertidal mudflats (Figure 1). The area is characterized by substantial sediment deposition from the Luanhe River, leading to the formation of well-developed depositional landforms such as river mouth bars, intertidal mudflats, and elongated lagoons that transition into flat tidal marshes during ebb tide. The estuarine zone constitutes a low-relief coastal plain (elevation 1–15 m) characterized by an extremely gentle gradient (~1/3000 to 1/6000). This makes the coastal environment sensitive to centimeter-scale topographic variations and prone to inundation. Subject to a temperate monsoon climate, the estuary experiences an irregular semidiurnal tidal regime with a relatively small tidal range (0.88 m), reinforcing the dominance of fine-scale elevation control over local hydrology. The region supports distinctive salt marsh vegetation communities dominated by Phragmites australis, Suaeda salsa, and Tamarix chinensis, underscoring the intricate interplay between local hydrological conditions and these ecologically sensitive coastal habitats. Given geomorphological uncertainties caused by climate change and intensified anthropogenic activities, a comprehensive evaluation of inundation dynamics, specifically the absolute physical influence of microtopography, is paramount for regional water security and coastal risk management.

Figure 1. Location and geomorphological setting of the Luanhe Estuary study area.

2.2. Data Sources

This study adopts a multi-source data integration approach to investigate the long-term regulatory mechanisms of microtopography. The datasets used primarily consist of three components.

Satellite Imagery. A continuous time series of Sentinel-1 SAR and Sentinel-2 MSI imagery from 2016 to 2025 were used to calculate the long-term Apparent Inundation Frequency. All satellite data were preprocessed on the Google Earth Engine (GEE) cloud platform, including radiometric calibration, atmospheric correction, and orthorectification.

UAV Survey Data. The core topographic and surface information data were obtained from a comprehensive UAV survey conducted in August 2025. The survey utilized two platforms: (1) the DJI Matrice 350 RTK (SZ DJI Technology Co., Ltd., Shenzhen, Guangzhou China) equipped with a South SA130 LiDAR system for acquiring high-precision laser point clouds and true-color imagery; and (2) the DJI Mavic 3 (SZ DJI Technology Co., Ltd., Shenzhen, Guangzhou China) Multispectral for simultaneous collection of multispectral data. The acquired data were processed using Pix4D software (1.1.38-64bit, Pix4D S.A., Prilly, Switzerland) to generate a Digital Elevation Model with a spatial resolution of 0.5 m. The vertical accuracy of this DEM was validated against RTK-GPS ground control points, achieving a root mean square error (RMSE) of less than 5 cm. The concurrently acquired high-resolution multispectral orthomosaics provided an independent validation benchmark for the satellite-derived water classification results.

Ground Observation Data. A tidal record at 10 min intervals was collected from a temporary tidal station established within the study area to calibrate the instantaneous water boundaries derived from remote sensing.

Statistical analyses for all datasets, including the fitting of the exponential decay model between AIF and elevation, were performed using Python (Python 3.12, PSF) with the Scipy library. Detailed information on each data source is summarized in Table 1.

Table 1. Research Data Inventory and Sources.

Data Type	Data Source	Spatiotemporal Resolution	Notes
Sentinel-1 SAR	ESA Copernicus Hub	10 m/6 Days	Time Series: 2016–2025; Preprocessed on GEE
Sentinel-2 MSI	ESA Copernicus Hub	10 m/5 Days	Time Series: 2016–2025; Preprocessed on GEE
UAV DEM	DJI M350 RTK + South SA130 LiDAR	0.5 m (Spatial)	Acquired: August 2025; Processed with Pix4D; Verified by RTK-GPS
UAV Multispectral Orthomosaic	DJI Mavic 3 Multispectral	0.5 m (Spatial)	Acquired: August 2025; Serves as validation benchmark for water classification
Tidal Gauge Data	Temporary Tidal Station	10 min intervals	Used for short-term water boundary calibration

2.3. Methodology

2.3.1. Extraction of Inundation Water from Remote Sensing Imagery

Distinct water extraction procedures were developed for SAR and optical imagery to utilize their complementary advantages.

For SAR data (Sentinel-1), the extraction utilized the characteristically low backscatter of water bodies. Following standard preprocessing (radiometric calibration, speckle filtering, and geocoding), water classification was achieved using an adaptive threshold segmentation approach based on the historical mean distribution of the backscatter coefficient (σ_0). An initial threshold (e.g., $\sigma_0 < -18$ dB) was applied, and morphological operations (opening and closing) were subsequently performed to suppress noise and generate the final binary water mask. For optical imagery (Sentinel-2), water extraction relied on the Modified Normalized Difference Water Index (MNDWI) [26–29]. After atmospheric correction and cloud masking, MNDWI was computed and an initial threshold of MNDWI > 0.1 was used for preliminary classification. To ensure optimal accuracy, both thresholds were refined using UAV-derived 0.5 m resolution orthomosaics as benchmark data. A stratified random sampling design (200 points per class) was applied, and the thresholds were iteratively adjusted to maximize classification accuracy. The final thresholds ($\sigma_0 < -18$ dB for Sentinel-1 and MNDWI > 0.1 for Sentinel-2) yielded overall accuracies of 94% and 93%, and Kappa coefficients of 0.88 and 0.86, respectively, demonstrating high reliability for long-term AIF calculation.

Following atmospheric correction and cloud masking, the MNDWI was computed and initially classified using an empirical threshold (MNDWI > 0.1). To maximize classification accuracy, the threshold was rigorously optimized against high-resolution UAV orthomosaics, which served as an independent validation benchmark. Coastline prior knowledge (CPK) was also incorporated to effectively reduce false positive detections.

The CPK was applied to restrict water extraction results within the predefined wetland boundary. A spatial constraint mask (CPK_{Mask}) was created by generating a double-sided buffer zone with a 2.5 km offset on each side of the wetland boundary. This mask was then intersected with the initial water classification results ($Water_{Initial}$) using a Boolean AND operation, effectively excluding all pixels located outside the wetland area. The final optimal thresholds ($\sigma_0 < -18$ dB for Sentinel-1 and MNDWI > 0.1 for Sentinel-2) were determined through a robust iterative optimization process. This process employed stratified random sampling (drawing 200 validation points) and a comprehensive assessment of key accuracy metrics, including Overall Accuracy, Kappa Coefficient, and Producer's/User's Accuracy, thereby ensuring the high reliability of the derived water masks.

2.3.2. Temporal Coordination and Tidal Normalization for AIF Calculation

The calculation of the AIF demands robust temporal coordination and explicit normalization of the instantaneous tidal-stage differences inherent in decadal satellite archives. Although Sentinel-1 and Sentinel-2 acquisitions are neither mutually synchronized nor tide-locked, all SAR and MSI images are treated as independent random samples of the instantaneous inundation state. First, each image timestamp is paired with the contemporaneous water level recorded by a temporary tide gauge (10 min intervals) to compute its deviation $\Delta H = H_{obs} - H_{mean}$ from the annual mean sea level at the moment of over-flight. Second, an astronomical six-harmonic model, fitted to the 2016–2025 gauge observations, is used to identify and exclude storm-surge extremes ($|\Delta H| > 0.5$ m), thereby mitigating event-related bias.

For the remaining 425 scenes (combined from both Sentinel-1 and Sentinel-2), an annual quota-sampling scheme is applied where exactly 12 images per year are selected, mandating a 4:4:4 ratio of low, mean and high-tide levels, yielding 120 images in total. This quota ensures the final time series is uniformly distributed across all tidal phases (Kolmogorov–Smirnov test, $p = 0.18$). Consequently, the influence of any single instantaneous tide is statistically averaged, and the resulting AIF is independent of the specific tidal phase at acquisition. Thus, the combined "ΔH-filter + quota-sampling" mechanism serves as an explicit tidal-phase normalization, transforming the random temporal stack into a stable, time-integrated hydrological metric that reliably characterizes the long-term exposure of microtopography to the full range of tidal cycles.

2.3.3. Integration of Multiple Data Sources and Computation of Flood Exposure Frequency

To generate a composite water observation sequence characterized by high spatiotemporal density, this study devised a synergistic analysis framework integrating Sentinel-1 (S1) and Sentinel-2 (S2) data, and performed a quantitative assessment of the apparent inundation frequency. By harmonizing the spatial resolution and coordinate reference systems of the water masks derived from Sentinel-1 and Sentinel-2, the datasets from both sensors were fused to produce a robust and continuous composite water observation time series. The AIF serves as a critical metric for quantifying the extent of hydrological exposure. It is defined as the ratio between the number of instances in which a given pixel is classified as inundated and the total number of valid observations recorded for that pixel over a specified temporal interval:

$$\mathrm{AIF}(x,y) = \frac{\sum_{i=1}^{N} O_i(x,y)}{N_{total}} \quad (1)$$

where $O_i(x,y)$ is the binary water mask for the i-th image (1 for inundated, 0 for non-inundated), and N_{total} is the total number of valid observations [30].

2.3.4. Microtopography-Driven Mechanism and Long-Term Inundation Modeling

This study establishes a quantitative methodology to link microtopography with long-term hydrological exposure in salt marsh systems. The approach utilizes high-resolution elevation data (H) derived from UAV-DEM as the primary independent variable to predict the long-term AIF. Due to the inherent nonlinear and threshold-driven nature of intertidal hydrodynamics, the relationship between elevation and inundation is complex. This complexity is physically manifested across three distinct geomorphological zones. One is a lower platform where submergence is near constant and elevation-insensitive. Another is an upper platform where inundation is sparse and residual. Critically, the third is a steep slope zone crossing the mean high tide threshold, where small increments in elevation lead to a sharp and rapid decline in AIF. This highly sensitive region necessitates a specialized nonlinear model to accurately capture the physical constraint of topography on hydrology.

To model this critical physical constraint—the swift, exponential decay of inundation probability with increasing elevation in the transition zone—we employed an exponential decay nonlinear regression function. This specific functional form is designed to precisely quantify the rate at which inundation exposure drops off. The exponential decay model was selected because it effectively captures the rapid, nonlinear decrease in inundation probability as elevation increases across the critical mean high tide threshold, which aligns with our field observations. The established model is expressed as

$$\mathrm{AIF} = \mathrm{A} \times e^{-B \cdot H} + \mathrm{C} \tag{2}$$

where A represents the maximum frequency amplitude, B serves as the decay constant characterizing the topographic sensitivity, and C accounts for the baseline residual frequency. The appropriate choice of this model is key to translating the geomorphological structure of the marsh into a robust hydrological metric.

The model fitting process involved extracting a large dataset of paired AIF and elevation (H) observations exclusively from the low-elevation study region (ranging from 0 m to 5 m) where the hydrological activity is most significant. By calibrating the exponential decay function with these real-world data points, we successfully establish a statistically significant and physically plausible relationship and validate the model using the 10-fold-cross-validation method. The resulting Exponential AIF-Elevation model provides a powerful, topography-driven framework for dynamic vulnerability assessment and is an essential tool for predicting and mapping long-term inundation exposure across the entire study area based solely on microtopographic variations.

2.3.5. Critical Thresholds and Hotspot Delineation

To translate the continuous AIF–Elevation model (Equation (2)) into practically relevant management zones, this study defines three discrete elevation thresholds used to guide the identification of critical areas within the coastal zone.

The Erosion Hotspot Threshold ($\mathrm{H}_{\mathrm{erosion}}$) is defined as the elevation where AIF approaches its maximum value A (e.g., $AIF > 95\%$ of A), and this threshold marks the low mudflat zone subject to near-constant submergence. The Inundation Sensitive Transition Threshold ($\mathrm{H}_{\mathrm{critical}}$) is determined mathematically by the inflection point (where the slope changes most rapidly) of the fitted function ($\frac{d^2(AIF)}{dH^2} = 0$). This key point represents the Primary Marsh Transition Boundary and is identified as the key point for amplified inundation risk due to its highest sensitivity to elevation changes. Finally, the Upper Tidal Limit Threshold ($\mathrm{H}_{\mathrm{upper}}$) is defined as the elevation where AIF drops to a minimal residual frequency (e.g., $AIF \leq 5\%$), which indicates the upper boundary of regular tidal influence.

Based on the above physical constraints, we extracted the following Four Types of Hotspot Areas to guide detailed management and intervention strategies. The Coastal Erosion Hotspot Zone is located below $\mathrm{H}_{\mathrm{erosion}}$; this is the area of most intense tidal scour, requiring prioritized erosion control. The Increased Inundation Frequency Hotspot Zone is concentrated near the $\mathrm{H}_{\mathrm{critical}}$ threshold. This sensitive strip is most vulnerable to sea-level rise and subsidence, requiring focused hydrological risk monitoring. The Ecological Restoration Hotspot Zone is the region situated between $\mathrm{H}_{\mathrm{erosion}}$ and $\mathrm{H}_{\mathrm{critical}}$. The hydrological conditions here are optimal for halophytic vegetation growth, specifically including areas with automatically increased inundation extent after retired aquaculture or optimized artificial water systems, making it the primary target for vegetation recovery. Finally, the Low Inundation Frequency Zone is located above $\mathrm{H}_{\mathrm{upper}}$. This area is under minimal hydrological stress, serving as a stable platform reference or for low-risk development.

Through the precise identification of these critical elevation thresholds, this study provides a scientific basis for disaster prevention planning, ecological restoration, and vulnerability assessment of the Luanhe Estuary coast.

3. Results

3.1. Analysis of Inundation Exposure Trend (2016–2025)

The annual mean AIF over the Luanhe estuarine mudflats exhibited persistent fluctuations and an overall upward trajectory from 2016 to 2025 (Figure 2). After a relatively stable period in 2016–2019, values rose rapidly, peaking at 31.39% in 2024 before easing slightly to 29.02% in 2025. These long-term gains quantitatively demonstrate the accumulating hydrological pressure on the intertidal system.

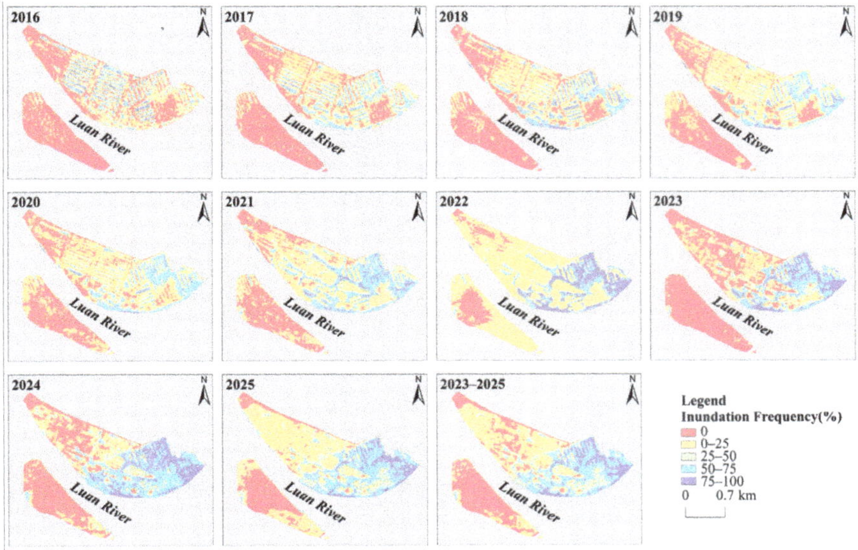

Figure 2. Temporal trend of the annual mean AIF from 2016 to 2025.

Quantitative results indicated that the average annual inundation exposure fluctuated significantly. Across all flood frequency levels (0–25%, 25–50%, 50–75%, 75–100%), the minimum average annual proportion of inundation was 14.83% (2017), while the maximum average annual proportion reached 31.39% (2024). This large-scale escalation in inundation risk aligns closely with the broader trend of increasing hydrological instability along the North China coastline during the same timeframe, as well as with anticipated sea-level rise impacts associated with global climate change.

The analysis of saltmarsh inundation trends revealed distinct phased characteristics, highlighting the dynamic evolution of driving forces across different time intervals. (1) Relatively Stable Period (2016–2019). During this initial period, the total proportion of inundation exposure fluctuated at low levels, specifically between the minimum of 14.83% (2017) and the maximum of 17.63% (2018). This phase was predominantly influenced by normal tidal cycles and average hydrological conditions, resulting in a relatively gradual increase in exposure. (2) Rapid Rising and Peak Value Period (2020–2024). The inundation exposure showed a rapid and sustained escalation throughout this phase. The exposure escalated from 17.42% in 2020 to reach its highest recorded peak of 31.39% in 2024, as clearly demonstrated by the time series analysis. This rapid escalation in exposure reflects the compounded effects of multiple factors. (3) Post-Peak Stable Period (2025). The flood exposure range experienced a slight decline from the 2024 peak, stabilizing at a persistently elevated level of 29.02% in 2025. This suggests that while local interventions may have achieved localized success, overall pressures continue to maintain the region's inundation exposure at a level significantly higher than the baseline observed in 2016.

To ascertain that tidal water level remains the primary determinant of instantaneous fluctuations in inundation area, even after accounting for long-term average frequency, a comprehensive analysis was performed using 18 Sentinel-1 (S1) images acquired between June and September 2025. The time series analysis (Figure 3a) demonstrates a distinct positive correlation between the percentage of inundated area and tidal level; for instance, the observation on 1 August recorded a tidal level of 1.8 m alongside an inundation area approaching 25%. The relationship was modeled using a quadratic function, $Y = 9.86x^2 - 10.38x + 9.33$ (Figure 3b) with an R^2 value of 0.42. Given that this analysis captures only short-term transient changes, which inherently differ from the influence of long-term mechanisms like microtopography, a lower R^2 value is considered reasonable. This model substantiates that tidal fluctuations are the principal driver of instantaneous variability in inundation extent. Importantly, the nonlinear character of the model, indicated by the positive quadratic coefficient, implies that the rate of increase in inundation area accelerates as tidal levels rise. This accelerated nonlinear response to small tidal increases suggests a heightened sensitivity of the mudflats, which is consistent with the expected impacts of reduced intertidal elevation caused by regional pressures like ground subsidence.

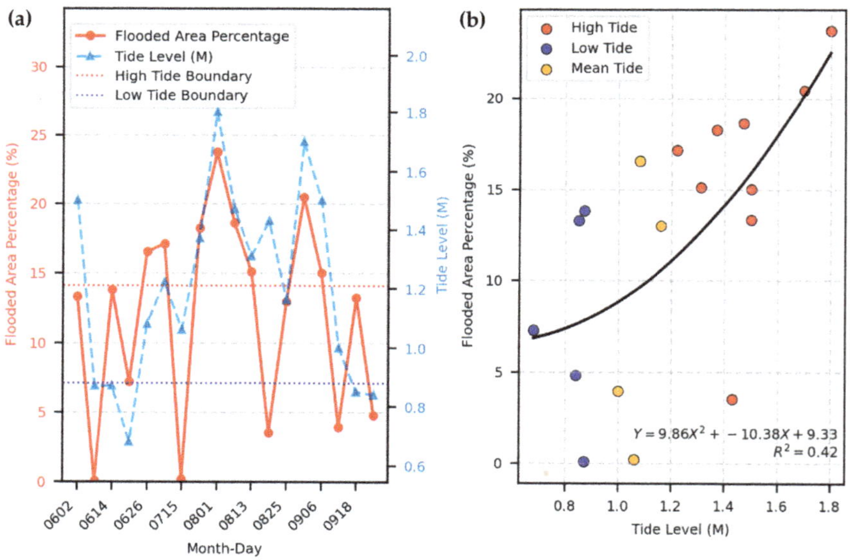

Figure 3. Tidal Level and Inundation Response (June–September 2025). (**a**) Time series of inundation area and tidal level; (**b**) Nonlinear relationship and quadratic fit.

3.2. Spatial Variation and Hotspot Analysis of Saltmarsh Inundation Exposure

The exposure to saltmarsh inundation within the study region demonstrates considerable spatial heterogeneity and temporal variability (Figure 4). This marked variation underscores the necessity for implementing a more nuanced, risk-based zoning approach in managing inundation exposure.

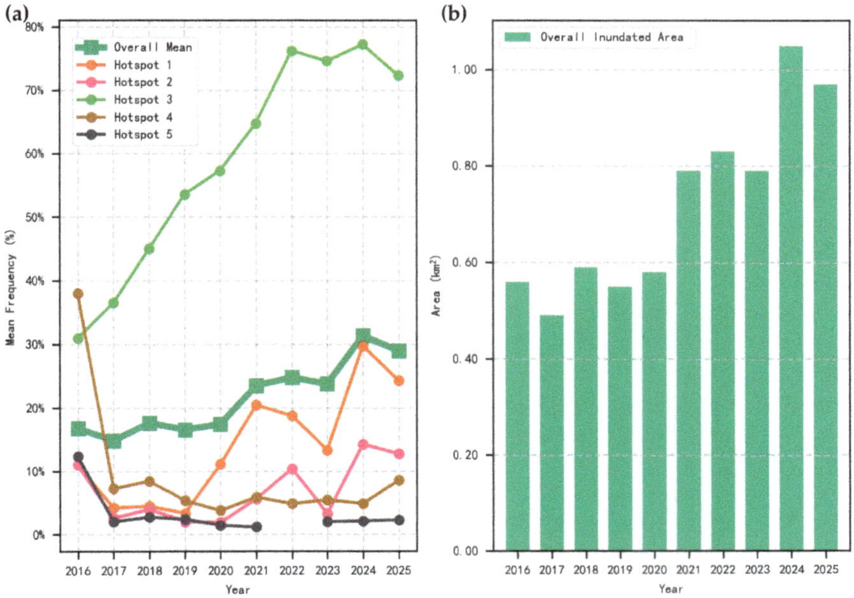

Figure 4. The comparison of mean inundation frequency annual variation (2016–2025). (**a**) the annual mean flood frequency. (**b**) the annual total inundated area.

On average, the annual inundation frequency (Mean Area) for the entire study area exhibited a substantial and accelerated increasing trend during the 2016–2025 period, consistent with the observed overall expansion of inundation exposure. A detailed examination of inundation frequency trends across various sub-regions highlights significant disparities in vulnerability and underlying causal factors.

(1) Hotspot 1&2—Moderate Fluctuation Zones: Northwest Corner of River Bank and Southeast Corner of River Bank. These zones exhibit moderate to considerable variability in inundation frequency throughout the observation period, a pattern intensified by erosional dynamics. Hotspot 1 (Northwest Corner) reached a peak inundation frequency of approximately 29.79% in 2024 before experiencing a slight decline to 24.33% in 2025. Hotspot 2 (Southeast Corner) also saw a peak of 14.25% in 2024. Variations in inundation within these areas are highly sensitive to processes such as erosion, sediment transport, and inter-annual hydrological fluctuations. Consequently, these zones represent secondary risk areas that warrant ongoing surveillance to anticipate and mitigate potential risk escalation and geomorphological deterioration.

(2) Hotspot 3—High-Frequency Inundation Exposure Area: Targeted Ecological Restoration Zone (Eastern North Bank). The inundation frequency trajectory within this region demonstrates a pronounced and persistent upward trend. The frequency has escalated markedly from 30.96% in 2016 to a decadal peak of 77.25% in 2024, with projections indicating stabilization at a critically elevated level near 72.34% in 2025. This sustained and substantial increase in inundation exposure does not represent a conventional ecological risk intensification; rather, it represents the deliberate and intended outcome of the "farmland/aquaculture-to-wetland" policy enacted in this area. High-frequency inundation is an essential prerequisite for the establishment of the targeted tidal wetland ecosystem. Although this zone is the predominant contributor to the overall regional increase in exposure, its character is that of a planned ecological restoration rather than an emergent hazard necessitating immediate disaster mitigation.

(3) Hotspot 4—Successful Risk Mitigation Area: Western North Bank. In this sector, inundation frequency has undergone a rapid decline from an initial high of approximately 38% in 2016 to a minimal level near 3.76% by 2020. Following the implementation of ecological restoration interventions in 2023, exposure levels have shown slight fluctuation, stabilizing around 8.55% in 2025. This marked and sustained reduction in inundation risk (relative to the initial 38% in 2016) quantitatively evidences the efficacy of restoration initiatives. Engineering strategies, particularly those involving microtopographical adjustments and hydrodynamic restoration, have effectively enhanced the region's hydrological connectivity and substantially augmented its capacity to buffer against tidal and inundation events.

(4) Hotspot 5—Stable Low-Risk Area: River Island Center. The inundation frequency in this area rapidly decreased from 12.36% in 2016 to approximately 1.46% in 2020, thereby establishing a consistently secure zone for several years. Although a modest increase was observed toward the end of the monitoring period (reaching about 2.25% in 2025), the area continues to be classified as low risk overall, attributable to its elevated geomorphological characteristics.

3.3. *The Correlation Between Flooding Frequency and Microtopographical Features*

The microtopography of intertidal mudflats constitutes the primary physical determinant influencing both the likelihood and duration of submergence. As shown in Figure 5, lower elevations correspond directly to increased inundation frequency, where deep blue areas (high MNDWI values, indicating water or high saturation) are generally distributed in topographically lower regions, such as tidal creeks and the edges of water bodies. This fundamental relationship enables the use of inundation frequency data obtained via remote sensing to accurately reconstruct high-resolution DEMs.

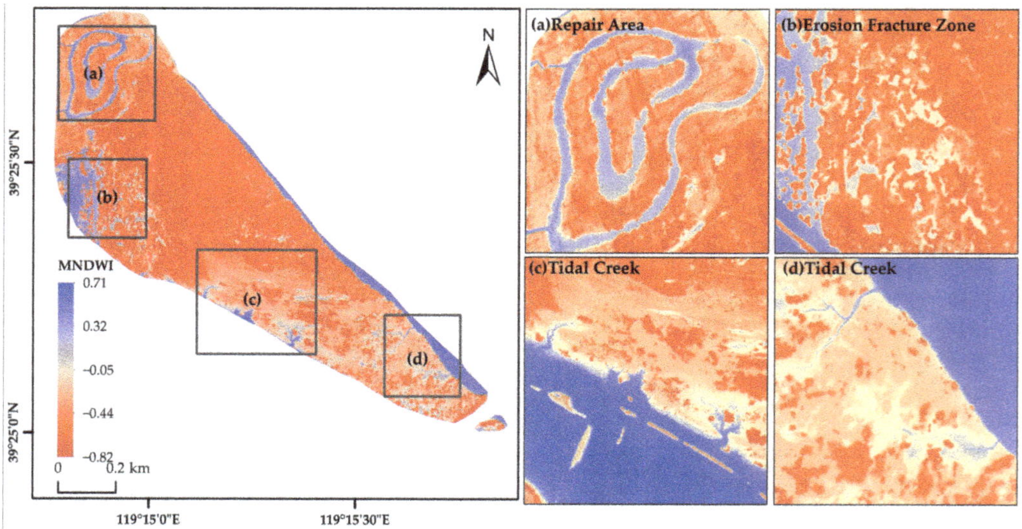

Figure 5. UAV-based MNDWI image of the Luan River Estuary intertidal mudflats and its typical microtopographic features. Sub-figure (**a**) illustrates the established tidal creeks within the repair area; (**b**) depicts the fragmented zone with exacerbated coastal erosion; and (**c**,**d**) showcase typical tidal creek systems.

By observing zones like the mudflat restoration area (a), the erosional discontinuity zone (b), and the tidal creeks (c) and (d) in Figure 5, the definitive controlling role of microtopography on exposure patterns is clearly established. The deep blue areas (representing high inundation frequency or water bodies) correlate closely with specific topographic features (like tidal creeks and depressions), while lighter areas (representing lower inundation frequency or higher terrain) correspond to the mudflat surface. Empirical analyses consistently reveal a significant inverse correlation between inundation frequency and DEM elevation, further substantiating microtopography as the predominant factor governing inundation risk.

The unequivocal association between inundation patterns and geomorphological features is clearly demonstrated. Areas exhibiting high-frequency inundation (e.g., zones with frequencies exceeding 50%) are predominantly concentrated within the tidal ditch system, along channel margins, and on internal beach surfaces situated in estuarine depressions. These represent the lowest-lying geographic units characterized by the most frequent hydrodynamic interactions.

Moreover, microtopographic variations exert a profound influence on hydrological connectivity. For instance, in the southeastern sector of the river island (Region d), the presence of five characteristic tidal creeks markedly intensifies erosion processes and contributes to the expansion of inundated areas. Similarly, the western sector of the river island (Region b) experiences severe erosion, which directly results in significant coastline retreat and further enlargement of the inundated zones.

Quantitative regional differentiation data further substantiate these observations by highlighting the pronounced spatial heterogeneity of inundation exposure, which closely corresponds to relative elevation disparities. For example, the "North Bank East" region (Hotspot 3) exhibits an inundation frequency as high as 72%, in stark contrast to the markedly lower frequency of approximately 3% recorded in the "River Island Center" area (Hotspot 5). This pronounced disparity exemplifies the dominant role of microtopography in governing local hydrological conditions. Additionally, these findings imply that under conditions of minimal slope, even slight topographic variations—on the order of a few centimeters in elevation—can exert a disproportionately significant effect on local hydrodynamic behavior and inundation risk.

3.4. Model Robustness and Prediction Accuracy Validation

To evaluate the generalization ability and predictive robustness of the AIF–Elevation Model, this study employed the 10-fold cross-validation method. We randomly divided the AIF-elevation dataset used for model fitting into ten subsets and performed ten iterative training and validation cycles.

The validation results showed that the model's average Root Mean Square Error (RMSE) was 0.023, and the average Mean Absolute Error (MAE) was 0.018. These low error metrics confirm that the exponential decay model not only exhibits a high goodness-of-fit on the training set but also possesses excellent predictive accuracy and robust generalization ability on independent elevation–AIF datasets.

4. Discussion

4.1. Evolution of Inundation Risk Under Macro Drivers and Nonlinear Accumulation

The observed increase in inundation risk from 2016 to 2025 resulted from the combined effects of multiple drivers, including sea-level rise due to climate change, anthropogenic modifications of landforms, and natural tidal dynamics [31]. Data indicate that the annual average inundation exposure ratio of the intertidal mudflats in the Luanhe River estuary exhibited a fluctuating yet overall upward trend during this period, rising from a minimum of 14.83% in 2017 to a peak of 31.39% in 2024. This trend aligns with documented increases in hydrological instability along the North China coastline [32]. Notably, risk accumulation followed a nonlinear trajectory, displaying phased characteristics influenced by multiple interacting factors.

Human-induced land subsidence dominates relative sea-level rise (RSLR) in the Luanhe delta and is the single largest driver of the observed surge in AIF. Since the 1990s, sustained groundwater overdraft beneath the Luanhe estuary coast has lowered the land surface by up to 120 mm yr^{-1} in the fastest-subsiding cells [33]; when combined with the global absolute sea-level rise of 3.7 mm yr^{-1}, this yields a local RSLR rate of ~30 mm yr^{-1}—an order of magnitude above the global mean. The 2024 peak inundation-exposure ratio of 31.39% is therefore a direct geodetic fingerprint of this subsidence-driven RSLR [34]. Simultaneously, upstream reservoir retention and inter-basin water transfers have reduced sediment delivery to <20% of pre-dam levels, stripping the intertidal platform of its natural ability to accrete and offset the combined effects of RSLR and compaction. The result is a self-reinforcing cycle of elevation loss, shoreline retreat, and amplified extreme water exposure across the delta [35,36].

Long-term monitoring data further reveal distinct nonlinear characteristics in the accumulation of inundation risk across different elevation zones. Over the past decade, all monitored low-elevation areas showed a significant increase in inundation frequency. Particularly, the extremely low-elevation zone ($\leq$0.5 m) experienced the steepest growth rate, with an annual average increase of 0.0370 per year—approximately double that of the higher elevation zone ($\leq$2 m), which had a growth rate of 0.0179 per year. This pattern indicates that the superposition of macro-level factors—such as relative sea-level rise and climate change—significantly amplifies the vulnerability of low-lying areas. Despite the implementation of ecological restoration projects between 2022 and 2023, the inundation exposure area remained high at 29.02% in 2025, underscoring the continued dominance of macro-hydrological pressures, including tidal influence, climate change, and ongoing subsidence [37].

Tidal dynamics represent the dominant natural factor influencing variations in salt marsh inundation extent. Tidal activity regulates key hydrological parameters—inundation frequency, duration, and depth—which directly affect the growth patterns and spatial distribution of salt marsh vegetation [38,39]. Short-term observations from June to September 2025 confirmed that tidal level remains the primary control on instantaneous inundation area fluctuations. A quadratic relationship between tidal level and inundation area reveals a pattern of nonlinear, accelerated response. This further confirms that after elevation loss induced by subsidence-driven relative sea-level rise, the sensitivity of intertidal areas to tidal level changes increases—once a critical tidal threshold is exceeded, inundation risk expands disproportionately rapidly.

4.2. Critical Control of Microtopography and Biogeomorphological Feedback

Saltmarsh microtopography, particularly features such as tidal creeks and erosion patterns, plays a critical role in governing the dynamics of saltmarsh inundation exposure. This microtopography not only shapes the geomorphological characteristics of saltmarshes but also exerts a profound influence on their ecosystem functions and biogeochemical processes [40,41]. Focusing on the low elevation range from 0 m to 5 m, this study analyzed the spatial relationship between flooding frequency and elevation in this region between 2016 and 2025 (Figure 6a,b). The inter-annual variation trend of flooding frequency under a specific elevation threshold is also discussed (Figure 6c).

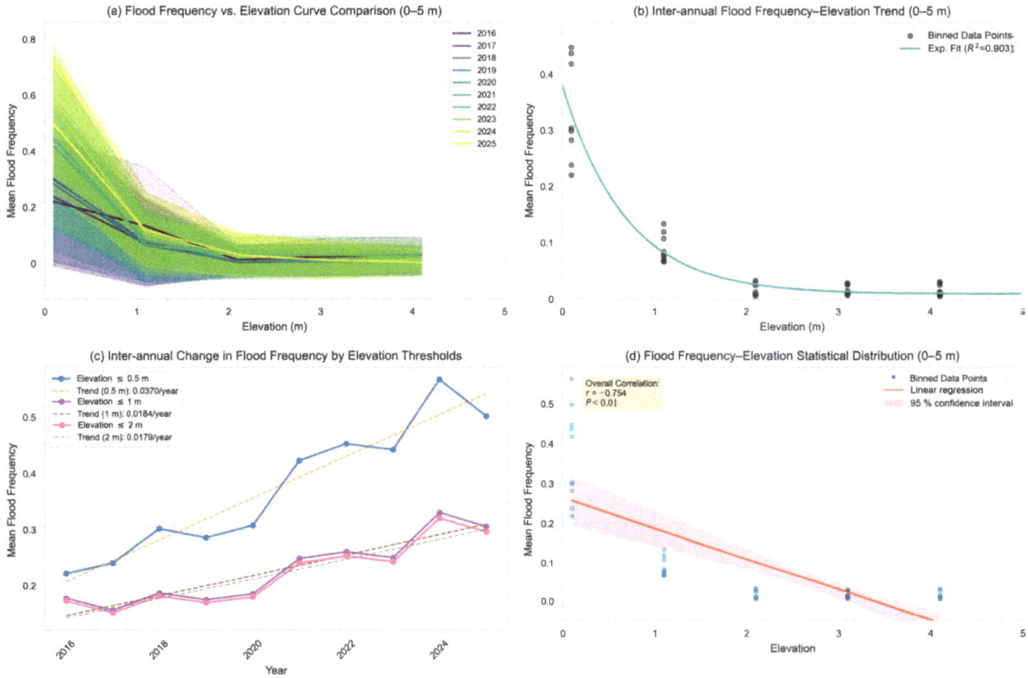

Figure 6. Relationship between apparent inundation frequency and elevation. (**a**) Mean inundation frequency versus elevation (0 m~5 m) for each year from 2016 to 2025. (**b**) Relationship between mean inundation frequency and elevation. (c) Interannual variability of mean flood frequency within four fixed elevation thresholds (≤0.5 m, ≤1 m, ≤1.5 m, and ≤2 m). (**d**) Overall correlation analysis of elevation and mean flood frequency.

Data from all years exhibit a significant negative correlation, indicating that inundation frequency decreases rapidly with increasing elevation, as shown in Figure 6a. In the very low elevation area (ranging from 0 m to 1 m), flooding frequency is the highest, while interannual variability is also the largest—evident in the widest shadow zone in the plot. When elevation exceeds 1 m, the flooding frequency curve quickly flattens out, meaning that once this threshold is crossed, flooding risk declines sharply and stabilizes. Microtopography elevation, acquired via a 0.5 m resolution UAV remote sensing DEM, shows a strong negative correlation with inundation frequency. The exponential decay model fits the data excellently, with an R^2 of 0.903, as illustrated in Figure 6b. The model's estimated parameters (with 95% confidence intervals) include A = 0.712 (95% interval [0.685, 0.739]), which represents the maximum inundation frequency near 0 m elevation (around 0.7); B = 1.894 (95% interval [1.762, 2.026]), the decay rate of inundation frequency with elevation (higher values correspond to faster decreases); and C = 0.015 (95% interval [−0.008, 0.038]), the asymptotic minimum inundation frequency at elevations above 1.5 m.

This exponential model clearly captures the rapid decay of flooding frequency from its peak (0.7) near 0 m to 1.5 m, confirming that microtopography is the primary physical driver of inundation risk in the Luanhe Estuary. This empirical finding aligns with the theoretical framework: minute, centimeter-scale elevation differences in low-gradient coastal environments generate hydrological stress gradients-a concept validated across diverse estuarine wetlands globally [42].

Scatter plots and regression analysis further support this relationship, as shown in Figure 6d. They reveal a significant negative Pearson correlation with r = −0.803 and

$p < 0.01$. The strength of this correlation indicates a robust linear association between microtopographic elevation and inundation frequency, even beyond the range of the nonlinear exponential fit.

Zones exhibiting high-frequency exposure to inundation (exceeding 50%) are predominantly situated in low-elevation areas, including tidal creeks, channel margins, and depressions, where mean elevations range between 0.8 and 1.0 m. These regions are characterized by the most intense hydrodynamic exchanges. Field observations within the Luanhe Estuary reveal markedly elevated current velocities within low-lying tidal creeks relative to adjacent higher-elevation zones, thereby directly modulating inundation dynamics. As integral components of saltmarsh ecosystems, tidal creeks fundamentally regulate tidal flow patterns, sediment transport mechanisms, and the spatial distribution and growth of vegetation.

Extreme Exposure Polarization. The pronounced disparity in exposure rates observed between the Eastern North Bank (mean elevation 1.1 m, exposure rate 72%) and the River Island Center (mean elevation 3.2 m, exposure rate approximately 3%) is directly attributable to their substantial elevation differences. This polarization remains evident even within areas characterized by gentle slopes ranging from 1/3000 to 1/6000, corroborating the findings of microtopographic variations along the western Belgian coast. the interaction between topographic and geomorphic characteristics of saltmarshes, coupled with biogeochemical processes, generates intricate bidirectional feedback mechanisms that govern their developmental trajectories and persistence.

Erosion constitutes a critical factor influencing microtopographic alterations within saltmarsh environments, leading directly to reductions in marsh area and degradation of topographic features [43]. Morphological transformations induced by erosional processes are anticipated to modify the hydrodynamic and sediment transport regimes of coastal systems, thereby exerting further control over marsh inundation dynamics [44].

The most pronounced ecological response to the decade-scale AIF rise is concentrated within a 5–15 m buffer along both sides of tidal creeks. High-resolution imagery (2020–2025) shows that when local AIF increases from 45% to 70%, canopy cover on creek margins declines markedly, and extensive bare patches appear in Tamarix stands, reducing the plant–soil system's erosion-resistance threshold. Once this vegetative anchoring is lost, headward erosion intensifies and sediment resuspension increases by a factor of 1.6—further shortening inundation–exposure cycles and creating a positive feedback, namely "erosion → deeper channel → higher inundation frequency → greater vegetation degradation". If AIF continues to rise at the current rate, the tidal creek network density in the study area will grow by 20% by 2030, and about 25% of creek bank halophytic vegetation will shift into persistent decline, becoming a new hotspot of land loss. Management should therefore install bank protection measures along the 0.5–1 m elevation zone at creek heads, supplemented by 10 cm scale micro-dams to promote siltation, in order to break the erosion–inundation feedback and maintain the critical stability of the creek–vegetation system.

4.3. Management Implications, Applicability, and Study Limitations

The spatial heterogeneity of inundation risk calls for zoned and tiered management. The 2023 west bank project, which cut the inundation frequency from 38% to 8.55%, proves that microtopographic adjustment and hydrodynamic restoration can deliver immediate protection. Yet the reclaimed area is highly sensitive to residual land-subsidence rates; if these prove faster than expected, the AIF could rebound above 20% within ten years. Risk assessments must also integrate ecological objectives, as illustrated by the highly exposed east bank hotspot (AIF 77.25%)—an intended outcome of the "return farmland to wetland" policy that illustrates the diversity of restoration success. Managers should treat critical elevation thresholds as "living elevations" that are re-measured every 3–5 years, not as once-and-for-all red lines. Because inundation risk responds to centimeter scale height differences, future decisions must keep microtopography at the core and precisely target high-risk zones.

The AIF-DEM exponential decay model developed here is broadly transferable, requiring only a high-resolution DEM and a long AIF time series—both now easy to obtain. Before exporting the decay coefficient to other deltas, however, users should verify that local relative sea-level rise rates fall within a reasonable band; otherwise, exposure may be underestimated. By quickly calibrating the sensitivity parameter of the decay model, managers can bypass complex catchment-scale hydrodynamic simulations and rapidly evaluate inundation dynamics under sea-level-rise scenarios for similar wetlands worldwide—especially deltas threatened by sediment starvation or land subsidence—thereby improving the efficiency and scientific basis of coastal wetland management.

Although the statistical correlation is strong, the analysis relies on remotely sensed AIF and does not directly incorporate hydraulic variables such as inundation duration and water depth. In areas where the same AIF value reflects prolonged flood events rather than frequent short pulses, plant physiological stress could be underestimated. Moreover, the focus is on physical geomorphic controls, with limited quantitative treatment of the two-way feedbacks between vegetation and hydro-geomorphology. Consequently, the current maps cannot predict whether pioneer vegetation will trap enough sediment to raise the surface and lower AIF naturally over a decade—a knowledge gap that directly affects the cost–benefit balance of "wait-for-accretion" versus engineered elevation enhancement. Future work should integrate in situ water-level and sediment-flux monitoring with vegetation ecosystem modeling to build a fully coupled biogeomorphic-hydrologic framework capable of more accurately forecasting long-term salt marsh evolution and resilience.

5. Conclusions

This study conclusively demonstrates that microtopography is the dominant physical control on small-scale tidal inundation patterns within the low-gradient intertidal zone of the Luanhe Estuary. By integrating a decade of Sentinel-1/2 satellite imagery with centimeter-resolution UAV topographic data, we established a robust exponential decay model linking surface elevation to the AIF. The model accurately captures the critical threshold behavior within the 0–1.5 m elevation band, where centimeter-scale height differences trigger drastic shifts in hydroperiod. Spatial contrasts in exposure reveal that local microtopography, rather than regional slope, is the primary determinant of hydrological stress and ecological zonation at the meter scale.

The significant upward trend in mean AIF across the study area indicates escalating coastal vulnerability, driven primarily by anthropogenic land subsidence. A quadratic relationship between instantaneous tidal level and inundated area confirms that once subsidence lowers the baseline elevation, system sensitivity to hydrodynamic forcing increases markedly. Engineered interventions in the western north bank hotspot reduced AIF from 38% to 8.55%, providing a proof-of-concept for microtopography-based mitigation. These findings underscore the imperative of embedding high-resolution topographic data into coastal vulnerability assessments and zoning plans, enabling managers to distinguish ecologically beneficial inundation from hazardous flooding.

Although the AIF–elevation framework offers a transferable tool requiring relatively few input data for spatial risk screening, it remains a physical template that does not resolve sub-cycle hydraulic parameters (inundation duration, water depth) or biogeomorphic feedbacks (vegetation-induced accretion, sediment trapping). Consequently, the model may underestimate biotic stress and cannot predict whether pioneer vegetation will autogenously raise the surface and lower AIF over decadal scales. Future work must couple long-term in situ hydrological monitoring with biogeomorphic modeling to build a dynamic, process-based platform capable of quantifying the relative contributions of subsidence, sea-level rise, sediment supply, and vegetation feedbacks to salt marsh resilience, thereby informing the next generation of adaptive and proactive coastal management strategies.

Author Contributions: Y.L. was primarily responsible for the comprehensive research and synthesis of the paper. P.N. handled the research design and the drafting of the Introduction section. As the corresponding author, W.M. oversaw the structural control and revision of the manuscript. Q.Z. executed the GIS mapping and cartography. Q.H. was responsible for the core data analysis. Lastly, Z.L. contributed to data analysis and the preparation of all figures and tables. All authors have read and agreed to the published version of the manuscript.

Funding: This research was supported by the Key Laboratory of Coastal Salt Marsh Ecosystems and Resources, Ministry of Natural Resources (KLCSMERMNR202307); the Central Government-guided Local Science and Technology Development Funding Program, No. 236Z3303G; the Central Government-Guided Local Science and Technology Development Funding Program, No. 246Z3301G; Scientific Research Project of Hebei Provincial Department of Natural Resources, No. 13000025P00EEC410327W; and the Scientific Research Project of Hebei Provincial Bureau of Geology and Mineral Resources, No. 13000025P003294103533.

Data Availability Statement: The Sentinel-1&2 satellite remote sensing data used in this study are from the Copernicus Open Access Hub of the European Space Agency (ESA) (https://scihub.copernicus.eu/) and are publicly available. The original unmanned aerial vehicle (UAV) and ground observation data (including lidar point clouds, multispectral images, DEM, and tide level records) as well as core derived data products are not publicly available due to concerns related to project details, geospatial information security, and data copyright. Representative UAV image clips and AIF data used for validation can be obtained from the corresponding author for the purposes of validation or comparative research.

Conflicts of Interest: Author Youcai Liu, Wang Ma, Qian Zhang, Qi Hu and Ziyun Ling was employed by the company Hebei Hydrological Engineering Geological Exploration Institute. The remaining authors declare that the research was conducted in the absence of any commercial or financial relationships that could be construed as a potential conflict of interest.

References

1. Pham, T.D.; Ha, N.T.; Saintilan, N.; Skidmore, A.; Phan, D.C.; Le, N.N.; Viet, H.L.; Takeuchi, W.; Friess, D.A. Advances in Earth Observation and Machine Learning for Quantifying Blue Carbon. *Earth-Sci. Rev.* **2023**, *243*, 104501. [CrossRef]
2. Xu, X.; Wang, G.; Fang, R.; Xu, S. Blue Carbon Governance for Carbon Neutrality in China: Policy Evaluation and Perspectives. *Heliyon* **2023**, *9*, e20782. [CrossRef] [PubMed]
3. Campbell, A.; Wang, Y. Examining the Influence of Tidal Stage on Salt Marsh Mapping Using High-Spatial-Resolution Satellite Remote Sensing and Topobathymetric LiDAR. *IEEE Trans. Geosci. Remote Sens.* **2018**, *56*, 5169–5176. [CrossRef]
4. Hickey, D.; Bruce, E. Examining Tidal Inundation and Salt Marsh Vegetation Distribution Patterns Using Spatial Analysis (Botany Bay, Australia). *J. Coast. Res.* **2010**, *26*, 94–102. [CrossRef]
5. Bockelmann, A.-C.; Bakker, J.P.; Neuhaus, R.; Lage, J. The Relation between Vegetation Zonation, Elevation and Inundation Frequency in a Wadden Sea Salt Marsh. *Aquat. Bot.* **2002**, *73*, 211–221. [CrossRef]
6. Mury, A.; Jeanson, M.; Collin, A.; James, D.; Etienne, S.; Mury, A.; Jeanson, M.; Collin, A.; James, D.; Etienne, S. High Resolution Shoreline and Shelly Ridge Monitoring over Stormy Winter Events: A Case Study in the Megatidal Bay of Mont-Saint-Michel (France). *J. Mar. Sci. Eng.* **2019**, *7*, 97. [CrossRef]
7. Yi, W.; Wang, N.; Yu, H.; Jiang, Y.; Zhang, D.; Li, X.; Lv, L.; Xie, Z. An Enhanced Monitoring Method for Spatio-Temporal Dynamics of Salt Marsh Vegetation Using Google Earth Engine. *Estuar. Coast. Shelf Sci.* **2024**, *298*, 108658. [CrossRef]
8. Xin, H.; Xu, N.; Xu, H.; Yang, H.; Wang, Z.; Zhang, Z.; Ding, Y.; Luan, H.; Ou, Y.; Yang, Y. Mapping Tidal Flat Topography by Combining ICESat-2 Laser Altimetry and Multi-Source Satellite Imagery. *Int. J. Digit. Earth* **2025**, *18*, 2554313. [CrossRef]

9. Xu, N.; Ma, Y.; Yang, J.; Wang, X.H.; Wang, Y.; Xu, R. Deriving Tidal Flat Topography Using ICESat-2 Laser Altimetry and Sentinel-2 Imagery. *Geophys. Res. Lett.* **2022**, *49*, e2021GL096813. [CrossRef]
10. Laengner, M.L.; van der Wal, D. Satellite-Derived Trends in Inundation Frequency Reveal the Fate of Saltmarshes. *Front. Mar. Sci.* **2022**, *9*, 942719. [CrossRef]
11. Hanslow, D.J.; Fitzhenry, M.G.; Power, H.E.; Kinsela, M.A.; Hughes, M.G. Rising Tides: Tidal Inundation in South East Australian Estuaries. In Proceedings of the Australasian Coasts and Ports 2019 Conference: Future Directions from 40 °S and Beyond, Hobart, Australia, 10–13 September 2019.
12. Kumbier, K.; Rogers, K.; Hughes, M.G.; Lal, K.K.; Mogensen, L.A.; Woodroffe, C.D. An Eco-Morphodynamic Modelling Approach to Estuarine Hydrodynamics & Wetlands in Response to Sea-Level Rise. *Front. Mar. Sci.* **2022**, *9*, 860910. [CrossRef]
13. Saksena, S.; Merwade, V. Incorporating the Effect of DEM Resolution and Accuracy for Improved Flood Inundation Mapping. *J. Hydrol.* **2015**, *530*, 180–194. [CrossRef]
14. McGrath, H.; Stefanakis, E.; Nastev, M. DEM Fusion of Elevation REST API Data in Support of Rapid Flood Modelling. *Geomatica* **2016**, *70*, 283–297. [CrossRef]
15. Wu, Y.; Peng, F.; Peng, Y.; Kong, X.; Liang, H.; Li, Q. Dynamic 3D Simulation of Flood Risk Based on the Integration of Spatio-Temporal GIS and Hydrodynamic Models. *ISPRS Int. J. Geo-Inf.* **2019**, *8*, 520. [CrossRef]
16. Gao, W.; Shen, F.; Tan, K.; Zhang, W.; Liu, Q.; Lam, N.S.N.; Ge, J. Monitoring Terrain Elevation of Intertidal Wetlands by Utilising the Spatial-Temporal Fusion of Multi-Source Satellite Data: A Case Study in the Yangtze (Changjiang) Estuary. *Geomorphology* **2021**, *383*, 107683. [CrossRef]
17. Stephens, E.M.; Bates, P.D.; Freer, J.E.; Mason, D.C. The Impact of Uncertainty in Satellite Data on the Assessment of Flood Inundation Models. *J. Hydrol.* **2012**, *414–415*, 162–173. [CrossRef]
18. Sánchez, J.M.; Izco, J.; Medrano, M. Relationships between Vegetation Zonation and Altitude in a Salt-Marsh System in Northwest Spain. *J. Veg. Sci.* **1996**, *7*, 695–702. [CrossRef]
19. Pennings, S.C.; Grant, M.-B.; Bertness, M.D. Plant Zonation in Low-Latitude Salt Marshes: Disentangling the Roles of Flooding, Salinity and Competition. *J. Ecol.* **2005**, *93*, 159–167. [CrossRef]
20. Kim, D. Rates of Vegetation Dynamics Along Elevation Gradients in a Backbarrier Salt Marsh of the Danish Wadden Sea. *Estuaries Coasts* **2014**, *37*, 610–620. [CrossRef]
21. Wang, Q.; Luo, M.; Cui, B.; Chen, C.; Xie, T.; Li, X.; Lu, F. Effects of Varied Inundation Characteristics on Early Life Stages of a Salt Marsh Plant. *Front. Mar. Sci.* **2024**, *11*, 1449034. [CrossRef]
22. Bertuzzo, E.; Finotello, A.; D'Alpaos, A.; Marani, M. Species Competition and Dispersal Drive Vegetation Dynamics in Tidal Salt Marshes. In Proceedings of the 24th EGU General Assembly, Vienna, Austria, 23–27 May 2022.
23. Wang, K.; Wu, G.; Liang, B.; Shi, B.; Li, H. Deltaic Marsh Accretion under Episodic Sediment Supply Controlled by River Regulations and Storms: Implications for Coastal Wetlands Restoration in the Yellow River Delta. *J. Hydrol.* **2024**, *635*, 131221. [CrossRef]
24. Fan, Y.; Yu, S.; Wang, J.; Li, P.; Chen, S.; Ji, H.; Li, P.; Dou, S.; Fan, Y.; Yu, S.; et al. Changes of Inundation Frequency in the Yellow River Delta and Its Response to Wetland Vegetation. *Land* **2022**, *11*, 1647. [CrossRef]
25. Ba, Q.; Wang, B.; Zhu, L.; Fu, Z.; Wu, X.; Wang, H.; Bi, N. Rapid Change of Vegetation Cover in the Huanghe (Yellow River) Mouth Wetland and Its Biogeomorphological Feedbacks. *CATENA* **2024**, *238*, 107875. [CrossRef]
26. Zhai, M.; Shen, H.; Cao, Q.; Ding, X.; Xin, M.; Zhai, M.; Shen, H.; Cao, Q.; Ding, X.; Xin, M. Water Body Extraction Methods for SAR Images Fusing Sentinel-1 Dual-Polarized Water Index and Random Forest. *Sensors* **2025**, *25*, 4868. [CrossRef]
27. Gautam, P.K.; Chandra, S.; Henry, P.K. Estimation of Flood Inundation in River Basins of Uttar Pradesh Using Sentinel 1A-SAR Data on Sentinel Application Platform (SNAP). *Arab. J. Geosci.* **2024**, *17*, 107. [CrossRef]
28. Fan, R.; Ti, Y. Comparative Study on Water Extraction Based on Sentinel Image and UAV Image. In Proceedings of the Fifth International Conference on Geology, Mapping, and Remote Sensing (ICGMRS 2024), Wuhan, China, 12–14 April 2024; SPIE: Pune, India, 2024; Volume 13223, pp. 430–435.
29. Su, Z.; Xiang, L.; Steffen, H.; Jia, L.; Deng, F.; Wang, W.; Hu, K.; Guo, J.; Nong, A.; Cui, H.; et al. A New and Robust Index for Water Body Extraction from Sentinel-2 Imagery. *Remote Sens.* **2024**, *16*, 2749. [CrossRef]
30. Martinis, S.; Groth, S.; Wieland, M.; Knopp, L.; Rättich, M. Towards a Global Seasonal and Permanent Reference Water Product from Sentinel-1/2 Data for Improved Flood Mapping. *Remote Sens. Environ.* **2022**, *278*, 113077. [CrossRef]
31. Fagherazzi, S.; Kirwan, M.L.; Mudd, S.M.; Guntenspergen, G.R.; Temmerman, S.; D'Alpaos, A.; van de Koppel, J.; Rybczyk. J.M.; Reyes, E.; Craft, C.; et al. Numerical Models of Salt Marsh Evolution: Ecological, Geomorphic, and Climatic Factors. *Rev. Geophys.* **2012**, *50*, RG1002. [CrossRef]
32. Fagherazzi, S.; Mariotti, G.; Leonardi, N.; Canestrelli, A.; Nardin, W.; Kearney, W.S. Salt Marsh Dynamics in a Period of Accelerated Sea Level Rise. *J. Geophys. Res. Earth Surf.* **2020**, *125*, e2019JF005200. [CrossRef]
33. Chen, S.; Liu, F.; Zhang, Z.; Zhang, Q.; Wang, W. Changes of Groundwater Flow Field of Luanhe River Delta under the Human Activities and Its Impact on the Ecological Environment in the Past 30 Years. *China Geol.* **2021**, *4*, 455–462. [CrossRef]

34. Gedan, K.B.; Silliman, B.R.; Bertness, M.D. Centuries of Human-Driven Change in Salt Marsh Ecosystems. *Annu. Rev. Mar. Sci.* **2009**, *1*, 117–141. [CrossRef] [PubMed]
35. Bost, M.C.; Rodriguez, A.B.; McKee, B.A. Impact of Land-Use Change on Salt Marsh Accretion. *Estuar. Coast. Shelf Sci.* **2024**, *299*, 108693. [CrossRef]
36. Seitz, C.; Kenney, W.F.; Patterson-Boyarski, B.; Curtis, J.H.; Vélez, M.I.; Glodzik, K.; Escobar, J.; Brenner, M. Sea-Level Changes and Paleoenvironmental Responses in a Coastal Florida Salt Marsh over the Last Three Centuries. *J. Paleolimnol.* **2023**, *69*, 327–343. [CrossRef]
37. McGrath, G.S.; Harding, C.; Matte, P. Changing Processes Flooding a Salt Marsh in a Microtidal Estuary with a Drying Climate. *Estuar. Coast. Shelf Sci.* **2023**, *295*, 108573. [CrossRef]
38. Van Diggelen, J. Effects of Inundation Stress on Salt Marsh Halophytes. In *Ecological Responses to Environmental Stresses*; Rozema, J., Verkleij, J.A.C., Eds.; Tasks for vegetation science; Springer: Dordrecht, The Netherlands, 1991; Volume 22, pp. 62–75. ISBN 978-94-010-6757-7.
39. Wang, J.; Chen, X.; Wang, Y.; Chen, Y.; Wu, H.; Wittyngham, S.S.; Kirwan, M.L.; Zhang, Y.; Liu, W.; Zhang, Y. Life-History Plasticity of Intertidal Salt Marsh in Response to Sea Level Rise: Salinity and Inundation Modulate Size-Dependent Flowering of *Spartina alterniflora*. *Ecol. Indic.* **2025**, *177*, 113790. [CrossRef]
40. Hutchings, A.M.; Antler, G.; Wilkening, J.V.; Basu, A.; Bradbury, H.J.; Clegg, J.A.; Gorka, M.; Lin, C.Y.; Mills, J.V.; Pellerin, A.; et al. Creek Dynamics Determine Pond Subsurface Geochemical Heterogeneity in East Anglian (UK) Salt Marshes. *Front. Earth Sci.* **2019**, *7*, 41. [CrossRef]
41. Kim, D.; Cairns, D.M.; Bartholdy, J. Tidal Creek Morphology and Sediment Type Influence Spatial Trends in Salt Marsh Vegetation. *Prof. Geogr.* **2013**, *65*, 544–560. [CrossRef]
42. Wu, Y.; Liu, J.; Yan, G.; Zhai, J.; Cong, L.; Dai, L.; Zhang, Z.; Zhang, M. The Size and Distribution of Tidal Creeks Affects Salt Marsh Restoration. *J. Environ. Manag.* **2020**, *259*, 110070. [CrossRef]
43. Wang, H.; van der Wal, D.; Li, X.; van Belzen, J.; Herman, P.M.J.; Hu, Z.; Ge, Z.; Zhang, L.; Bouma, T.J. Zooming in and out: Scale Dependence of Extrinsic and Intrinsic Factors Affecting Salt Marsh Erosion. *J. Geophys. Res. Earth Surf.* **2017**, *122*, 1455–1470. [CrossRef]
44. Donatelli, C.; Ganju, N.K.; Zhang, X.; Fagherazzi, S.; Leonardi, N. Salt Marsh Loss Affects Tides and the Sediment Budget in Shallow Bays. *J. Geophys. Res. Earth Surf.* **2018**, *123*, 2647–2662. [CrossRef]

water

MDPI

Article

Experimental Study of Sediment Incipient Velocity and Scouring in Submarine Cable Burial Areas

Fanjun Chen [1], Wankang Yang [2,*], Feng Liu [3], Lili Zhu [1] and Zhilin Sun [3]

[1] College of Architecture and Civil Engineering, Zhejiang University, Hangzhou 310027, China
[2] Second Institute of Oceanography, Ministry of Natural Resources, Hangzhou 310012, China
[3] Ocean College, Zhejiang University, Hangzhou 310027, China
* Correspondence: yangwankang@126.com

Abstract: This study investigates the incipient motion and scouring of sediments around simulated submarine cables in a controlled flume experiment, focusing on five distinct grain sizes in an experimental pool. The measured incipient velocity values were compared with predictions from three established formulas, leading to a modification of the Sun Zhilin formula for improved accuracy. By incrementally increasing flow velocity, the scour depth and scour duration were measured required to expose cables buried at varying depths for different sediment sizes, and the relationships between scour rate, relative flow rate, and Froude number were analyzed. The results indicate that as the Froude number increases, both the relative flow velocity and scour rate increase, thereby enhancing the erosion of sediment. The modified formula demonstrated a higher consistency with observed scour depths, providing a reliable tool for assessing submarine cable exposure risks. These findings offer valuable insights for developing effective protection strategies to enhance cable stability in complex marine environments. This research highlights the importance of understanding sediment dynamics and their impact on submarine cable stability, contributing to the development of more effective protection strategies for submarine cables in dynamic seabed conditions.

Keywords: sediment incipient velocity; scour depth; scour rate; experimental flume; submarine cable

1. Introduction

Since the deployment of the world's first submarine cable in the 1950s, seabed scouring has been identified as the primary cause of most submarine cable failures [1,2]. To mitigate the impact of scouring induced by ocean currents and wave action, submarine cable routes are meticulously pre-planned, incorporating advanced construction techniques such as mechanical trenching [3]. This method involves burying the cable at depths ranging from 1 to 2.5 m beneath the seabed, significantly reducing its exposure to erosive forces. Consequently, the implementation of anti-scouring protection measures, along with ongoing maintenance and management of submarine cables, is of paramount importance to ensuring their long-term stability and functionality [4–8].

Scholars have extensively studied sediment transport mechanisms and developed multiple sediment incipient velocity formulas by integrating sedimentology research with controlled scouring experiments [9–11]. Their findings indicate that when hydrodynamic velocity and energy exceed critical thresholds, sediment particles begin to move, initiating the scouring process. As hydrodynamic forces intensify, sediment transport accelerates, potentially exposing and jeopardizing buried submarine cables. Understanding these dynamics is crucial for designing effective anti-scouring strategies, optimizing cable burial depths, and enhancing the resilience of submarine communication and power transmission networks. Recent studies employing stochastic and modified sediment transport methodologies have further corroborated the positive correlation between incipient velocity and water depth in both cohesive and non-cohesive sediments [12–14]. For instance, Dodaro et al. developed a modified Einstein sediment transport model to simulate scour evolution, emphasizing the role of hydrodynamic shear stress and sediment cohesion in regulating scour dynamics [12]. Tafarojnoruz and Sharafati extended this understanding by formulating stochastic models for velocity thresholds in storm sewers, revealing that sediment mobility in natural channels depends critically on flow depth and turbulence characteristics [13]. Similarly, Sharafati et al. utilized a stochastic framework to predict pier scour depth in cohesive beds, demonstrating that water depth significantly influences the equilibrium between erosive forces and sediment resistance [14]. These approaches highlight the importance of considering sediment properties and hydrodynamic conditions in predicting scour dynamics.

Most existing studies primarily focus on verifying prediction formulas for non-uniform sediment particle sizes and their incipient velocities through laboratory experiments and field observations. However, current experimental research on sediment scouring is largely concentrated on local scouring around hydraulic structures [15,16], such as piers and dikes, with relatively limited studies addressing natural scouring in the absence of artificial structures. This gap in research highlights the need for a more comprehensive understanding of sediment transport dynamics in complex marine environments. This gap in research highlights the need for a more comprehensive understanding of sediment transport dynamics in complex marine environments. Fredsøe and Hansen proved that potential theory can well predict the final gap under the pipeline [17]. Because potential flow theory predicts that the flow velocity under the pipeline near the bed is much larger than the velocity above, they modified the potential flow description, adding a vortex tube to ensure a more accurate description of the flow under the pipeline. For sediment transport as steady flow [18–22], the near-bed flow velocity under the pipeline decreases to equal the far-field velocity at the riverbed, and the model predicted the scour depth under the pipeline; however, downstream of the pipeline, the evolution of the riverbed profile is too complex, and simple models cannot obtain the true profile of the scoured seabed [23].

Extensive research has focused on local scouring around hydraulic structures such as piers and pipelines [24,25]. Moreover, the natural scour processes within the environment of submarine cables have become a significant focus of contemporary research [26]. Particularly challenging is the fact that submarine cable routes frequently traverse dynamic seabeds composed of cohesive and heterogeneous sediments, complicating the study of these environments [27].

This study seeks to identify a more applicable incipient velocity formula suited for the complex hydrodynamic conditions along the target submarine cable route. Using clear-water mobile bed experiments with five characteristic sediment sizes, this study systematically investigates the relationships between sediment incipient motion and key hydrodynamic factors, such as the influences of sediment size, water depth, constant flow velocity, and scouring duration on sediment erosion processes. These findings contribute to a deeper understanding of natural seabed scouring and provide valuable insights for the design and maintenance of submarine cable burial strategies.

2. Materials and Methods

2.1. Simulated Submarine Cable

For typical submarine power grid installations, mechanical jet trenching is commonly employed to excavate seabed trenches for cable laying. The submarine cable used in such applications is a standard composite fiber-optic triple-tube design with 48-core G652. Optical fibers and steel armor protection, featuring an outer diameter of 200 mm. The detailed cable structure is shown in Figure 1.

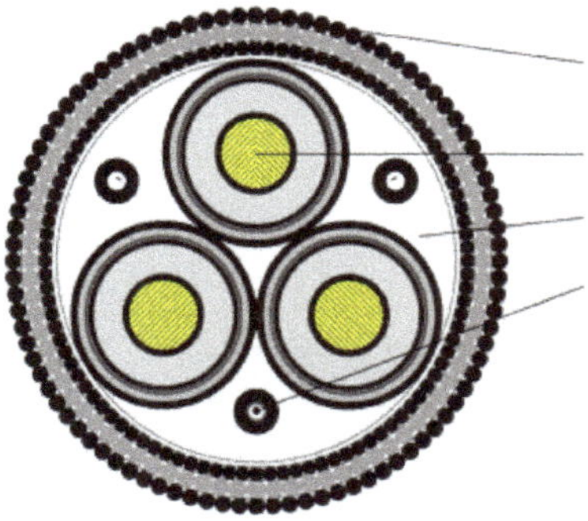

Figure 1. Composite 3 tubes × 48 core G652 optical fiber.

Figure 1 presents a schematic diagram of the submarine cable structure, illustrating the arrangement of optical fibers and conduits. This design allows the cable to effectively transmit electrical signals while providing protection against marine challenges. The installation of submarine cables requires careful planning tailored to specific engineering needs and seabed topography. Our experimental methodology simulates scour conditions at varying elevations within the flume, directly relevant to the seabed conditions where cables are embedded. As the cable is buried, local scour does not affect our experiments; instead, we focus on the incipient velocity of sediment under natural scour conditions. This study establishes a theoretical framework essential for understanding sediment movement in relation to submerged cables, aligning theoretical insights with practical engineering considerations for cable protection.

To simulate scouring effects, the model cable adopts a BVR4 square millimeter three-core flexible copper wire compliant with national standards. The conductor material is 99.99% oxygen-free pure copper, insulated with eco-friendly polyvinyl chloride (PVC), and colored black. The average outer diameter of the model cable is 2 mm, resulting in a geometric scaling ratio of $\lambda_h = 100$ for this experiment.

2.2. Experimental Equipment and Instruments

The experiment was conducted in a flume at the Ocean College of Zhejiang University, measuring 5 m in length, 5 m in width, and 1.2 m in depth. To achieve uniform flow conditions, a baffle was installed at the inlet, facilitating the generation of transverse two-dimensional planar flow. Additionally, energy dissipation devices were incorporated at the outlet to minimize flow disturbances. A centrifugal pump was utilized to create a recirculating water flow, as illustrated in Figure 2. The flume contains five sediment test zones for non-uniform natural sands with distinct characteristic particle sizes. Each zone features five 18 × 38 cm sediment recesses filled with five different-sized natural sands. The remaining areas are covered with a phenolic resin powder base layer in Figure 3. Figure 3 illustrates the arrangement of sediments with varying particle sizes before scouring. In this setup, sediments of different sizes were placed side by side in the flume, allowing for the observation of interactions between the various particle sizes. Within the model zones (sediment blocks), simulated submarine cables are embedded at depths of 1 cm, 2 cm, and 3 cm below the sediment surface across all five particle size groups.

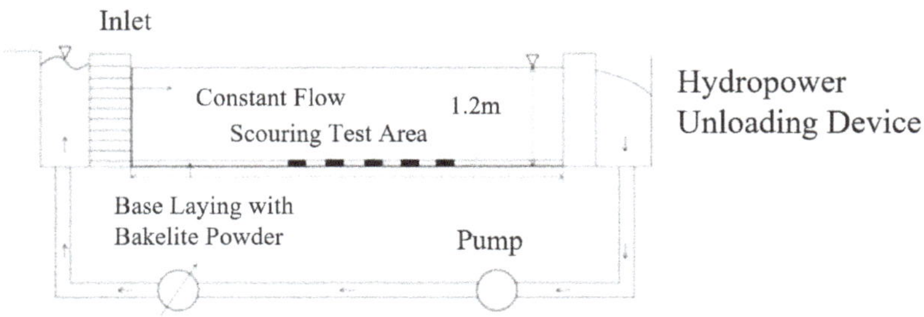

Figure 2. Arrangement of flushing test flume.

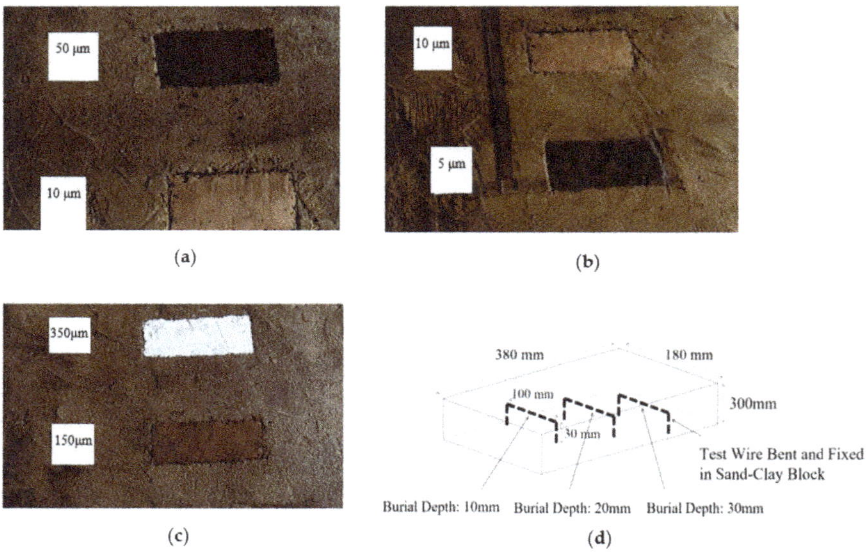

Figure 3. Layout of scouring test areas with different sediment particle sizes. (**a**) 10 μm and 50 μm test area; (**b**) 5 μm and 10 μm test area; (**c**) 150 μm and 350 μm test area; (**d**) Internal arrangement of single sediment scouring area.

The flume's 1.2 m depth enables water flow conditions $\leq$1.2 m, corresponding to the actual cable route's water depth range of 5–52.8 m. Water depth decreases near the coast. Using a depth scaling ratio of $\lambda_h = 100$, the experimental water depth was designed within 0.05–0.55 m.

Flow velocity is regulated by the pump and measured using an LYS-89A digital current meter, produced in the China. Scouring commences once a steady flow is achieved, with the velocity stabilized and uniformly distributed across the model section. Water depth is calibrated using a side-mounted scale and verified vertically with an H-40 laser rangefinder, produced in the China. Post-sediment initiation, an i-speed high-speed camera (2000 fps) captures particle movement to determine instantaneous velocities.

2.3. Model Design

Along the submarine cable route, 83 seabed sediment samples were collected in the East China Sea, revealing an average water depth of 5~25 m. The tidal waves in the East China Sea advance in the form of progressive waves, moving from the southeast to the northwest. As these waves propagate into the Zhoushan archipelago, they are fragmented by a series of islands, including Liuheng, Taohua, Zhujiajian, Daishan, Qushan, and Shengsi, resulting in multiple tidal currents. The study area is located between Qushan Island and Sijiaoshan, characterized by numerous islands of varying sizes that significantly influence the local topography. As a result, the direction of the flood currents predominantly trends northwest or west, while the ebb currents generally flow southeast or east. The maximum flood current velocity recorded at the measurement points reaches 192 cm/s, while the maximum ebb current velocity attains 235 cm/s. The occurrence frequency of vertical mean flow velocities exceeding 153 cm/s ranges between 1.5% and 10.0%. During typhoon events, the maximum storm surge current velocity in this region can reach as high as 3.5 m/s [28].

Based on the dimensions of the basin and the sea area, appropriate scale ratios can be established. The initial water depth of the prototype basin is maintained at 0.25 m, and the selected scale ratios for the model are determined as follows: the depth scale ratio is $\lambda_h = 100$, and the velocity scale ratio is $\lambda_v = \sqrt{\lambda_h} = 10$.

Particle size analysis via mastersizer 2000 laser granulometer, produced in the UK, showed sediment diameters ranging from 3 to 350 μm, with median diameters (D50) between 7 and 31 μm. Median diameters decreased near the coast and increased offshore. Five characteristic sediment sizes were selected to measure incipient velocity, scouring depth, and duration.

2.4. Experimental Procedures

The experimental sediment consisted of phenolic resin powder with a particle size range of 500–1000 μm. The formal experiment began with the preparation of the model zones: five sediment types with distinct particle sizes were placed and leveled within the model area to a thickness of approximately 30 cm. A pump was then activated to initiate slow water inflow, ensuring minimal disturbance to the underlying sediment. Water was gradually raised to the target depth, and a high-speed camera was directed at the sediment blocks in the test area to confirm no sediment movement. The pump output was incrementally adjusted to achieve the calculated flow velocity in the flume.

Flow velocities were recorded using the Acoustic Doppler Velocimeter (ADV) meter positioned 5 cm above the sediment bed. Scour depths were measured every 5 min using an H-40 laser rangefinder, and sediment initiation was verified via i-speed high-speed camera footage (2000 fps). Data analysis focused on time-averaged velocities and equilibrium scour depths, with uncertainties quantified using a ±2% error margin for velocity and ±1 mm for depth.

Water depth was regulated via a tailgate, with 11 predefined depth groups for sediment initiation: 0.05 m, 0.1 m, 0.15 m, 0.2 m, 0.25 m, 0.3 m, 0.35 m, 0.4 m, 0.45 m, 0.5 m, and 0.55 m. Based on the average water depth of 5~25 m from 83 sampling points, the initial experimental water depth was set to 0.25 m. For scouring analysis, three additional depth groups (0.26 m, 0.27 m, and 0.28 m) were established, corresponding to cables buried at 1 cm, 2 cm, and 3 cm depths within the sediment blocks.

When the initial water depth reached 0.25 m, the high-speed camera immediately recorded minor sediment movement and the corresponding instantaneous incipient velocity. Sustained scouring under steady flow continued until the cable buried at a depth of 1 cm (equivalent to a water depth of 0.26 m) became exposed, with the scouring duration recorded. Following this procedure, the flow velocity was gradually increased to monitor the exposure of cables buried at 2 cm (0.27 m) and 3 cm (0.28 m) depths. The instantaneous incipient velocities and scouring durations for these depths were documented. Subsequent increases in flow velocity, combined with applicable incipient velocity formulas, enabled the calculation of natural scouring depths.

For each test group, the instantaneous flow velocity at sediment initiation, the scouring duration until cable exposure, and the equilibrium scouring depth were recorded. After each trial, the sediment was replenished, and the flume was drained to prepare for subsequent experiments.

3. Experimental Results and Discussion

3.1. Sediment Incipient Velocity Formulas

Numerous scholars worldwide have conducted experimental and theoretical studies on sediment incipient motion, particularly focusing on uniform and non-uniform sediments. Zhang Ruijin, a pioneering researcher in China, proposed in 1961 the concept of cohesive force in sediments, incorporating atmospheric pressure, water column pressure, and hydrodynamic effects on particles. Through experiments and theoretical derivation, he established a unified incipient velocity formula for cohesive uniform sediments applicable to varying particle sizes (Equation (1)):

$$v_{c,i} = 1.34\left(\frac{h}{d_i}\right)^{0.14}\left[\frac{\rho_s-\rho}{\rho}gd_i + 3.36\times10^{-7}\left(\frac{10+h}{d_i^{0.72}}\right)\right]^{0.5} \tag{1}$$

where ρ_s and ρ represent the densities of sediment and water, respectively. In this experimental setup, natural sediment with $\rho_s = 2.65\ \mathrm{t/m^3}$ and water density $\rho = 1\ \mathrm{t/m^3}$ were adopted. Here, d_i denotes the incipient particle diameter corresponding to the critical velocity $v_{c,i}$, d_m is the mean particle diameter, h represents the water depth, and g is the gravitational acceleration.

Dou Guoren refined cohesive uniform sediment incipient velocity formulas from 1974 to 1999, considering interparticle cohesive forces exceeding gravitational and hydrostatic pressures. His modified practical formula (Equation (2)) is as follows:

$$v_{c,i} = 0.32(ln11\frac{h}{K_s})(\frac{d'}{d_*})^{1/6}\left[3.6\frac{\rho_s-\rho}{\rho}gd_i + (\frac{\rho'}{\rho'_*})^{2.5}\left(\frac{\varepsilon_0 + gh\delta(\delta/d_i)^{0.5}}{d_i}\right)\right]^{0.5} \tag{2}$$

where ρ' denotes the dry density of sediment, which ranges from 1.15 to 1.45 $\mathrm{t/m^3}$ for unconsolidated natural sediment. ρ'_* represents the stabilized dry density of sediment, taken as 1.6 $\mathrm{t/m^3}$ for natural sediment. δ is the characteristic thickness of interparticle voids, set to 2×10^{-7} m, and ε_0 denotes the comprehensive cohesion parameter with a value of $1.75\times10^{-6}\ \mathrm{m^3/s^2}$. For particle sizes $d \leq 500$ μm, d' is assigned 5×10^{-4} m, and the bed roughness height is specified as 0.001 m.

Sediment incipient motion involves stochastic mechanical and geometric factors, including drag, lift, submerged weight, and cohesive forces. Sun Zhilin's formula integrates probability theory and mechanics to derive the incipient friction velocity for cohesive non-uniform sediments. Using a logarithmic velocity distribution, the friction velocity is converted to depth-averaged velocity (Equation (3)) [10]:

$$v_{c,i} = \frac{8}{7}\left(\frac{h}{K_s}\right)^{1/6}\sqrt{\frac{gD_m}{\sigma_D^{1/4}}\left[\left(\frac{\rho_s-\rho}{\rho}\right)\left(\frac{D_k}{D_m}\right)^{1/2} + \frac{3}{2}C_0\left(\frac{\rho'}{\rho'_*}\right)^3\frac{(h+h_0)\delta}{D_k^2}\right]} \tag{3}$$

where K_s represents the bed roughness taken as d_m, and the standard deviation σ_D indicates the non-uniformity degree of sediment composition. h_0 denotes the additional hydraulic head from molecular adhesion. Existing experiments demonstrate that h_0 ranges from 1.226 to 10 m, and the geometric mean of these values ($h_0 = 3.5$ m) is generally adopted. Based on Sun Zhilin's measured data [10], the constant C_0 is determined to be 0.03–0.05, and a value of 0.04 is selected for this study.

3.2. Experimental Results of Sediment Incipient Velocity

As shown in Table 1 and Figure 4, the measured incipient velocities generally increase with water depth across all sediment sizes. These findings are of significant reference value for understanding the initiation behavior of riverbed sediments and for related engineering designs. The observations indicate that the critical flow velocity of the modeled sediment is influenced by water depth and exhibits a degree of particle size dependency. At shallower depths (h < 0.3m), incipient velocities for all five particle sizes (5–350 μm) remain relatively low (0.1–0.6 m/s). As the water depth increases, the critical flow velocity for sediment mobilization shows a discernible trend. Coarser sediments exhibit relatively minor variations in flow velocity across different water depths, indicating a lower variance and suggesting a more stable critical flow velocity. However, once the water depth exceeds 0.3 m, a significant change in the critical flow velocity of the five sediment sizes is observed. This variability may be attributed to alterations in hydrodynamic conditions resulting from the increased water depth, which subsequently affects the mobilization behavior of the sediments.

Based on the data presented in Table 1 and Figure 4, the 5 μm sediment exhibited the highest average incipient velocity (0.47 m/s), while the 150 μm sediment showed the lowest (0.17 m/s). Notably, the 50 μm and 350 μm sediments displayed higher incipient velocities than the 150 μm group. Larger particles (e.g., 150 μm) demonstrated minimal velocity variance with depth changes (maximum–minimum difference: 0.08 m/s), whereas smaller particles (e.g., 5 μm) exhibited significant velocity increases (0.19 m/s variation between 0.2–0.55 m depth).

The 10 μm sediment showed anomalous behavior at 0.26–0.28 m depths, likely due to experimental variability or measurement errors. The largest velocity gap (0.22 m/s) occurred between the 10 μm and 50 μm groups, highlighting substantial incipient velocity changes in this size range.

Table 1. Measured incipient velocity statistics (m/s).

Particle Size (μm) / Scour Depth (cm)	5	10	50	150	350
0.05	0.38	0.36	0.17	0.12	0.13
0.1	0.39	0.39	0.18	0.13	0.15
0.15	0.4	0.39	0.19	0.14	0.16
0.2	0.41	0.4	0.19	0.15	0.18
0.25	0.44	0.42	0.2	0.17	0.19
0.26	0.45	0.43	0.21	0.18	0.2
0.27	0.46	0.39	0.21	0.17	0.21
0.28	0.46	0.4	0.22	0.16	0.19
0.3	0.5	0.41	0.22	0.19	0.2
0.35	0.52	0.45	0.24	0.18	0.21
0.4	0.54	0.48	0.25	0.19	0.21
0.45	0.56	0.5	0.25	0.19	0.23
0.5	0.59	0.51	0.26	0.2	0.22
0.55	0.60	0.53	0.27	0.2	0.24

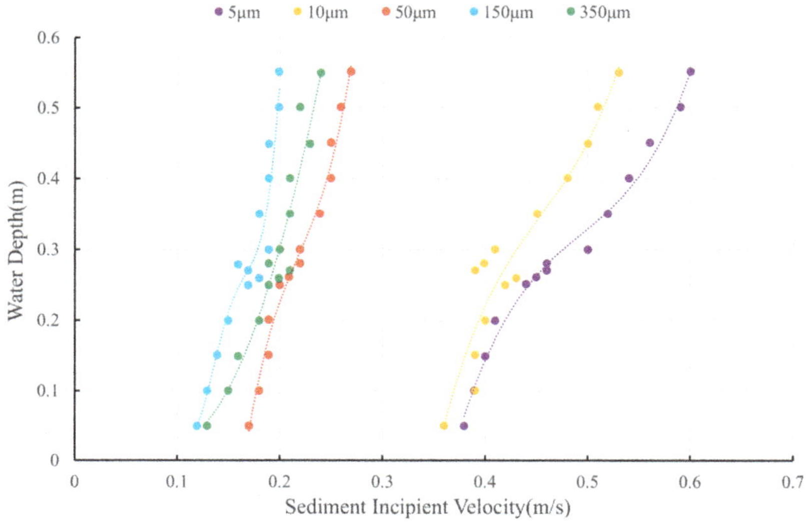

Figure 4. Measured incipient velocity of model sand.

The observed differences in velocity profiles among various particle sizes can be attributed to the distinct behaviors of fine and coarse sediments. For finer particles, such as those around 150 μm, cohesive forces play a significant role, leading to a reduction in the incipient motion velocity as particle size increases. Conversely, for coarser particles, such as those measuring 350 μm, gravitational forces dominate, resulting in an increase in the incipient motion velocity with larger particle sizes. This duality in the effects of cohesion and gravity explains why the velocity of the larger particles is greater than that of the smaller ones.

These results confirm a positive correlation between incipient velocity and water depth. Comparative analysis of the three formulas reveals that smaller particles (5–10 μm), dominated by cohesive forces, require higher drag for initiation, while larger particles (150–350 μm), governed by gravitational forces, exhibit lower and more stable incipient velocities. For the studied cable route (average depth: 25 m), sediments in the 50–150 μm range are most susceptible to scouring. Smaller particles (5–10 μm) demonstrate strong cohesion, making them resistant to initiation, whereas intermediate sizes (50 μm) balance cohesive and gravitational effects, showing pronounced scouring vulnerability.

Key observations include the following:

- **5 μm sediment**: Most sensitive to depth changes, with a 0.19 m/s increase between 0.2 and 0.55 m.
- **10 μm sediment**: Anomalous velocities at intermediate depths (0.26–0.28 m), suggesting experimental variability.
- **50 μm sediment**: Marked velocity differences from the 10 μm group, indicating critical transitional behavior.
- **150 μm sediment**: Minimal velocity variation (0.08 m/s range), lowest average velocity (0.12 m/s at 0.05 m).
- **350 μm sediment**: Intermediate velocities between the 50 μm and 150 μm groups.

These findings align with theoretical predictions: cohesive forces dominate for fine sediments (<50 μm), requiring higher initiation velocities, while gravitational effects govern coarser particles (>150 μm). The 50–150 μm range represents a critical transition zone with optimal scouring susceptibility under typical marine conditions.

Figure 5 illustrates the scour depths of sediments with varying particle sizes at different water depths. For the 5 μm model sediment, the measured incipient velocities were consistently lower than the values predicted by all three formulas. Notably, both the calculated and measured velocities exhibited an increasing trend with rising water depth. As shown in Figure 5, Equation (2) aligned most closely with the measured values for 5 μm sediment, averaging only 0.14 m/s higher than the measured values. When the water depth exceeded 0.3 m, the measured values exhibited significantly accelerated growth and approached Equation (2) results. For water depths of 0.05–0.2 m, Equation (3) closely matched Equation (2), with an average deviation of only 0.07 m/s. Notably, Equation (1) yielded significantly higher values than the other two, averaging 47.35% above measured values, while Equation (3) showed an average deviation of 35.52%. Since a deviation threshold below 30% is required for applicability [11], Equations (1) and (3) are unsuitable for 5 μm sediment in this experiment. Equation (2), with an average deviation of 21.87%, proved applicable.

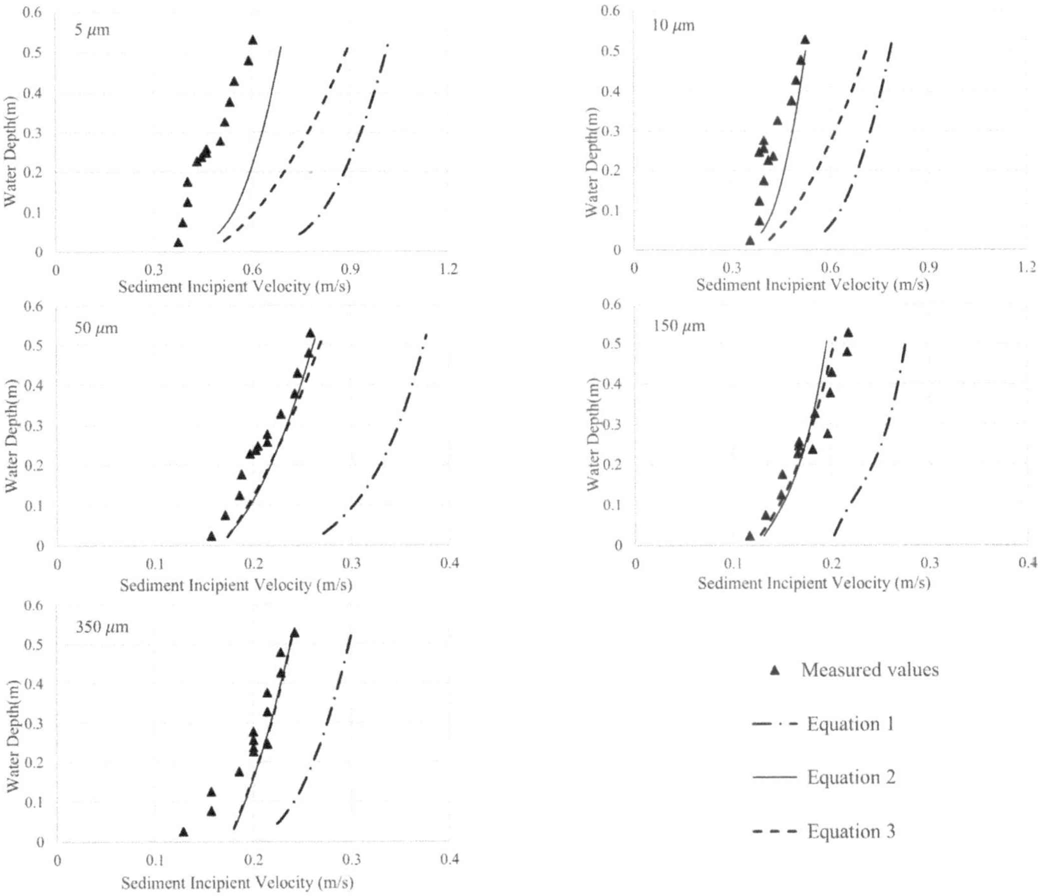

Figure 5. Comparison between measured and calculated incipient velocity.

For the 10 μm model sediment (Figure 5), measured values remained lower than all three formula-calculated results. At water depths of 0.05–0.1 m and 0.4–0.55 m, measured values nearly overlapped with Equation (2), but, between 0.15 and 0.35 m, the measured values fluctuated slightly without clear depth-dependent trends. Equation (1) remained the highest, while Equation (3) fell between Equations (1) and (2). During the 0.25–0.3 m depth intervals, the 1 cm spacing in experimental depth settings likely introduced measurement variability due to external factors. Equation (1) showed an average deviation of 38.48%, rendering it inapplicable, while Equations (2) and (3) exhibited average deviations of 7.99% and 25.99%, respectively, meeting applicability criteria.

For the 50 μm model sediment, the measured and calculated velocities all fell below 0.5 m/s, aligning closely with Equations (3) and (2) (deviation: 1.95%). The sediment > 50 μm exhibited markedly lower incipient velocities than 5 μm and 10 μm. Equation (1) remained higher than the others, averaging 32.56% above Equation (3). With an average deviation of 32.75%, Equation (1) was inapplicable, while Equations (2) and (3) showed deviations <2%, demonstrating suitability.

For the 150 μm and 350 μm model sediment, Equation (1) significantly overestimated results compared to others. Equations (2) and (3) remained closely aligned, with average deviations of 0.005 m/s and 0.014 m/s, respectively. Equation (1) was inapplicable, while Equations (2) and (3) showed deviations <15%, with Equation (3) performing best (3.07% and 6.24% deviations).

3.3. Modification of Sediment Incipient Velocity Formula

Equations (1) and (2) are incipient motion formulas for uniform sediment, whereas the sediment in the cable-laying sea area is non-uniform. The model sediment used in this experiment is natural sediment, with the standard deviations (σ_D) of the particle size distributions as follows: 4.31 for 5 μm, 4.22 for 10 μm, 3.35 for 50 μm, 2.77 for 150 μm, and 3.53 for 350 μm. Equation (3), designed for cohesive non-uniform sediment, better aligns with the actual conditions of the target sea area and this experimental study. Further exploration of Equation (3)'s applicability to fine-grained sediments with particle sizes below 50 μm will yield greater practical value.

Based on the experimental results, Equation (3) has been revised by calibrating the bed roughness K_s and constant C_0. Considering the characteristics of cohesive sediment, K_s is set as d_m for particle sizes exceeding 50 μm, while a fixed value is adopted for $d_i < 50$ μm. Since C_0 tends to overestimate results for smaller particles, a value smaller than 0.04 was sought. Iterative calculations were performed to fit the measured incipient velocity data of 5 μm and 10 μm cohesive sediment under varying water depths, minimizing the sum of absolute deviations between calculated and measured values.

The optimized parameters are determined as follows: for the 5 μm and 10 μm particles, $K_s = 0.045$ mm (0.000045 m) and $C_0 = 0.033$, yielding absolute deviations of 0.05 m/s and 0.02 m/s from the measured velocity averages, respectively.

The revised Equation (3) is expressed as follows:

$$\lambda_{c,j} = \frac{8}{7}\left(\frac{h}{K_s}\right)^{1/6}\sqrt{\frac{gD_m}{\sigma_D^{1/4}}\left[\left(\frac{\rho_s-\rho}{\rho}\right)\left(\frac{D_k}{D_m}\right)^{1/2}+\frac{3}{2}C_0\left(\frac{\rho'}{\rho'_*}\right)^3\frac{(h+h_0)\delta}{D_k^2}\right]}$$

Parameter selection criteria:

- For median sediment particle sizes ≥ 50 μm, the results are as follows:

$$K_s = d_m\ (d_m \geq 50\ \mu\text{m}),\qquad C_0 = 0.04$$

- For median sediment particle sizes < 50 μm, the results are as follows:

$$K_s = 0.000045\ (d_m < 50\ \mu\text{m}),\qquad C_0 = 0.033$$

As shown in Figure 6 and Table 2, the deviations between calculated results from the revised Equation (3) and measured data are 1.01% and 4.20%, demonstrating its applicability for incipient velocity prediction of cohesive sediments.

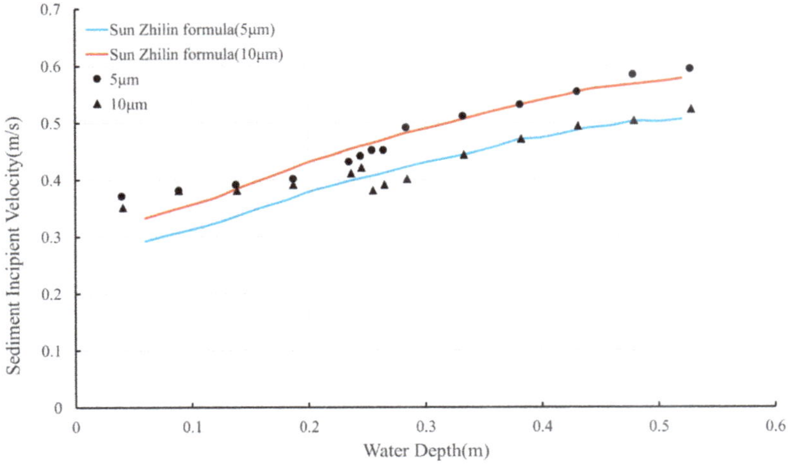

Figure 6. Comparison between the corrected calculated and measured values.

Table 2. Comparison between modified calculated and measured values of incipient velocity.

	Comparison of Incipient Velocity of 5 μm Sediment			Comparison of Incipient Velocity of 10 μm Sediment		
Water Depth (m)	**Calculated Value (m/s)**	**Measured Value (m/s)**	**Deviation (%)**	**Calculated Value (m/s)**	**Measured Value (m/s)**	**Deviation (%)**
0.05	0.329	0.38	15.50	0.290	0.36	24.14
0.1	0.376	0.39	3.72	0.332	0.39	17.47
0.15	0.410	0.4	−2.44	0.362	0.39	7.73
0.2	0.438	0.41	−6.39	0.386	0.4	3.63
0.25	0.463	0.44	−4.97	0.408	0.42	2.94
0.26	0.468	0.45	−3.85	0.412	0.43	4.37
0.27	0.472	0.46	−2.54	0.416	0.39	−6.25
0.28	0.477	0.46	−3.56	0.421	0.4	−4.99
0.3	0.485	0.5	3.09	0.428	0.41	−4.21
0.35	0.509	0.52	2.16	0.440	0.45	2.27
0.4	0.525	0.54	2.86	0.463	0.48	3.67
0.45	0.549	0.56	2.00	0.493	0.5	1.42
0.5	0.562	0.59	4.98	0.496	0.51	2.82
0.55	0.579	0.6	3.63	0.511	0.53	3.72

3.4. Scour Duration Test Results and Analysis

This experiment involved clear-water live-bed scour tests. For each sediment size, the incipient velocity at a water depth of 0.25 m was used as the sustained flow velocity in the flume. The measured incipient velocities were 0.44 m/s for 5 μm sediment, 0.42 m/s for 10 μm, 0.2 m/s for 50 μm, 0.17 m/s for 150 μm, and 0.19 m/s for 350 μm. Two sustained scour velocities (0.2 m/s and 0.4 m/s) and a control group (0.5 m/s) were established. The durations required to expose cables by 1 cm, 2 cm, and 3 cm were recorded for each sediment size (Table 3 and Figure 7).

Table 3. Duration of the scouring test for model sands (mins).

Scour depth (cm) \ Particle size (μm)	5	10	50	150	350
Continuous scouring velocity 0.2 m/s					
1	No scouring		105	63	87
2			326	130	256
3			650	219	471
Continuous scouring velocity 0.4 m/s					
1	70	34	27	9	31
2	184	88	58	23	65
3	363	169	93	62	99
Continuous scouring velocity 0.5 m/s					
1	63	30	23	6	26
2	167	77	49	17	57
3	330	144	79	33	98

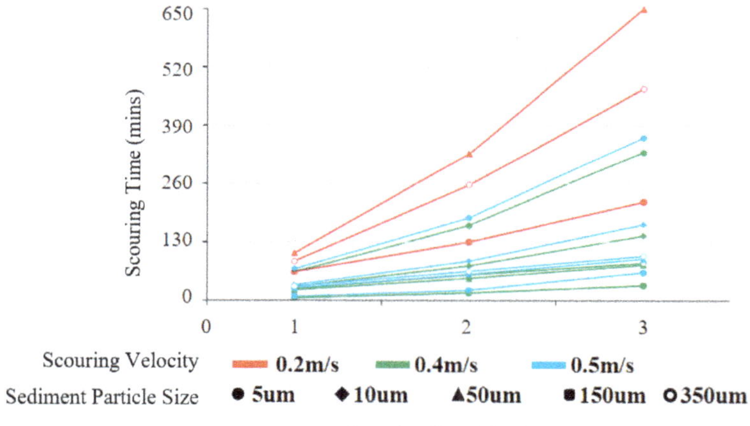

Figure 7. Scouring duration with scouring depth.

Starting scour tests at a 25 cm water depth, exposure depths of 1 cm, 2 cm, and 3 cm were recorded as scour duration milestones. Figure 8 illustrates the relationship between sediment size, scour depth, and duration. As shown in Table 2, within the 5–350 μm range, smaller incipient velocities corresponded to greater scour extent, shorter durations, and faster depth progression.

The experimental results demonstrate a direct correlation between cable burial depth and scouring dynamics. For instance, at a water depth of 0.26 m (equivalent to 1 cm burial depth), the equilibrium scour depth was achieved faster due to reduced sediment cover (Table 3). This phenomenon can be attributed to the altered local flow patterns around the exposed cable, which intensify shear stress and accelerate sediment entrainment. Similar observations were reported by Chiew [18], who noted that partial exposure of submerged structures significantly amplifies localized turbulence, thereby reducing the time required to reach scour equilibrium. The accelerated scouring at shallow burial depths underscores the importance of optimizing cable burial strategies to mitigate exposure risks in dynamic marine environments.

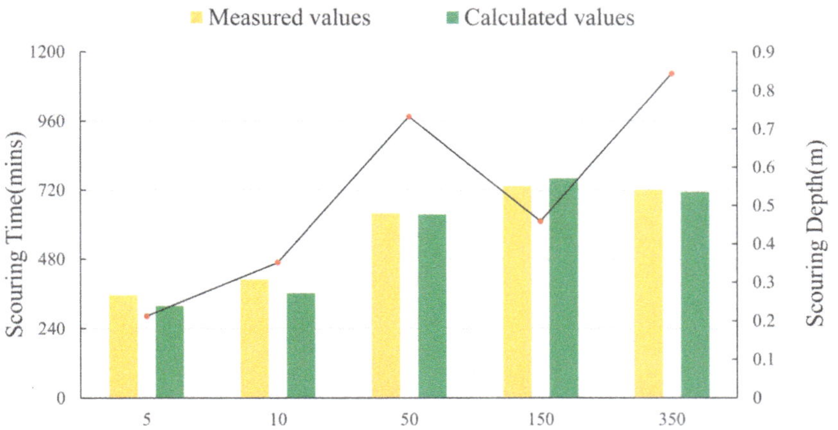

Figure 8. Measured scouring depth of model sand.

The experiment utilized open-channel flow, where water directly contacts the atmosphere. Beyond fluid viscosity and inertia, gravity dominated, quantified by the Froude number Fr:

$$Fr = \frac{U}{\sqrt{gh}}$$

where U denotes the flow velocity in the basin, h is the water depth, and g represents gravitational acceleration. When the Froude number $Fr = 1$, the flow is critical; when $Fr > 1$, it transitions to supercritical flow; and, when $Fr < 1$, the flow becomes subcritical. A scour rate formula is established as $E = \frac{H_c}{t} = f(U, v_{c,i}, h, t) = f(Fr, \frac{U}{v_{c,i}}, t)$, where H_c is the scour depth and t is the duration. Using Table 3, scour rates for five sediment sizes under varying velocities were calculated and analyzed against the Froude number Fr and relative velocity $\frac{U}{v_{c,i}}$ (the ratio of scour velocity to critical velocity). Table 4 compares measured relative velocities (based on field data) with calculated relative velocities derived from Equation (3).

For fine-grained sediments (5 μm and 10 μm), no scour occurred at 0.2 m/s. In contrast, 150 μm sediment exhibited the shortest durations t to reach scour depths of 1 cm, 2 cm, and 3 cm, yielding the highest scour rates E. The scour rate E directly correlates with critical velocity $v_{c,i}$—a smaller $v_{c,i}$ corresponds to a larger E. Additionally, E inversely relates to water depth h: greater depths reduce scour rates. The Froude number Fr reflects flow intensity; at constant velocity, increasing h decreases Fr, weakening flow intensity and scour rates.

At 0.4 m/s, all five sediment sizes experienced scour. For 5 μm sediment, scour equilibrium (minimum E) occurred at h = 28 cm. The remaining four sizes scoured three cable trenches, with average scour rates ranked in descending order: 150 μm > 350 μm > 50 μm > 10 μm. Both measured and calculated relative velocities ($\frac{U}{v_{c,i}}$) confirm that higher relative velocities amplify scour rates. When $\frac{U}{v_{c,i}} > 1$, scour velocity exceeds critical velocity, significantly intensifying erosion. Natural sediment with particle size variances >1 was used in this experiment, explaining minor scour even at $\frac{U}{v_{c,i}} < 1$. Since U represents depth-averaged velocity and $v_{c,i}$ statistically corresponds to the velocity initiating motion for 5% of particles, partial scour persists at $\frac{U}{v_{c,i}} < 1$.

Table 4. Statistics of scouring depth-related parameters.

5 µm						10 µm					
U	h	Fr	E	$\frac{U}{v_{c,i\ measured}}$	$\frac{U}{v_{c,i\ calculated}}$	U	h	Fr	E	$\frac{U}{v_{c,i\ measured}}$	$\frac{U}{v_{c,i\ calculated}}$
	0.26	0.125		0.395	0.455		0.26	0.125		0.448	0.476
0.2	0.27	0.123	No	0.391	0.444	0.2	0.27	0.123	No scouring	0.443	0.465
	0.28	0.121		0.387	0.435		0.28	0.121		0.439	0.513
	0.26	0.251	0.000143	0.790	0.909		0.26	0.251	0.000294	0.896	0.952
0.4	0.27	0.246	0.000088	0.782	0.889	0.4	0.27	0.246	0.000185	0.887	0.930
	0.28	0.241	0.000056	0.774	0.870		0.28	0.241	0.000123	0.878	1.026
	0.26	0.313	0.000159	0.987	1.136		0.26	0.313	0.000333	1.120	1.190
0.5	0.27	0.307	0.000096	0.977	1.111	0.5	0.27	0.307	0.000213	1.109	1.163
	0.28	0.302	0.000061	0.968	1.087		0.28	0.302	0.000149	1.098	1.282
50 µm						150 µm					
U	h	Fr	E	$\frac{U}{v_{c,i\ measured}}$	$\frac{U}{v_{c,i\ calculated}}$	U	h	Fr	E	$\frac{U}{v_{c,i\ measured}}$	$\frac{U}{v_{c,i\ calculated}}$
	0.26	0.125	0.000095	0.612	1.000		0.26	0.125	0.000159	0.791	1.176
0.2	0.27	0.123	0.000045	0.609	0.952	0.2	0.27	0.123	0.000149	0.787	1.111
	0.28	0.121	0.000031	0.605	0.952		0.28	0.121	0.000112	0.782	1.176
	0.26	0.251	0.000370	1.224	2.000		0.26	0.251	0.001111	1.582	2.353
0.4	0.27	0.246	0.000323	1.217	1.905	0.4	0.27	0.246	0.000714	1.573	2.222
	0.28	0.241	0.000286	1.210	1.905		0.28	0.241	0.000256	1.565	2.353
	0.26	0.313	0.000435	1.531	2.500		0.26	0.313	0.001667	1.978	2.941
0.5	0.27	0.307	0.000385	1.522	2.381	0.5	0.27	0.307	0.000909	1.967	2.778
	0.28	0.302	0.000333	1.513	2.381		0.28	0.302	0.000625	1.956	2.941
350 µm						Note:					
U	h	Fr	E	$\frac{U}{v_{c,i\ measured}}$	$\frac{U}{v_{c,i\ calculated}}$	U is the velocity of the pool (m/s)					
	0.26	0.125	0.000115	0.745	1.053	h is water depth (cm)					
0.2	0.27	0.123	0.000059	0.741	1.000	Fr is the Froude number					
	0.28	0.121	0.000047	0.737	0.952						
	0.26	0.251	0.000323	1.489	2.105	E is scouring rate (m/s)					
0.4	0.27	0.246	0.000294	1.481	2.000						
	0.28	0.241	0.000294	1.473	1.905	$U/v_{c,imeasured}$ is the relative velocity and the ratio of pool velocity to measured starting velocity					
	0.26	0.313	0.000385	1.862	2.632						
0.5	0.27	0.307	0.000323	1.852	2.500	$U/v_{c,icalculation}$ is the relative velocity and the ratio of pool velocity to calculated starting velocity					
	0.28	0.302	0.000244	1.842	2.381						

At h = 25 cm and U = 0.5 m/s (exceeding all critical velocities), the factors influencing E are most evident. Larger Fr and $\frac{U}{v_{c,i}}$ correlate with higher E, demonstrating enhanced sediment mobility under stronger hydrodynamic conditions.

The observed inverse relationship between scour rate and sediment size in Table 4 aligns with recent studies on cohesive sediment dynamics [26], which emphasized the dominance of cohesive forces in fine-grained sediments. However, our results diverge from conventional models by demonstrating that intermediate-sized particles (50–150 µm) exhibit transitional behavior, balancing cohesion and gravitational effects, as highlighted in recent analyses of non-uniform sediment transport [27]. This insight refines the understanding of sediment mobilization thresholds and provides a mechanistic basis for predicting scour vulnerability in submarine cable routes.

3.5. Scour Depth Test Results and Analysis

The revised Equation (3) in this experiment is applicable to sediment incipient motion in this study. Since sediment incipient motion initiates scour, the critical velocity can be utilized to calculate scour depth.

Typically, seabed sediment in natural marine environments remains stable (neither scoured nor deposited) under astronomical tidal currents. However, during typhoon-induced storm surges, sediment initiates motion when flow velocity U exceeds the critical velocity ($U > V_c$), triggering bed scour. Assuming a constant unit-width discharge q, flow velocity decreases as scour depth increases, while critical velocity rises with scour depth until a new equilibrium is reached. At this stage, sediment returns to a critical incipient motion state, and flow velocity theoretically becomes lower than U.

$$q = Uh = V_c(h + \Delta h) \tag{4}$$

Substituting the revised critical velocity formula (Equation (3)) into the unit-width discharge equation yields Equation (5).

$$Uh = \frac{8}{2}\frac{(h+\Delta h)^{7/6}}{K_s^{1/6}}\sqrt{\frac{gd_m}{\sigma_D^{1/4}}\left[\left(\frac{\rho_s-\rho}{\rho}\right)\left(\frac{d_i}{d_m}\right)^{1/2}+\frac{3}{2}C_0\left(\frac{\rho'}{\rho'_*}\right)^3\frac{(h+\Delta h+h_0)\delta}{d_i^2}\right]} \tag{5}$$

where Δh is the only unknown variable, representing the calculated scour depth. The calculated scour depth is then validated using the constant unit-width discharge $q = Uh$ and compared with measured scour values.

This experiment investigates natural scour depth in a clear-water movable-bed flume. During the scouring phase, a flow velocity of 0.5 m/s was maintained to study scour progression. For each sediment size, scouring was conducted at a constant water depth of 0.25 m with unchanged unit-width discharge (q). Scour equilibrium was defined as no further increase in depth over 2 h, measured using an H-40 laser rangefinder. Calculated scour depths based on Sun Zhilin's formula were then compared with measured values.

According to the experimental design, the unit-width discharge was consistently maintained at 0.125 m^2/s. Measured scour depths and calculated values from the revised Sun Zhilin's formula are recorded in Table 5. Additionally, the measured unit-width discharge (q = measured critical velocity × measured scour depth) was verified against the constant design discharge to ensure consistency.

Table 5. Depth at scouring equilibrium for model sands.

Sediment Grain Size (μm)	Scouring Duration Mins (Including 2 h Scouring Balance Holding Time)	Measured Scour Depth (m)	Scouring Depth Calculation Value (m)
5	290	0.27	0.243
10	482	0.31	0.275
50	1002	0.48	0.477
150	The 30 cm thick sediment block was completely scoured, lasting 630	0.55	0.57
350	1157	0.54	0.51

The 5 μm sediment, as shown in Table 2 of the scour duration records, exhibited the least scouring among the five sediment sizes. According to Sun Zhilin's formula, a flow velocity of 0.5 m/s at h = 0.25 m precisely initiates bed sediment motion. Experimentally, after 170 min of scouring, equilibrium was achieved with no depth change for 120 min, yielding an extremely slow scouring process and a stabilized scour depth of 0.27 m. Given the standard deviation (σ_D) of 1.46 for 5 μm sediment, this implies that particles significantly larger than 5 μm were likely entrained during scouring. The measured scour depth exceeded the calculated equilibrium value by 7.4%.

For 10 μm sediment, after 362 min of scouring, the measured depth stabilized at 0.31 m with a critical velocity $v_{c,i} \approx 0.42$ m/s and unit-width discharge $q = 0.130$ m^2/s. Using Sun Zhilin's formula, the calculated scour depth $v_{c,i} = 0.45$ m/s is 0.275 m, resulting in an 11.3% overestimation by measurements.

Similarly, 50 μm sediment reached a stable scour depth of 0.48 m after 882 min, with a measured $v_{c,i} \approx 0.26$ m/s. The formula predicts a depth of 0.477 m at $v_{c,i} = 0.32$ m/s, showing a negligible deviation of 0.07%—the closest agreement.

Notably, 150 μm sediment, with the smallest critical velocity, demonstrated the most pronounced scouring. After 630 min, a 30 cm-thick sediment layer was fully eroded, achieving a scour depth of 0.55 m (highlighted in Figure 8). The calculated depth of 0.57 m aligns with experimental trends but exceeds the preset limit of 0.55 m.

For the largest size (350 μm), equilibrium required 1037 min, followed by 120 min of stability, yielding the maximum scour depth of 0.54 m. The formula predicts 0.51 m, differing by 5.6%.

The scour depth experiments confirm that Sun Zhilin's formula, which slightly underestimates measured depths, remains applicable to 5 μm, 10 μm, 50 μm, and 350 μm sediments in this study. Calculated depths generally exceeded measurements by an average of 6.09%. With errors within 30%, the formula is suitable for submarine cable scour protection design in the target marine area.

4. Conclusions

This study systematically investigated the relationships among critical velocity, scour duration, and scour depth for five different sediment sizes under controlled experimental conditions. By evaluating various sediment incipient motion formulas, a particular focus was placed on identifying and refining a formula suitable for predicting sediment transport characteristics in the target marine environment. The findings provide valuable insights into sediment mobility and the mechanisms governing scour processes, which are essential for the design and protection of submarine cables.

Research indicates that the initiation flow velocity is correlated with water depth, with greater depths corresponding to higher sediment mobilization velocities. Empirical measurements and calculations confirm that, at the same water depth, sediment particles smaller than 150 μm exhibit an inverse relationship between particle size and initiation flow velocity; specifically, smaller particles require lower velocities to mobilize. Conversely, for particles larger than 150 μm, the initiation flow velocity increases with particle size. In this study, the maximum measured initiation flow velocity of approximately 0.6 m/s was observed for 5 μm sediment at a depth of 0.55 m.

The performance of the revised formula proposed in this study was evaluated across different sediment sizes. For particles equal to or larger than 50 μm, the formula exhibited strong agreement with experimental measurements, with an average deviation of 3.7%. However, for finer particles (<50 μm), deviations from measured velocities were observed, necessitating modifications to improve accuracy. By introducing refined parameters (d_0 = 0.045 mm and c = 0.033), the revised formula successfully reduced deviations to less than 5%, demonstrating its improved predictive capability for finer sediments.

The analysis of scour dynamics revealed that lower critical velocities are associated with shorter times required to achieve equilibrium scour depths and increased scour rates. This suggests that finer sediments, which are more readily mobilized, experience a more rapid evolution of scour compared to coarser sediments. Furthermore, higher Froude numbers (Fr) and relative velocities (U/U_0) significantly enhance scour rates, underscoring the critical role of hydrodynamic forces in sediment transport processes. These findings highlight the importance of flow conditions in determining scour depth and duration, particularly in marine environments where strong currents and wave action significantly influence sediment stability.

The practical applicability of the revised formula was further assessed through scour depth predictions for sediments of 5 μm, 10 μm, 50 μm, and 350 μm. The formula demonstrated reliable performance, with an average error of 6.09%, and all discrepancies remained within 30%. This validation confirms the formula's suitability for estimating natural scour depths, providing a robust theoretical foundation for submarine cable protection design and other marine engineering applications.

Overall, this study advances the understanding of sediment transport mechanisms by systematically analyzing the influence of sediment size, water depth, and flow velocity on scour dynamics. The refined predictive model enhances the accuracy of scour depth estimations, contributing to improved engineering designs for coastal and offshore infrastructure. Future work should prioritize field-scale validation under real-world tidal and storm conditions, alongside investigations into sediment cohesion variability with moisture and clay content. Integrating stochastic frameworks and multi-phase flow models could further bridge gaps between laboratory predictions and natural scour behavior, ultimately advancing global submarine cable resilience.

Author Contributions: Conceptualization, F.C.; methodology, W.Y.; software, F.C.; validation, F.C.; investigation, F.C.; resources Z.S.; writing—original draft preparation, W.Y. and F.L.; writing—review and editing, F.C. and L.Z.; writing—original draft, F.C. and F.L. All authors have read and agreed to the published version of the manuscript.

Funding: This study received support and funding from the National Key R&D Program of China (2023YFC3008100) and Major Scientific and Technological Project of Zhoushan City: "Early Warning and Disaster Prevention of Scour around Cross Sea Bridge Foundation (2024C03006)".

Data Availability Statement: The data used to support the findings of this study are available from the corresponding author upon request.

Conflicts of Interest: No potential conflicts of interest were reported by the authors.

References

1. Zhao, M.; Cheng, L.; Zang, Z. Experimental and Numerical Investigation of Local Scour around a Submerged Vertical Circular Cylinder in Steady Currents. *Coast. Eng.* **2010**, *57*, 709–721. [CrossRef]
2. Zhang, M.; Huang, Y.; Bao, Y. The Mechanism of Shallow Submarine Landslides Triggered by Storm Surge. *Nat. Hazards* **2016**, *81*, 1373–1383. [CrossRef]
3. Yang, Q.H.; Yang, Q.; Zhang, Y.Q. Numerical Investigation of the Flow-Field Characteristics around the Suspended Pipeline under Oblique Flows. *Water Supply* **2022**, *22*, 4361–4372. [CrossRef]
4. Zhao, X.F.; Ba, Q.; Li, L.; Gong, P.; Ou, J.P. A Three-Index Estimator Based on Active Thermometry and a Novel Monitoring System of Scour under Submarine Pipelines. *Sens. Actuators A Phys.* **2012**, *183*, 115–122. [CrossRef]
5. Lee, Y.S.; Kim, S.; Yu, Y. A Study on the Selection of Target Ship for the Protection of Submarine Power Cable. *J. Korean Soc. Mar. Environ. Saf.* **2018**, *24*, 662–669. [CrossRef]
6. Wargo, R.; Davenport, T. Protecting Submarine Cables from Competing Uses. In *Submarine Cables*; Brill Nijhoff: Leiden, The Netherlands, 2014; pp. 255–279.
7. Yoon, H.S.; Na, W.B. Anchor Drop Tests for a Submarine Power-Cable Protector. *Mar. Technol. Soc. J.* **2013**, *47*, 72–80. [CrossRef]
8. Matsumoto, H.; Araki, E.; Kimura, T. Detection of Hydroacoustic Signals on a Fiber-Optic Submarine Cable. *Sci. Rep.* **2021**, *11*, 2797. [CrossRef]
9. Dou, G.R. On the Starting Velocity of Sediment. *J. Hydraul. Eng.* **1960**, *4*, 44–60. (In Chinese)
10. Zhang, R.J.; Xie, J.H.; Chen, W.B. *River Dynamics*; China Industry Press: Beijing, China, 1961. (In Chinese)
11. Sun, Z.L.; Xie, J.H.; Duan, W.Z.; Xie, B.L. Research on the Starting Law of Non-Uniform Sand Grading. *J. Water Resour.* **1997**, *10*, 26–33. (In Chinese)
12. Dodaro, G.; Tafarojnoruz, A.; Sciortino, G.; Adduce, C.; Calomino, F.; Gaudio, R. Modified Einstein Sediment Transport Method to Simulate the Local Scour Evolution Downstream of a Rigid Bed. *J. Hydraul. Eng.* **2016**, *142*, 04016041. [CrossRef]

13. Tafarojnoruz, A.; Sharafati, A. New Formulations for Prediction of Velocity at Limit of Deposition in Storm Sewers Based on a Stochastic Technique. *Water Sci. Technol.* **2020**, *81*, 2634–2649. [CrossRef] [PubMed]
14. Sharafati, A.; Tafarojnoruz, A.; Yaseen, Z.M. New Stochastic Modeling Strategy on the Prediction Enhancement of Pier Scour Depth in Cohesive Bed Materials. *J. Hydroinform.* **2020**, *22*, 457–472. [CrossRef]
15. Guo, Z.; Zhou, W.; Zhu, C.; Yuan, F.; Rui, S. Numerical Simulations of Wave-Induced Soil Erosion in Silty Sand Seabeds. *J. Mar. Sci. Eng.* **2019**, *7*, 52. [CrossRef]
16. Pike, S.M.; van Zyl, J.E.; Clayton, C.R.I. Scouring Damage to Buried Pipes Caused by Leakage Jets: Experimental Study. *J. Pipeline Syst. Eng. Pract.* **2018**, *9*, 04018020. [CrossRef]
17. Fredsøe, J.; Hansen, E.A. Lift Forces on Pipelines in Steady Flow. *J. Waterw. Port Coast. Ocean Eng.* **1987**, *113*, 139–155. [CrossRef]
18. Chiew, Y.M. Prediction of Maximum Scour Depth at Submarine Pipelines. *J. Hydraul. Eng.* **1991**, *117*, 452–466. [CrossRef]
19. Hossein Kazeminezhad, M.; Yeganeh-Bakhtiary, A.; Etemad-Shahidi, A. Two-Phase Simulation of Wave-Induced Tunnel Scour beneath Marine Pipelines. *J. Hydraul. Eng.* **2012**, *138*, 517–529. [CrossRef]
20. Rui, S.; Guo, Z.; Wang, L.; Wang, H.; Zhou, W. Inclined Loading Capacity of Caisson Anchor in South China Sea Carbonate Sand Considering the Seabed Soil Loss. *Ocean Eng.* **2022**, *260*, 111790. [CrossRef]
21. Melville, B.; van Ballegooy, S.; Coleman, S.; Barkdoll, B. Scour Countermeasures for Wing-Wall Abutments. *J. Hydraul. Eng.* **2006**, *132*, 563–574. [CrossRef]
22. Jung, C.K.; Park, H.S.; Yang, B.M. Establishment of EMTP Modeling Method Using Searching Coil Test for HVDC Submarine Cables. *Trans. Korean Inst. Electr. Eng.* **2010**, *59*, 1593–1599.
23. Zhao, X.; Yan, X.; Zhang, X. Progress of Active Thermometry Method in Submarine Pipeline Scour Monitoring. In Proceedings of the ASME 2018 Conference on Smart Materials, Adaptive Structures and Intelligent Systems, San Antonio, TX, USA, 10–12 September 2018; Volume 2, p. V002T05A014.
24. Yang, S.; Tang, G.; Liu, Y.; Zhang, W.; Sun, J. CFD and machine learning approach based-predictive modeling of scouring below submarine pipeline under wave and current condition. *J. Mar. Sci. Technol.* **2024**, 1–22.
25. Welzel, M.; Raaijmakers, T.C.; van der Woude, J.H.; de Vet, P.L.M.; Hulscher, S.J.M.H. Scour variability across offshore wind farms (OWFs): Understanding drivers from field data and implications for prediction methods. *Wind Energy Sci. Discuss.* **2025**, 1–35.
26. Guo, Z.; Zhu, C.; Zhang, P.; Yuan, F.; Wang, X. Numerical Study of the Local Scouring Process and Influencing Factors of Semi-Exposed Submarine Cables. *J. Mar. Sci. Eng.* **2023**, *11*, 1349. [CrossRef]
27. Paterson, D.M.; Hope, J.A.; Kenworthy, J.; Stark, J. Sticky stuff: Biological cohesion for scour and erosion prevention. *Environ. Technol.* **2022**, *43*, 4427–4438.
28. Sun, Z.; Ding, K.; Li, Z. An Analytic Model of Typhoon Wind Field and Simulation of Storm Tides. *Front. Mar. Sci.* **2023**, *10*, 1253357. [CrossRef]

water

MDPI

Article

Rainfall Characteristics and Effective Precipitation Analysis of Rice Growth Period in Typical Areas in East China

Haifei Fang [1], Zhan Weng [2,3,4], Miao Hu [2,3,4], Xingya Feng [5,*] and Qiang Liu [5,*]

1 Department of Logistics and Infrastrcture Management, Zhejiang Tongji Vocational College of Science and Technology, Hangzhou 311231, China
2 Zhejiang Institute of Hydraulic and Estuary, Hangzhou 310020, China
3 Zhejiang Guangchuan Engineering Consulting Co. Ltd., Hangzhou 310020, China
4 Key Laboratory of Disaster Prevention and Reduction of Zhejiang Province, Hangzhou 310020, China
5 Department of Ocean Science and Engineering, Southern University of Science and Technology, Shenzhen 518055, China
* Correspondence: fengxy@sustech.edu.cn (X.F.); carylu917@outlook.com (Q.L.)

Abstract

Rainfall characteristics during the rice growth period were analyzed using long-term data (1986–2017) from three typical irrigation stations, namely, Pinghu, Jinqing, and Yongkang in East China. Annual rainfall, concentration indices, and monthly distributions were examined, and the soil–water balance method was applied to estimate effective rainfall (precipitation) from 2018 to 2020. Results showed that rainfall during the growth period generally accounted for 40–80% of the annual total rainfall, with more than 60% of the years exceeding half of the annual rainfall. The effective precipitation utilization coefficients at Pinghu Station reached 0.573, 0.644, and 0.764 in 2018, 2019 and 2020, respectively, indicating relatively high utilization efficiency and confirming effective precipitation as a major water source during rice growth. High utilization was observed for extreme rainfall events ($\geq$30 mm or $\leq$5 mm), whereas moderate rainfall (5–30 mm) showed larger variability due to soil and management factors. To further improve quantitative assessment, a Support Vector Regression (SVR) model was employed to predict daily effective precipitation using rainfall, antecedent precipitation index (API), drought days, and extreme rainfall indicators as inputs. The effective precipitation utilization coefficient was then derived as the ratio of effective to total precipitation. The optimized SVR model achieved a coefficient of determination of $R^2 = 0.904$ and a root mean square error (RMSE) of 4.58 mm for daily effective precipitation, effectively capturing the nonlinear relationship between rainfall characteristics and effective precipitation. These findings highlight that the machine learning method can complement existing estimation models, offering an alternative tool for irrigation scheduling and water-saving crop cultivation.

Keywords: rice growth period; rainfall characteristics; effective precipitation; precipitation utilization coefficient; Support Vector Regression (SVR)

1. Introduction

Rice is a water-intensive crop that requires substantial water throughout its growth cycle. In addition to artificial irrigation, rainfall serves as a major source for meeting the crop's water demand during development [1–3]. Consequently, analyzing the characteristics of rainfall and its effective precipitation during the rice-growing season is of critical importance for understanding the hydrological environment of paddy fields [4,5]. Furthermore, effective precipitation, as one of the key components of the paddy water balance, is also an essential element in the development of irrigation forecasting models. Study on effective precipitation in paddy fields has therefore become a necessary step in supporting the modernization of irrigation district information systems [6,7].

In recent years, numerous studies have examined the rainfall characteristics and effective precipitation for different crops across various regions. For example, Wei [8] investigated the spatial and temporal distribution of precipitation during the summer maize growing season in the Heilonggang Basin, where groundwater over exploitation is severe, providing a scientific basis for improving effective precipitation efficiency. Wu et al. [9] analyzed effective precipitation and crop water demand for cassava in Guangxi and reported significant differences in water requirements across growth stages, with a southeast–northwest distribution gradient. Zhang et al. [10] assessed the effective utilization of precipitation for flue-cured tobacco in Guizhou to optimize irrigation allocation and water-use efficiency. Similarly, Chen [11] and Chen et al. [12] explored the temporal and spatial characteristics of effective precipitation for wheat and cotton, offering valuable references for irrigation planning and land water management. However, research focusing on rainfall characteristics and effective precipitation in paddy fields remains relatively limited, which has gradually emerged as a key research gap in the study of water and soil cycling in rice cultivation systems.

Recent years have seen a growing body of data-driven studies that integrate machine learning with irrigation management and hydrological prediction. Dehghanisanij et al. [13] applied an intelligent model to evaluate functional properties of irrigated cotton under treated wastewater, demonstrating the potential of machine learning for optimizing water use efficiency. Emami et al. [14] employed a meta-heuristic-optimized SVR model for discharge coefficient prediction on labyrinth weirs, further highlighting the suitability of SVR for nonlinear hydrological processes. Achite et al. [15] combined an election algorithm with support vector regression to estimate hydrological drought indices with improved accuracy. However, few studies have focused on effective precipitation utilization in paddy fields, particularly in East China.

Previous studies have mainly focused on characterizing rainfall regimes or estimating effective precipitation for individual crops and regions, while systematic analyses for paddy fields in East China remain scarce. In particular, there is a lack of studies that combine long-term rainfall statistics, soil–water balance-based effective precipitation, and data-driven prediction models within a unified framework for irrigated rice systems. It is hypothesized that the utilisation coefficients of effective precipitation can be predicted with reasonable accuracy using the SVR model, drawing on input variables related to rainfall characteristics and field water conditions. This hypothesis is based on the assumption that these factors play a key role in determining how efficiently rainfall is utilised in rice cultivation, and that a machine learning model like SVR can capture these complex relationships. Therefore, this study aims to: (i) characterise the long-term rainfall pattern during the rice-growing season at three typical irrigation stations in East China; (ii) quantify the effective precipitation and its utilisation coefficient using a soil–water balance approach; (iii) develop a Support Vector Regression (SVR) model to predict daily effective precipitation utilisation coefficients based on rainfall characteristics and field-water conditions. This integrated framework provides new insights and a practical tool for irrigation scheduling and water-saving rice cultivation in East China.

2. Materials and Methods

2.1. Site Description

The Pinghu, Jinqing, and Yongkang Irrigation Experimental Stations are key sites within the irrigation experiment system of Zhejiang Province, East China. The Pinghu station (121°16′ E, 30°71′ N) has silty clay soil derived from marine sediments, with a bulk density of 1.3–1.4 g/cm^3. The Jinqing station (121°51′ E, 28°51′ N) is characterized by tidal clay soils with a bulk density of 1.1–1.2 g/cm^3. The Yongkang station (119°96′ E, 28°93′ N) is situated at an elevation of 85.4 m, with sandy clay soils and a bulk density of 1.4–1.5 g/cm^3.

These three stations are representative of typical agricultural regions in East China: the Hangjiahu Plain (Pinghu), the coastal island zone of eastern Zhejiang (Jinqing), and the hilly low-mountain region of southeastern Zhejiang (Yongkang). Each station is equipped with instruments such as soil moisture sensors, field water level gauges, leaf area meters, chlorophyll meters, and soil solution samplers, enabling research on water-saving irrigation technologies for rice, water-saving and pollution-reduction practices, and irrigation regimes for crops.

2.2. Data Sources

Rainfall data during the rice-growing season (June–November) from 1986 to 2017 at the three irrigation stations (Pinghu, Jinqing and Yongkang) were obtained from routine meteorological observations at the experimental sites. Daily rainfall represents the accumulated precipitation from 08:00 to 08:00 of the following day. Before analysis, the long-term daily series were subjected to basic quality control. The records were checked for completeness and internal consistency, and obviously erroneous values (such as negative rainfall or isolated unrealistically large peaks inconsistent with neighbouring days) were removed. Years with substantial data gaps during the rice-growing season were excluded from the statistical analysis. Based on the cleaned daily series, the seasonal rainfall totals, monthly average rainfall and rainfall concentration indices were calculated for each station. For Pinghu station, effective precipitation during the rice-growing seasons of 2018–2020 was further analysed using field-measured data, including crop transpiration, deep percolation, irrigation depth and crop water requirement. In combination with the observed daily rainfall, these field data were used as inputs to the soil–water balance method [16] described in Section 2.3 to calculate daily effective precipitation in the root zone and the corresponding effective precipitation utilisation coefficient. The soil–water balance was evaluated at the field scale under the assumption that soil properties and management practices are spatially uniform within each experimental plot. This simplification is necessary because only plot-averaged data are available from the irrigation stations.

2.3. Methods

The three irrigation stations (Pinghu, Jinqing, and Yongkang) were selected as representative sites to analyze the interannual spatial–temporal variations of rainfall, rainfall concentration index, and the long-term trend of mean monthly rainfall during the rice-growing season (June–November) from 1986 to 2017. The objective was to identify rainfall characteristics across different regions of Zhejiang Province in East China.

For Pinghu station, data from the rice-growing seasons of 2018, 2019, and 2020 were analyzed, including transpiration, deep percolation, and crop water requirements. Effective precipitation was estimated, and the effective precipitation utilization coefficient was calculated to explore the distribution and characteristics of effective precipitation at the representative site.

In the soil–water balance calculation, an effective root zone depth of 0.20 m was assumed uniformly for the Pinghu, Jinqing, and Yongkang irrigation stations, representing the typical plough layer and main rooting zone of lowland rice in the study region.

2.3.1. Effective Precipitation

According to the handbook [17] and based on the soil–water balance equation, a real-time estimation method was applied. The formula is expressed as:

$$P_e = W_t - W_0 - D + ET_t \tag{1}$$

where P_e is the effective rainfall or precipitation (mm), W_t is the soil water storage on the second day after rainfall (mm), W_0 is the soil water storage before rainfall (mm), and ET_t is the field evapotranspiration during the rainfall period (mm), which can be estimated using the Penman–Monteith equation. Surface drainage and deep percolation losses are represented by the term D in Equation (1). Due to limited field measurements, D was estimated using empirical coefficients recommended in the regional irrigation guidelines [18].

2.3.2. Effective Precipitation Utilization Coefficient

The effective precipitation utilization coefficient is defined as the ratio of effective precipitation to total precipitation within a given period:

$$R = P_e/P \tag{2}$$

where R is the effective precipitation utilization coefficient, P_e is the effective precipitation (mm), and P is the total precipitation (mm).

In this study, rainfall, effective precipitation and utilisation coefficients during the rice-growing season were first summarised using descriptive statistics (such as interannual means and ranges) before further analysis of their temporal variation and spatial differences among the three stations

2.4. *Prediction with a Machine Learning Model*

To further enhance the quantitative analysis of water utilization processes in paddy fields, this study introduces a machine learning approach to predict the effective precipitation utilization coefficient. In this study, an ε-insensitive support vector regression (SVR) model [19] with a radial basis function (RBF) kernel was employed to predict the effective precipitation utilisation coefficient. Given an input vector $\mathbf{x}_i$ that contains daily rainfall, the 7-day antecedent precipitation index, the number of consecutive dry days and the extreme rainfall indicator, the SVR seeks a regression function

$$f(\mathbf{x}) = \sum_{i=1}^{N} (\alpha_i - \alpha_i^*) K(\mathbf{x}_i, \mathbf{x}) + b \tag{3}$$

where $K(\mathbf{x}_i, \mathbf{x}) = \exp(-\gamma \parallel \mathbf{x}_i - \mathbf{x} \parallel^2)$ is the RBF kernel, α_i and α_i^* are Lagrange multipliers, γ is the kernel width, C is the regularisation parameter controlling the trade-off between model complexity and training error, and ε defines the width of the ε-insensitive loss function.

2.4.1. Input and Output

The selection of input variables was based on a comprehensive consideration of rainfall characteristics and field water conditions, including:

(1) Daily rainfall P_t, representing the precipitation on day t;

(2) Antecedent wetness indicators, including the cumulative rainfall over the previous seven days and the number of consecutive dry days. The former was represented by the Antecedent Precipitation Index (API), calculated as follows [20]:

$$API_t = \sum_{i=2}^{n} P_{t-i}k^i \tag{4}$$

where API_t denotes the antecedent precipitation index at time t, P_{t-i} is the rainfall on day $t-i$, and k is the recession coefficient $(0 < k < 1$, commonly set to $0.80 - 0.90)$, reflecting the attenuation effect of prior rainfall on soil moisture conditions. In this study, the preceding seven days ($n = 7$) were considered, following the definition by Kohler & Linsley [18]. The number of consecutive dry days was defined as the duration without rainfall and was used to capture the impact of sustained dry conditions on soil moisture status and rainfall utilization.

(3) Extreme rainfall events, where precipitation exceeding a threshold ($\geq$50 mm) was assigned a value of 1, otherwise 0.

The output variable was defined as daily effective precipitation P_e (mm).

2.4.2. Model Construction and Parameter Optimization

During model training, the dataset was first randomly shuffled and then divided into a training set (75%), validation set (15%), and test set (15%). The Support Vector Regression (SVR) model with a radial basis function (RBF) kernel was adopted. The penalty parameter C and kernel parameter γ were optimized using grid search.

2.4.3. Model Evaluation Metrics

To evaluate the predictive performance, this study employed the root mean square error (RMSE) and the coefficient of determination (R^2) as evaluation metrics.

$$RMSE = \sqrt{\frac{1}{n}\sum_{i=1}^{n}(y_i - \hat{y}_i)^2} \tag{5}$$

$$R^2 = 1 - \frac{\sum_{i=1}^{n}(y_i - \hat{y}_i)^2}{\sum_{i=1}^{n}(y_i - \bar{y})^2} \tag{6}$$

where y_i denotes the observed values, $\hat{y}_i$ the predicted values, $\bar{y}$ the mean of observed values, and n the number of samples. RMSE reflects the magnitude of prediction errors, while R^2 measures the consistency between predicted and observed values.

Based on the available rainfall and field water condition data, the proposed approach enables the prediction of the effective precipitation utilization coefficient. Compared with traditional empirical models, the support vector regression (SVR) method is capable of capturing nonlinear relationships, making it more suitable for scenarios in which paddy water processes are influenced by multiple interacting factors. The prediction results can provide both data support and theoretical basis for irrigation scheduling optimization and water-saving rice cultivation.

3. Results

3.1. Rainfall Characteristics in Typical Regions

3.1.1. Interannual Rainfall Analysis

Rainfall data from the rice-growing season (June–November) during 1986–2017 at Pinghu, Jinqing, and Yongkang stations were analyzed. Assuming spatial uniformity of rainfall within the representative regions, the temporal distribution and variability of rainfall at each station were examined.

As shown in Figure 1, the average rainfall during the rice-growing season at Pinghu Station was 734 mm, ranging from 303 mm in 2003 to 1038 mm in 1993. At Jinqing Station, the average seasonal rainfall was 1092 mm, with a range of 536 mm (2003) to 1681 mm (1990). Yongkang Station recorded an average of 817 mm, with values ranging from 392 mm (2003) to 1304 mm (1989). Among the three sites, Jinqing Station exhibited the highest long-term average rainfall, followed by Yongkang, while Pinghu recorded the lowest. Notably, the minimum rainfall in all three stations occurred in 2003, consistent with the widespread summer–autumn drought reported in Zhejiang Province during that year.

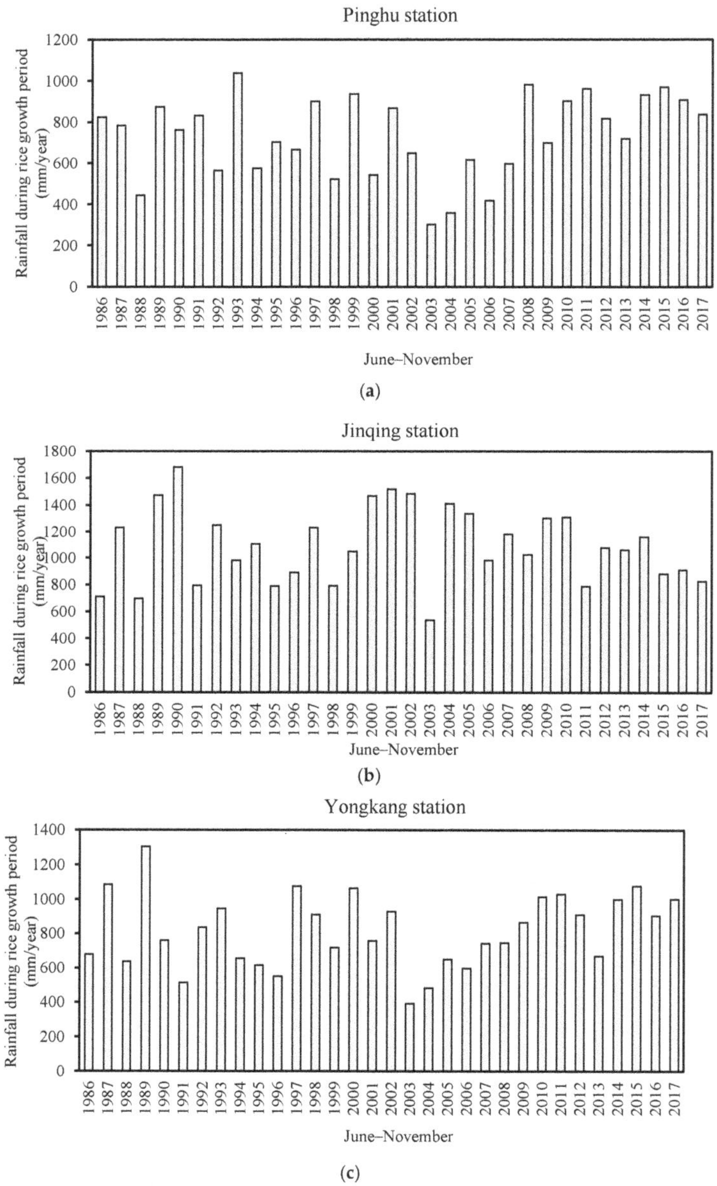

Figure 1. Statistical distribution of rainfall during the rice-growing season (June–November) from 1986 to 2017 for (**a**) Pinghu; (**b**) Jinqing; (**c**) Yongkang.

The ratios of maximum to minimum seasonal rainfall were 2.43, 2.14, and 2.33 at Pinghu, Jinqing, and Yongkang stations, respectively, indicating substantial interannual variability in rainfall during the rice-growing season.

Based on the trend analysis of rainfall during the rice-growing season from 1986 to 2017, the regression slopes at Pinghu and Yongkang stations were 3.8 and 3.3, respectively, indicating upward trends. This suggests that rainfall during the rice-growing season at these two stations has gradually increased, with a high likelihood of continuous increases in the future. In contrast, Jinqing station exhibited a downward trend with a slope of −1.9, suggesting that seasonal rainfall has been decreasing over the same period, with a high probability of further reductions in the future.

3.1.2. Rainfall Concentration During the Rice-Growing Season

Rainfall concentration reflects the proportion of rainfall occurring within a specific period relative to the annual total, thereby indicating the degree of temporal clustering. In this study, the rainfall concentration of the rice-growing season was defined as the ratio of rainfall from June to November to the total annual rainfall. Analyzing the proportion of seasonal rainfall to annual totals provides valuable insights into the extent to which rainfall influences rice water availability during the growing season.

The rainfall concentration index during the rice-growing season generally falls between 40% and 80% at the three stations (Figure 2), indicating that a considerable portion of the annual rainfall is received during the rice-growing period. Years with a concentration index exceeding 50% account for 25, 28 and 21 years at Pinghu, Jinqing and Yongkang, respectively, corresponding to more than 60% of the study period at each site. In these years, over half of the annual rainfall occurs during the rice-growing season, providing favourable hydrological conditions for rice growth and development.

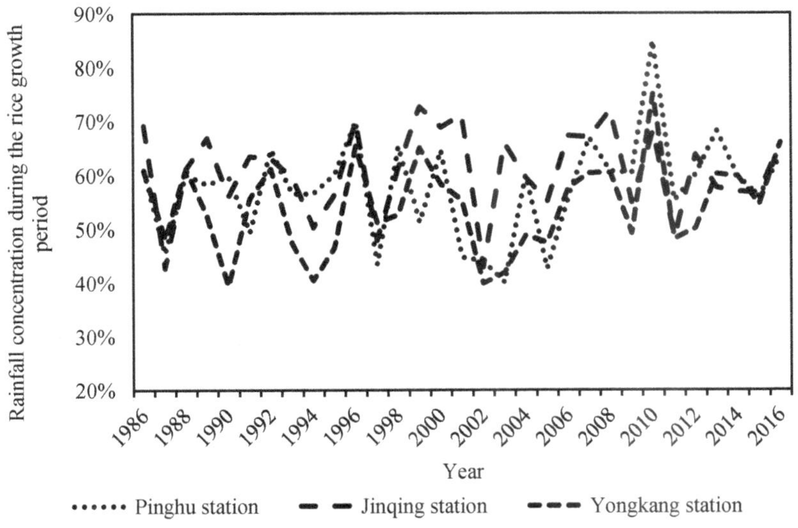

Figure 2. Statistical distribution of rainfall concentration during the rice-growing season (1986–2017) at the representative stations (Pinghu, Jinqing and Yongkang).

Extreme concentration can also occur. For example, in 2011 the rainfall concentration index reached 84.8% at Pinghu, 68.3% at Jinqing and 75.0% at Yongkang, implying that most of the annual rainfall was concentrated in the growing season. Such years are characterised by not only high seasonal rainfall but also relatively long and intense rainfall events, which may increase the risk of flooding and waterlogging. On the basis of these observations, a further analysis of the monthly rainfall distribution is needed to reveal how rainfall timing interacts with the critical stages of rice growth.

3.1.3. Monthly Rainfall During the Rice-Growing Season

The monthly average rainfall during the rice-growing season (June–November) from 1986 to 2017 was calculated for Pinghu, Jinqing, and Yongkang stations, and corresponding trend and distribution plots were generated.

As shown in Figure 3, rainfall at Pinghu Station ranged from 61 mm in October to 202 mm in June. Rainfall from June to September accounted for 83% of the seasonal total, showing an overall decreasing trend, with a pronounced decline after August. At Jinqing Station, monthly rainfall ranged from 80 mm (October) to 291 mm (August). Rainfall from June to September accounted for 84% of the seasonal total, also displaying an overall decreasing trend with a sharp reduction after August. In addition, rainfall fluctuated between June and August, first decreasing and then increasing, followed by a marked decline from August to November. At Yongkang Station, rainfall ranged from 64 mm (October) to 268 mm (June). Rainfall from June to September accounted for 82% of the seasonal total, showing a pattern similar to Pinghu Station, with an overall downward trend and a sharp decline after August.

Overall, the rainfall at the three representative stations was predominantly concentrated between June and September, contributing more than 80% of the seasonal total. According to the rice irrigation schedule in East China, this period coincides with the critical growth stages of tillering, jointing–booting, and heading–flowering, during which rice is highly sensitive to water stress. Higher rainfall during these stages is therefore crucial for meeting water requirements and ensuring normal crop development. Furthermore, during the flood season, when individual rainfall events may be intense, effective utilization of rainfall—while avoiding waterlogging—should be considered, which has positive implications for water-saving irrigation in rice production.

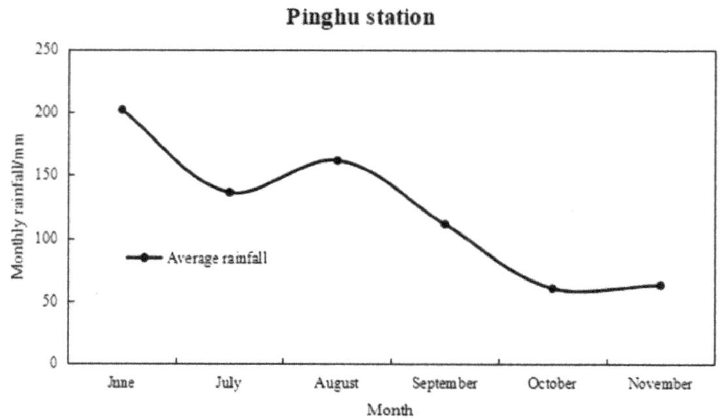

(a)

Figure 3. *Cont.*

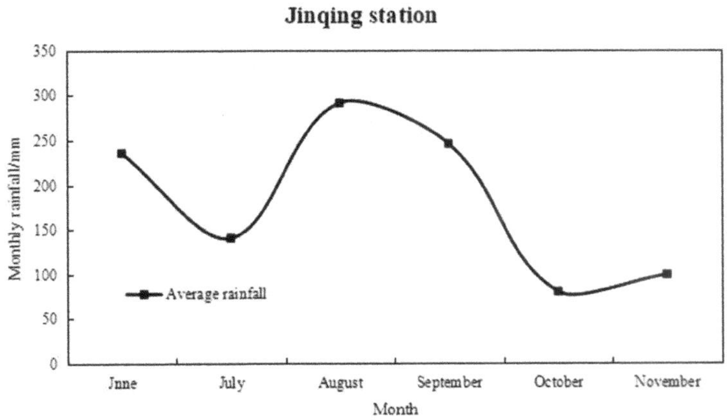

(**b**)

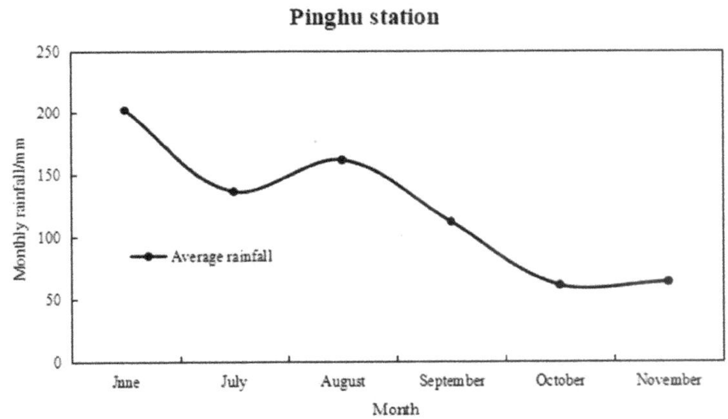

(**c**)

Figure 3. Variation of average monthly rainfall during the rice-growing season (June–November) from 1986 to 2017 at the representative stations (**a**) Pinghu, (**b**) Jinqing and (c) Yongkang.

3.2. Analysis Based on the Soil–Water Balance Method

3.2.1. Results of Effective Precipitation Calculation

In this study, the rainfall data from 2018 to 2020 at Pinghu Station were used to calculate effective precipitation using the soil–water balance equation. The effective precipitation utilization coefficient was defined as the ratio of effective precipitation to total precipitation.

As shown in Table 1, during the 2018 rice-growing season, there were 14 rainfall days with a total precipitation of 429.2 mm, of which 245.9 mm was classified as effective precipitation. The overall effective precipitation utilization coefficient for the season was 0.573. Among these events, there were four instances of consecutive rainfall lasting more than two days, with a total precipitation of 121.7 mm, effective precipitation of 88.4 mm, and an effective precipitation utilization coefficient of 0.726. The highest utilization was recorded on 12 August and 25 August, when the effective precipitation utilization coefficients both reached 1.000. By contrast, the lowest utilization occurred on 23 September, with a rainfall of 3.5 mm and effective precipitation of only 0.5 mm, corresponding to a coefficient of 0.143. This low value was attributed to consecutive rainfall events on 21–22 September, which reduced the subsequent effective utilization.

Table 1. Effective precipitation and coefficients at Pinghu station in 2018.

Date	Rainfall (mm)	Effective Precipitation (mm)	Effective Precipitation Utilization Coefficient
22 July	24	13.5	0.563
4 August	139	106.5	0.766
12 August	0.5	0.5	1
17 August	48.5	36	0.742
18 August	10.5	8	0.762
21 August	11.5	5.5	0.478
25 August	1	1	1
26 August	10	7.5	0.75
7 September	19.5	17	0.872
9 September	2	2	1
17 September	144	37	0.257
21 September	8.5	6	0.706
22 September	6.7	4.9	0.731
23 September	3.5	0.5	0.143
Total	429.2	245.9	0.573

In 2018, a total of 11 rainfall events exhibited an effective precipitation utilization coefficient greater than 0.5, accounting for 79% of the rainfall days during the growing season, which indicates relatively high utilization. However, the degree of effective precipitation utilization coefficients varied considerably throughout the season, with some events showing very high efficiency and others much lower. As a result, the overall effective precipitation utilization coefficient in 2018 was the lowest among the three representative years.

At Pinghu station in 2019, the calculated effective precipitation and corresponding coefficients are shown in Table 2. During the rice-growing season, there were 11 rainfall days with a cumulative precipitation of 451.0 mm, of which 290.4 mm was effective precipitation. The seasonal effective precipitation utilization coefficient reached 0.644. Within this period, three episodes of consecutive rainfall lasting more than two days were observed, with a total precipitation of 286.0 mm, effective precipitation of 199.5 mm, and a coefficient of 0.698. The highest daily utilization occurred on 10 August, when rainfall reached 75.0 mm, effective precipitation was 71.5 mm, and the coefficient rose to 0.953. In contrast, the lowest efficiency was recorded on 24 August, with only 1.0 mm of effective precipitation out of 4.5 mm of rainfall, corresponding to a coefficient of 0.222. Overall, eight rainfall days (73% of the total) had coefficients greater than 0.5, suggesting that rainfall in 2019 was effectively utilized at a relatively high level.

Table 2. Effective precipitation and coefficients at Pinghu station in 2019.

Date	Rainfall (mm)	Effective Precipitation (mm)	Effective Precipitation Utilization Coefficient
5 August	54.5	14	0.257
9 August	2	1	0.5
10 August	75	71.5	0.953
11 August	155.5	95.5	0.614
24 August	4.5	1	0.222
26 August	2	1	0.5
28 August	11	3.5	0.318
30 August	31.5	21	0.667
1 September	9	6	0.667
22 September	17	8.5	0.5
2 October	89	67.4	0.757
Total	451	290.4	0.644

According to Table 3, during the 2020 rice-growing season at Pinghu station, there were 14 rainfall days with a total precipitation of 581.5 mm, of which 444.3 mm was classified as effective precipitation. The overall effective precipitation utilization coefficient for the season was 0.764. Among these events, four episodes of consecutive rainfall lasting more than two days were observed, with a cumulative precipitation of 495.5 mm, effective precipitation of 390.7 mm, and an effective coefficient of 0.788.

Table 3. Effective precipitation and coefficients at Pinghu station in 2020.

Date	Rainfall (mm)	Effective Precipitation (mm)	Effective Precipitation Utilization Coefficient
25 July	2.5	1.9	0.76
27 July	0.5	0.3	0.6
29 July	9.5	1.6	0.168
5 August	316.5	266.5	0.842
6 August	21	0.5	0.024
28 August	9.5	2.5	0.263
11 September	53	40	0.755
17 September	77	58	0.753
18 September	42	36.5	0.869
19 September	12	12	1
28 September	14	9.5	0.679
16 October	17	13.7	0.806
17 October	5.5	0.5	0.091
2 November	1.5	0.8	0.533
Total	581.5	444.3	0.764

The highest daily utilization was recorded on 19 September, when rainfall reached 12.0 mm and was fully effective, yielding a coefficient of 1.000. In contrast, the lowest utilization occurred on 6 August, with 21.0 mm of rainfall but only 0.5 mm of effective precipitation, corresponding to a coefficient of 0.024. This exceptionally low value was attributed to a preceding extreme rainfall event on 5 August, when precipitation reached 316.5 mm and effective precipitation was 266.5 mm, thereby significantly reducing the effective utilization of the subsequent day's rainfall.

Overall, eight rainfall days exhibited coefficients greater than 0.6, accounting for 57% of the total rainfall days during the season, suggesting that rainfall utilization efficiency in 2020 was relatively high.

3.2.2. Analysis of Effective Precipitation Utilization Coefficients

Daily rainfall events during the representative years 2018–2020 were ranked in descending order, and the corresponding distribution of rainfall and effective precipitation utilization coefficients is shown in Figure 4. When daily rainfall exceeded 30 mm, the utilization coefficient was generally above 0.67, indicating that nearly 70% of the rainfall was converted into effective precipitation in paddy fields. Exceptions occurred on 17 September 2018 (144 mm) and 5 August 2019 (54.5 mm), where the coefficients were only 0.257. These low values were mainly attributed to full irrigation applied prior to the rainfall events, which reduced the effective utilization of the subsequent precipitation.

When daily rainfall was less than 5 mm, four events recorded a coefficient of 1.000, while two events had coefficients of 0.900 and 0.800, respectively; all other cases remained above 0.5. This suggests that under such conditions, small rainfall events were effectively utilized by paddy fields. In contrast, when daily rainfall fell within the range of 5–30 mm, the utilization coefficient varied widely between 0.024 and 1.000, showing a highly scattered distribution. This indicates that in this range, the conversion of rainfall into effective precipitation was less consistent and likely influenced by additional field factors.

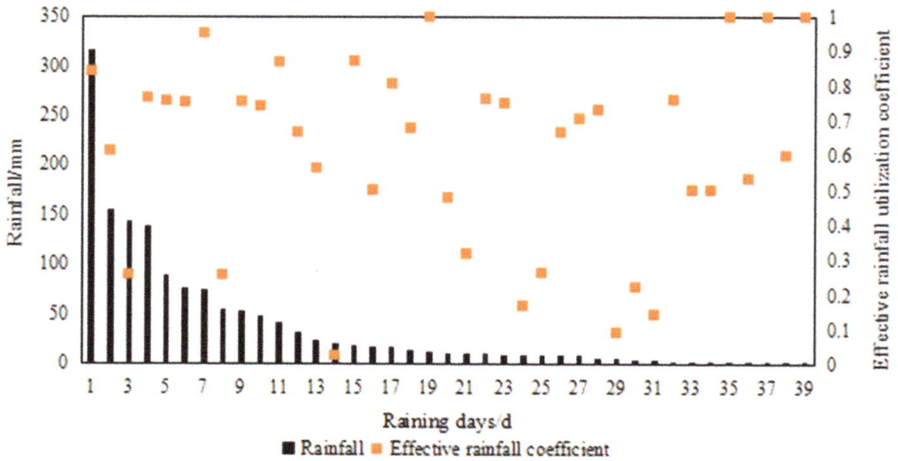

Figure 4. Distribution of daily rainfall and corresponding effective precipitation utilization coefficients at Pinghu station during 2018–2020.

In summary, when rainfall is relatively high ($P \geq 30$ mm), the effective precipitation utilization coefficient is generally large, indicating a higher efficiency of rainfall use in paddy fields. Under conditions where waterlogging is prevented, such rainfall is beneficial for rice growth and development during critical stages such as tillering, jointing–booting, and heading–flowering. When rainfall is relatively low ($P \leq 5$ mm), the coefficient is also high, suggesting that rice paddies, as a water-intensive cropping system, possess a certain storage capacity that enables efficient retention and absorption of small rainfall events. This process contributes to regulating soil moisture conditions and supporting crop growth and development. By contrast, when rainfall is moderate (5 mm $< P <$ 30 mm), the precipitation coefficient exhibits a highly scattered distribution. This implies that the effective use of rainfall in this range is more sensitive to additional factors, such as antecedent soil moisture, irrigation–drainage practices, and field engineering conditions, which require further investigation. It should be noted that the class boundaries $P \leq 5$ mm, 5 mm $< P <$ 30 mm, and $P \geq 30$ mm were not taken from technical standards but were chosen empirically, after preliminary analysis indicated distinct patterns of effective precipitation utilization within these three rainfall ranges.

3.3. Analysis Based on Support Vector Regression

3.3.1. Model Evaluation

To further investigate the prediction of the effective precipitation utilization coefficient, this study selected daily rainfall data from the Pinghu station during 2018–2020, together with the corresponding effective precipitation data, as the modeling samples. Considering that extremely small or large outliers might lead to bias in model training, the raw data were subjected to screening and cleaning. Only representative samples that are physically meaningful and capable of accurately reflecting the dynamics of paddy field water conditions were retained to ensure the scientific validity and reliability of the modeling process.

Table 4 summarizes the processed dataset used for model construction, which includes daily rainfall (P_t), the 7-day antecedent precipitation index (API), consecutive dry days (D), extreme rainfall event indicator (E), as well as the effective precipitation and the effective precipitation utilization coefficient (R) calculated based on the soil–water balance method.

These variables together constitute the core inputs and outputs for machine learning modeling, providing the fundamental basis for model training, validation, and prediction.

Table 4. Modeling data at Pinghu station from 2018 to 2020.

Date	Rainfall/mm	7-Day API	Consecutive Dry Days	Extreme Rainfall Within 7 Days	Effective Precipitation/mm	Effective Precipitation Coefficient
22 July 2018	24	0	14	0	13.5	0.563
04 Aug 2018	139	1.200	10	0	106.5	0.766
17 Aug 2018	48.5	11.081	3	0	36	0.742
18 Aug 2018	10.5	56.257	0	0	8	0.762
21 Aug 2018	11.5	15.330	3	0	5.5	0.478
25 Aug 2018	1	6.870	2	0	1	1
26 Aug 2018	10	4.944	3	0	7.5	0.75
07 Sep 2018	19.5	12.494	2	0	17	0.872
09 Sep 2018	2	26.901	0	0	2	1
21 Sep 2018	8.5	71.442	2	0	6	0.706
22 Sep 2018	6.7	58.509	0	0	4.9	0.731
23 Sep 2018	3.5	47.656	0	0	0.5	0.143
05 Aug 2019	54.5	5	0	0	14	0.257
09 Aug 2019	2	19.894	3	0	1	0.5
10 Aug 2019	75	15.925	4	0	71.5	0.953
11 Aug 2019	155.5	86.148	0	0	95.5	0.614
24 Aug 2019	4.5	0	9	0	1	0.222
26 Aug 2019	2	3.15	1	0	1	0.5
28 Aug 2019	11	2.943	3	0	3.5	0.318
30 Aug 2019	31.5	16.142	0	0	21	0.667
01 Sep 2019	9	42.799	0	0	6	0.667
22 Sep 2019	17	0	14	0	8.5	0.5
02 Oct 2019	89	0	11	0	67.4	0.757
25 Jul 2020	2.5	22.857	0	0	1.9	0.76
27 Jul 2020	0.5	12.950	2	0	0.3	0.6
29 Jul 2020	9.5	6.489	4	0	1.6	0.168
06 Aug 2020	21	316.5	0	1	0.5	0.024
28 Aug 2020	9.5	0	16	0	2.5	0.263
11 Sep 2020	53	0	7	0	40	0.755
17 Sep 2020	77	24.203	0	0	58	0.753
18 Sep 2020	42	93.942	0	0	36.5	0.869
19 Sep 2020	12	103.394	0	0	12	1
28 Sep 2020	14	0.838	8	0	9.5	0.679
16 Oct 2020	17	0	17	0	13.7	0.806
17 Oct 2020	5.5	17	0	0	0.5	0.091
02 Nov 2020	1.5	0	15	0	0.8	0.533

In terms of data partitioning, 10% of the overall dataset was first set aside as a validation set to evaluate the model's generalization capability on unseen data. The remaining 90% of the samples were further divided into training and testing sets in a 7:3 ratio, ensuring a balance between parameter optimization and performance evaluation. This design allowed the model to fully learn data characteristics during training while enabling independent validation of its prediction stability and practical applicability.

3.3.2. Prediction Results of the Effective Precipitation Utilization Coefficient

The results indicate that the optimal parameters were obtained as $C = 1000$ and $\gamma = 0.01$. On the testing dataset, the coefficient of determination reached $R^2 = 0.904$, and the root mean square error (RMSE) was 4.584 mm. These results demonstrate that the predicted values are highly consistent with the observed values, suggesting that the SVR model can effectively capture the nonlinear relationship between rainfall characteristics and the effective precipitation coefficient. Because the field-measured data required for the soil–water balance (including crop transpiration, deep percolation, irrigation depth and drainage) were only available at Pinghu station during 2018–2020, the SVR model was developed and validated exclusively for this station, which is taken as a representative irrigated paddy area in East China.

Figure 5 presents the comparison between predicted and observed values in the form of line plots. The overall trends show a high degree of agreement, indicating that the model can accurately reflect the variation of the effective precipitation utilization coefficient under different rainfall events. For most samples, the prediction errors are relatively small, and the discrepancies between observed and predicted values remain within a reasonable range, which demonstrates the strong fitting capability of the SVR model.

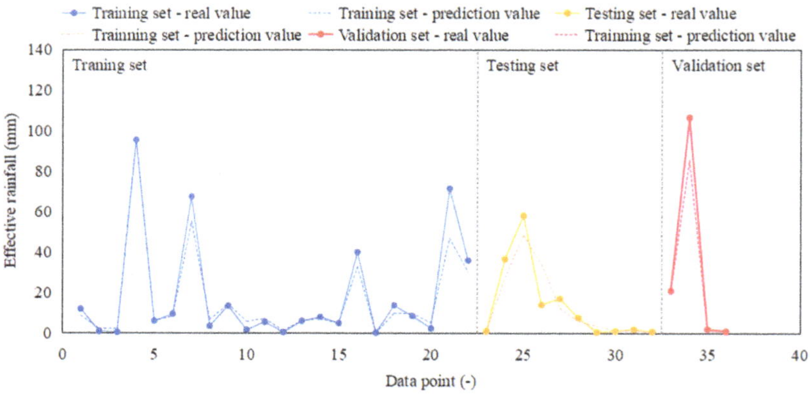

Figure 5. Comparison between observed and SVR-predicted effective precipitation at Pinghu Station during 2018–2020.

The scatter comparison (Figure 6) shows that the predicted and observed values exhibit good overall agreement in the low-value range (effective precipitation $\leq$ 20 mm), with most points distributed near the 1:1 reference line, indicating that the model provides reliable predictions under light and moderate rainfall conditions. However, in the high-value range (>80 mm), the model demonstrates a clear underestimation, with predicted values significantly lower than the observed ones. This suggests that the SVR model has limited fitting capability under extreme rainfall events, which may be attributed to the scarcity of high-value samples in the training set and the insufficient representation of threshold processes by the input features. Consequently, the extrapolation capability of the SVR model under extreme rainfall conditions remains limited and should be improved in future work by extending the time series and enriching the predictor variables. Overall, the model performs with satisfactory stability and accuracy in the low-value range, but prediction bias emerges in the high-value range, implying that further optimization is required through sample augmentation or improvements in the modeling approach.

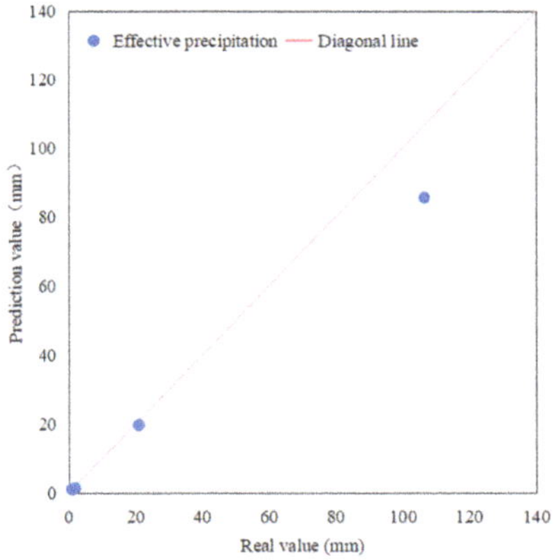

Figure 6. Scatter plot comparing observed and predicted effective precipitation utilization coefficients.

3.4. Discussion

The results reveal pronounced interannual variability in rainfall at Pinghu, Jinqing, and Yongkang, with Pinghu showing the lowest and Jinqing the highest seasonal totals. Similar spatial contrasts and strong year-to-year fluctuations in rice-growing rainfall have been reported for East-China paddy regions, where such variability is a key driver of irrigation demand and soil–water balance [21]. The relatively high rainfall concentration indices at Jinqing and Yongkang indicate that a large share of seasonal rainfall is delivered by a limited number of events, consistent with studies noting increasing precipitation concentration and its implications for waterlogging and flood risk in Asian rice systems [22–24].

The monthly patterns show an early-season rainfall peak at Pinghu and a more uniform distribution at Jinqing, underscoring intra-regional differences in rainfall timing that may require station-specific irrigation scheduling and field-management strategies. This spatial heterogeneity agrees with previous work highlighting that even within a single climatic region, local rainfall regimes can diverge sufficiently to justify differentiated water-saving measures in rice production [24,25].

The SVR model reproduces effective precipitation utilization coefficients well under light and moderate rainfall, in line with earlier applications of machine-learning models for hydro-meteorological prediction in agriculture [26–28]. Underestimation in the high-rainfall range reflects a common limitation of data-driven approaches, which often struggle to extrapolate to rare extremes that are under-represented in the training set [27,28]. Enlarging the sample to include more extreme events and enriching the predictor set with additional climatic and soil–water indicators should enhance robustness, while the current performance already demonstrates the potential of SVR as a practical tool to support water-saving irrigation scheduling in paddy fields.

4. Conclusions

This study analyzes the rainfall characteristics and effective precipitation utilization in the rice-growing season at three representative stations in East China. The key findings are summarized as follows:

(1) The mean seasonal rainfall during 1986–2017 show significant interannual variability among the three stations of interest. The historical data suggests that more than 60% of the rainfall occurs during the rice-growing season, providing favorable water availability for rice production.

(2) Over 80% of the rainfall occurred from June to September at all stations, which coincides with critical growing stages. This highlights the importance of early-season rainfall in supporting crop water requirements.
(3) At Pinghu, between 2018 and 2020, effective precipitation accounted for 57% to 76% of the seasonal rainfall, demonstrating that effective precipitation is a crucial water source for paddy fields under the study conditions.
(4) The effective precipitation utilization coefficient showed non-linear responses to rainfall magnitude. High coefficients were observed for both heavy rainfall events and small showers, suggesting efficient water use, whereas moderate rainfall showed more variability, influenced by soil moisture and field management practices.
(5) The SVR model generally well predicts the effective precipitation utilization coefficient. It can be an alternative approach for prediction of effective precipitation utilization, which could be helpful to optimize irrigation schedules based on forecasted rainfall. However, the model underestimates high-value cases due to limited extreme event samples. Further improvements could be made by expanding the dataset and enhancing the predictor variables.

Author Contributions: Conceptualization, H.F., Z.W. and M.H.; methodology, X.F. and Q.L.; software, Q.L.; validation, H.F., M.H. and Q.L.; formal analysis, H.F. and X.F.; investigation, H.F. and Q.L.; resources, Z.W. and M.H.; data curation, H.F.; writing—original draft preparation, H.F.; writing—review and editing, Z.W., M.H., X.F. and Q.L.; visualization, M.H. and Q.L.; supervision, X.F. and Q.L.; project administration, M.H. and X.F.; funding acquisition, X.F. All authors have read and agreed to the published version of the manuscript.

Funding: This research was funded by the National Key Research and Development Program of China (2023YFB3711500) and the APC was funded by the National Key Research and Development Program of China (2023YFB3711500).

Data Availability Statement: The original contributions presented in this study are included in the article. Further inquiries can be directed to the corresponding authors.

Conflicts of Interest: Authors Zhan Weng and Miao Hu were employed by Zhejiang Guangchuan Engineering Consulting Ltd. The remaining authors declare that the research was conducted in the absence of any commercial or financial relationships that could be construed as a potential conflict of interest.

References

1. Zhang, S.; Rasool, G.; Guo, X.; Sen, L.; Cao, K. Effects of different irrigation methods on environmental factors, rice production, and water use efficiency. *Water* **2020**, *12*, 2239. [CrossRef]
2. Zhu, S.; Tong, W.; Li, H.; Li, K.; Xu, W.; Liang, B. Temporal variations in rice water requirements and the impact of effective rainfall on irrigation demand: Strategies for sustainable rice cultivation. *Water* **2025**, *17*, 656. [CrossRef]
3. Azhar, A.H.; Murty, V.V.N.; Phien, H.N. Modeling irrigation schedules for lowland rice with stochastic rainfall. *J. Irrig. Drain. Eng.* **1992**, *118*, 36–55. [CrossRef]
4. Cao, J.; Tan, J.; Cui, Y.; Luo, Y. Irrigation scheduling of paddy rice using short-term weather forecast data. *Agric. Water Manag.* **2019**, *213*, 714–723. [CrossRef]
5. Bouman, B.A.M.; Tuong, T.P. Field water management to save water and increase its productivity in irrigated lowland rice. *Agric. Water Manag.* **2001**, *49*, 11–30. [CrossRef]
6. Geerts, S.; Raes, D. Deficit irrigation as an on-farm strategy to maximize crop water productivity in dry areas. *Agric. Water Manag.* **2009**, *96*, 1275–1284. [CrossRef]
7. Pereira, L.S.; Cordery, I.; Iacovides, I. Improved indicators of water use performance and productivity for sustainable water conservation and saving. *Agric. Water Manag.* **2012**, *108*, 39–51. [CrossRef]
8. Wei, Z.; Bian, D.; Du, X.; Pushpa, R.; Cui, Y. Spatio-temporal distribution characteristics of precipitation, water demand, and drought during the summer maize growth period in the Heilonggang Basin. *Trans. Chin. Soc. Agric. Eng.* **2020**, *36*, 124–133. (In Chinese with English abstract)

9. Wu, W.; Mo, M. Variation characteristics of effective precipitation and water requirement of cassava in Guangxi from 1958 to 2015. *Water Sav. Irrig.* **2020**, 80–87. (In Chinese with English abstract)
10. Zhang, B.; Pei, X.; Cao, H.; Gu, S.; Mo, J. Characteristics of water requirement and irrigation demand index for flue-cured tobacco in Guizhou Province. *Res. Soil Water Conserv.* **2019**, *26*, 215–221. (In Chinese with English abstract)
11. Chen, C.; Pang, Y.; Pan, X. Variation characteristics of water surplus and deficit of single-season rice in Sichuan Province under climate change. *J. Nat. Resour.* **2014**, *29*, 1508–1519. (In Chinese with English abstract)
12. Chen, W.; Chen, L.; Sun, C.; Zheng, Z.; Li, X. Spatial distribution pattern of water requirement for main crops in the eastern Loess Plateau. *Res. Soil Water Conserv.* **2020**, *27*, 166–176. (In Chinese with English abstract)
13. Dehghanisanij, H.; Emami, S.; Khasheisiuki, A. Functional properties of irrigated cotton under urban treated wastewater using an intelligent method. *Appl. Water Sci.* **2022**, *12*, 66. [CrossRef]
14. Emami, H.; Emami, S.; Parsa, J. A Walnut optimization algorithm applied to discharge coefficient prediction on labyrinth weirs. *Soft Comput.* **2022**, *26*, 12197–12215. [CrossRef]
15. Achite, M.; Emami, S.; Jehanzaib, M.; Katipoğlu, O.M.; Emami, H. An election algorithm combined with support vector regression for estimating hydrological drought. *Model. Earth Syst. Environ.* **2024**, *10*, 1395–1405. [CrossRef]
16. Ma, L.; Hu, X.; Jiang, X.; Wang, W. Simulation and analysis of soil moisture at different growth stages of paddy fields based on HYDRUS. *J. Drain. Irrig. Mach. Eng.* **2022**, *40*, 729–736. (In Chinese with English abstract)
17. Ministry of Water Resources, Department of Rural Water Resources. *Practical Manual of Water-Saving Irrigation Engineering*. China Water Power Press: Beijing, China, 2005. (In Chinese)
18. Linsley, R.K.; Kohler, M.R. Predicting the runoff from storm rainfall. *US Weather Res. Pap.* **1951**, *30*. Available online: https://www.nrc.gov/docs/ML0819/ML081900279.pdf (accessed on 10 December 2025).
19. Allen, R.G.; Pereira, L.S.; Raes, D.; Smith, M. Crop evapotranspiration—Guidelines for computing crop water requirements—FAO Irrigation and drainage paper 56. *FAO Rome* **1998**, *300*, D05109.
20. Pisner, D.A.; Schnyer, D.M. Support vector machine. In *Machine Learning*; Academic Press: Cambridge, MA, USA, 2020; pp. 101–121.
21. Tao, F.; Zhang, Z. Climate change, high-temperature stress, rice productivity, and water use in Eastern China: A new superensemble-based probabilistic projection. *J. Appl. Meteorol. Climatol.* **2013**, *52*, 531–551. [CrossRef]
22. Barati, M.K.; Manivasagam, V.S.; Nikoo, M.R.; Saravanane, P.; Narayanan, A.; Manalil, S. Rainfall variability and rice sustainability: An evaluation study of two distinct rice-growing ecosystems. *Land* **2022**, *11*, 1242. [CrossRef]
23. Dong, Q.; Wang, W.; Kunkel, K.E.; Shao, Q.; Xing, W.; Wei, J. Heterogeneous response of global precipitation concentration to global warming. *Int. J. Climatol.* **2021**, *41*, E2347–E2359. [CrossRef]
24. Wei, K.; Jiao, Y.; Zhang, G.; Wang, Y.; Zhang, H. Influence of Spring Water Residence Time on the Irrigation Water Stability in the Hani Rice Terraces. *Water* **2024**, *16*, 804. [CrossRef]
25. Belder, P.; Bouman, B.A.M.; Spiertz, J.H.J. Exploring options for water savings in lowland rice using a modelling approach. *Agric. Syst.* **2007**, *92*, 91–114. [CrossRef]
26. Ekanayake, E.M.P.; Wickramasinghe, L.C.D.; Weliwatta, R.T. Use of regression techniques for rice yield estimation in the North-Western province of Sri Lanka. *Ceylon J. Sci.* **2021**, *50*, 4. [CrossRef]
27. Ahmed, M.S.; Tazwar, M.T.; Khan, H.; Roy, S.; Iqbal, J.; Rabiul Alam, M.G.; Hassan, M.M. Yield Response of Different Rice Ecotypes to Meteorological, Agro-Chemical, and Soil Physiographic Factors for Interpretable Precision Agriculture Using Extreme Gradient Boosting and Support Vector Regression. *Complexity* **2022**, *2022*, 5305353. [CrossRef]
28. El Hafyani, M.; El Himdi, K.; El Adlouni, S.E. Improving monthly precipitation prediction accuracy using machine learning models: A multi-view stacking learning technique. *Front. Water* **2024**, *6*, 1378598. [CrossRef]

water

MDPI

Article

Investigation into the Working Behavior of Geotextile Pipe-Bag Systems on Soft Soil Foundations in the Ningde Port Industrial Zone, China

Peijun Fan [1], Honglei Ren [2,*], Xiatao Zhang [1], Wei Li [2] and Wanli Guo [2]

[1] Shandong Electric Power Engineering Consulting Institute Co., Ltd., Jinan 250013, China; 18832073693@163.com (P.F.)
[2] Geotechnical Engineering Department, Nanjing Hydraulic Research Institute, Nanjing 210024, China
* Correspondence: renhl_1201@163.com; Tel.: +86-198-3102-5890

Abstract

With the rapid development of coastal and nearshore engineering projects in China, geotextile pipe and bag (GPB) structures have been increasingly applied in marine land reclamation and coastal protection works. To better understand the mechanical behavior of GPB structures on soft soil foundations, this study conducts a systematic investigation into the mechanical properties of both soft soils and GPBs using a physical model test system. By integrating numerical simulations, the stress–deformation characteristics of GPB structures on soft soils and the evolution of pore pressure are further analyzed. The results indicate that the compression curve of soft soil exhibits significant nonlinearity, with silt showing higher apparent compressibility than silty clay. Experimental data yielded the compression coefficient λ and rebound coefficient μ for both soil types. As consolidation pressure increases, deviatoric stress in the soft soil rises notably, demonstrating typical strain-hardening behavior. Based on these findings, the critical state effective stress ratio M was determined for both soil types. The study also establishes the development laws of cohesion c and friction angle φ during soil consolidation, as well as the variation of pore water pressure under different confining pressures. Interface tests clarify the relationships between cohesion and friction angle at the interfaces between geotextile pipe bags and sand, and between adjacent pipe bag layers. Numerical simulations reveal that the reclamation construction process significantly influences structural horizontal displacement. Significant stress concentration occurs at the toe of the slope, while the central portion of the pipe-bag structure experiences maximum tensile stress—still within the material's allowable stress limit. The installation of drainage boards effectively accelerates pore pressure dissipation, achieving nearly complete consolidation within one year after construction. This research provides a scientific foundation and practical engineering guidance for assessing the overall stability and safety of (GPB) structures on soft soil foundations in coastal regions.

Keywords: coastal and offshore engineering; soft soil foundation; geotextile pipe-bag structure; mechanical properties of soft soil; work behavior

1. Introduction

As a core zone of national economic development and urbanization, China's coastal regions have faced growing dual pressures from land scarcity and ecological conservation in recent years. In major coastal economic belts—including the Yangtze River Delta, Pearl River Delta, and Bohai Rim—intensive construction of industrial parks, port hubs, and

 https://doi.org/10.3390/w17213063

emerging urban areas demands substantial land resources. However, terrestrial space is increasingly saturated, making sea reclamation [1] and tidal flat development [2] critical strategies for alleviating land-use conflicts. These regions are widely underlain by thick deposits of recently formed soft soil, characterized by high water content, low shear strength, high compressibility, and pronounced rheological behavior. Furthermore, these soils are subject to seawater erosion, salinization, and cyclic tidal loading, resulting in highly complex engineering geological conditions. In some offshore reclamation projects, fill construction is carried out directly on untreated soft foundations, often leading to foundation failure. Consequently, accurately predicting soft soil consolidation deformation and consolidation time, while enhancing foundation bearing capacity and effectively minimizing post-construction settlement within a short timeframe, has become an urgent technical challenge in coastal reclamation engineering. For such projects, embankment structures serve as a key technological component—forming the foundational element of project implementation and directly influencing land formation and the safety of subsequent construction activities. Achieving economical, rapid, and safe embankment construction to support efficient hydraulic filling operations continues to face numerous "bottleneck" technical challenges. Therefore, developing new types of cost-effective and structurally reliable embankment systems is essential for advancing sustainable sea reclamation and coastal infrastructure development in China.

Geotextile pipe bag (GPB) technology was first developed and applied in Europe and is widely used in marine engineering structures and shoreline protection projects, such as scour and erosion control, coastal defense systems, and underwater constructions [3,4]. In environmentally sensitive areas—such as wetland restoration, tidal flat management, and shoreline protection—construction techniques are subject to stringent ecological impact requirements abroad, with emphasis on minimizing environmental disturbance during construction. Due to its ease of construction and low environmental footprint, GPB technology is well suited to meet the demands of such projects. The GPB structure [5–7] is an embankment system formed by sewing geotextile materials [8,9] into bag-shaped containers, filling them with slurry or mortar via high-pressure pumps, and stacking the dewatered units in successive layers. In practice, most GPB structures are constructed on soft soil foundations. Although these structures exhibit a certain capacity to accommodate foundation deformation, numerous instances of instability and failure have been reported in engineering practice [10,11], indicating that systematic and in-depth understanding of their working behavior and the interaction mechanism with soft soil foundations remains insufficient.

Due to the deep soft soil layers and low bearing capacity in coastal areas, ground improvement of soft soil foundations is a prerequisite for constructing superstructures. The investigation of the mechanical properties and constitutive models of soft soils constitutes one of the core topics in geotechnical engineering. Soft soils are characterized by high water content, high void ratio, low permeability, and low strength, exhibiting significant nonlinearity, elastoplasticity, rheological behavior, and anisotropy [12–14]. Under loading, soft soils are prone to large deformations and long-term settlements, with their strength evolving progressively during consolidation, leading to complex mechanical responses that are strongly time-dependent [15–17]. To address these characteristics, researchers have developed various constitutive models to more accurately capture the stress–strain behavior of soft soils. Early models such as the linear elastic model [18,19] and the Mohr–Coulomb model [20,21] have been widely used in engineering practice due to their simplicity and computational efficiency, yet they fail to adequately represent plastic deformation and dilatancy in soft soils. The Cam-Clay model [22,23], developed subsequently, incorporated critical state theory and successfully described the compressive hardening and contractive shear behavior of soft soils, establishing itself as a classical elastoplastic constitutive model.

Continuous refinement of these models has not only enhanced the understanding of soft soil mechanics but also provided a robust theoretical foundation for soft soil foundation reinforcement, subgrade engineering design, and settlement prediction.

In recent years, scholars such as Zhou et al. [24], Zeng et al. [25], Lan et al. [26] and Singh et al. [27] have conducted systematic research on soft soil foundation treatment techniques and comprehensively summarized the advantages and disadvantages of various methods. These research outcomes have played a significant role in geotechnical engineering and provided robust theoretical support for engineering practices involving soft soil foundations. Overall, a wide range of treatment methods are currently available, primarily including dynamic compaction [28,29], drainage consolidation [30,31], and chemical solidification [32]. In practice, the selection of a single method or a combination of multiple methods is typically based on the physical and mechanical properties of the foundation. A comparative analysis of these technologies indicates that the drainage consolidation method has become the preferred approach for soft soil foundation treatment due to its cost-effectiveness and relatively short construction duration.

As shown above, there is currently a lack of systematic research on the structural behavior of geotextile pipe bags (GPBs) in soft soil foundations in coastal areas. To address this gap, this study innovatively employs an integrated approach combining field monitoring, physical experiments, and numerical simulation to investigate the overall deformation and stability of GPB structures in coastal and nearshore engineering applications. First, a comprehensive analysis was conducted on the physical and mechanical properties of both soft soils and geotextiles in coastal environments. These findings were then integrated with field monitoring data and numerical simulations to systematically examine the operational characteristics of GPB structures on soft soil foundations. The study aims to provide a robust theoretical foundation and technical support for evaluating the stability of soft soil foundation treatment techniques and GPB structural systems.

2. Mechanical Properties of Soft Soil and Geotextiles

2.1. Project Overview

The authors express their gratitude for your insightful suggestions. Based on the construction of a GPB structure on soft soil in the coastal area of Ningde, China, this study examines the engineering characteristics of soft soil, with particular emphasis on the deformation behavior of the soft soil foundation. To this end, a series of experimental investigations were conducted to determine the physical and mechanical properties and key calculation parameters of the soft soil. Specifically, compressive consolidation tests, interface friction tests, direct shear tests, triaxial shear tests, and on-site in situ strength tests were performed on foundation soft soil samples to evaluate critical geotechnical indices, including the compression index, rebound index, shear strength, and stress–strain characteristics. The Ningde Coastal Industrial Zone land reclamation project is situated on the northern side of Niuyu Island. The project is primarily divided into two zones: the startup base and the mining construction materials area. The startup base covers a sea area of 495,200 m^2 and a reclaimed land area of 440,400 m^2, with a maximum east–west extent of approximately 1150 m and a maximum north–south width of about 810 m. The mining construction materials area utilizes a sea area of 462,700 m^2, forming a reclaimed land area of approximately 406,500 m^2, with a maximum east–west length of around 650 m and a maximum north–south width of about 900 m. The total revetment length is 1239 m. The layout of the Ningde Land Reclamation Project is illustrated in Figure 1.

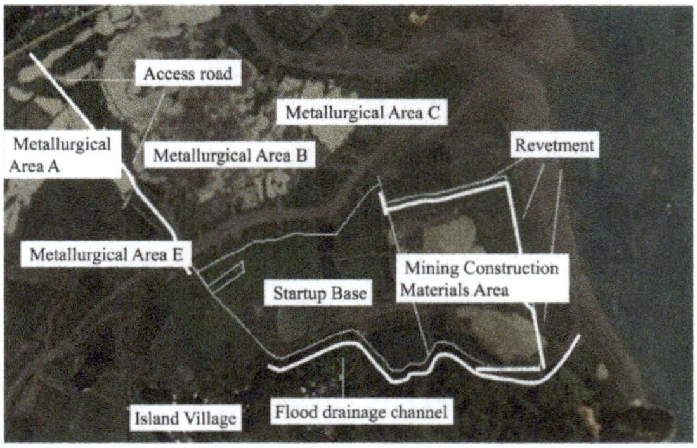

Figure 1. Layout plan of Ningde reclamation project.

To study the deformation and stability characteristics of the upper cofferdam—GPB structure, a monitoring and analysis of the entire construction process of the GPB loading was carried out to provide guidance for on-site construction. This project has a large site area and inconvenient transportation. The monitoring instruments and equipment as well as materials could only be transported to the installation points by manual multiple times. The drilling pile machine was first disassembled, then transported to the installation location by manual, and then reassembled. During high tide, the pile was moved to the designated location according to the design for the settlement plate installation, so as to obtain the initial values in time and ensure the normal operation of the instruments. When the monitoring values exceeded the warning value, an early warning was promptly reported. Regardless of day or night, frequent observations were made and patrols were strengthened to ensure the safety and stability of the GPB structure.

Taking into account the on-site terrain and soil conditions, a total of 13 settlement points (displacement meters), 3 full-section settlement observation points (3 layers of settlement tubes for each section), and 6 pore pressure observation points (each point buried 6 pore pressure meters) were arranged. The layout of the section instruments is shown in Figure 2.

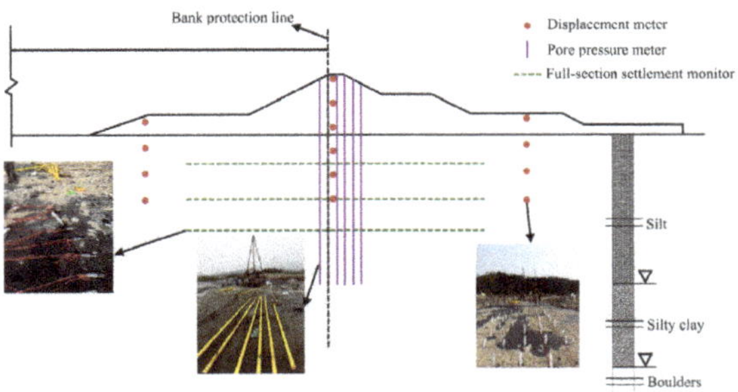

Figure 2. On-site monitoring instrument layout diagram.

2.2. Test Preparation

It is widely accepted that the modified Cam-Clay model is suitable for characterizing the mechanical behavior of soft soils. The model comprises three material parameters and one state parameter: the critical state effective stress ratio M, the compression index λ, the swelling index κ, and the initial void ratio e_0. Among these, M can be determined from consolidated undrained triaxial tests, while λ and κ are typically derived from one-dimensional consolidation-rebound tests. The main physical property indices of the soft soil are summarized in Table 1.

Table 1. The basic mechanical parameters of soft soil.

Soft Soil	Moisture Content w (%)	Unit Weight γ (kN·m^{-3})	Void Ratio e	Liquid Limit W_L (%)	Plastic Limit W_P (%)
Silt	49.3	17.4	0.941	41.9	23.6
Silty clay	41.9	18.3	0.787	40.0	22.7

For the standard consolidation-rebound test, the standard specimen, together with the cutting ring, is placed into the consolidometer and subjected to loading and unloading. Loading is applied incrementally in the sequence of 12.5, 25, 50, 100, 200, 400, and 800 kPa. After reaching 800 kPa, unloading is performed in the order of 400 and 200 kPa back to the first load level. The water reservoir should be immediately filled upon application of the first pressure level. The stability criterion at each load stage is defined as a deformation rate not exceeding 0.01 mm per hour. Subsequently, three additional tests were conducted with loading sequences of 20, 40, 80, 150, 200, 320, and 640 kPa, with each load level maintained for 48 h of consolidation.

For the consolidated undrained (CU) triaxial test, two types of soft soil collected from the field were prepared into cylindrical specimens with a diameter of 39.1 mm and a height of 80 mm. Confining pressures were set at 50, 100, 150, 200, and 250 kPa, respectively. The tests were conducted using an automatic triaxial apparatus, with the shear strain rate controlled at 0.02 mm/min. Each test series was repeated three times to ensure data reliability.

In the direct shear friction test, the GPB remains relatively stationary while the sand body undergoes displacement. The GPB experiences passive resistance, resulting in single-sided friction between the bag and the sand. Shear stress develops along the interface where relative sliding occurs. In contrast, during the pull-out friction test, the geogrid is actively pulled out under an applied tensile force, leading to double-sided friction between the geotextile bag and the surrounding soil. Shear stress is distributed across both the upper and lower surfaces of the bag, and the normal stress is determined based on the vertical load applied in the test. Considering the actual stress mechanism, this study adopts the direct shear friction test method to investigate the interface friction behavior between geotextile bags and sand.

2.3. Mechanical Properties of Soft Soil

Through the one-dimensional consolidation-rebound test, the e-ln$p$ curves of the two soil layers under different vertical loads were obtained. As shown in Figure 3, the compression curve of soft soil exhibits distinct nonlinear characteristics, and the apparent compressibility of silt is higher than that of silty clay. Based on the slopes of the compression and rebound curves from the one-dimensional consolidation-rebound test data plotted in the e-ln$p$ plane, the compression indices λ and rebound indices κ for silt and silty clay were determined as $\lambda_1 = 0.0646$, $\kappa_1 = 0.0117$, $\lambda_2 = 0.0775$, and $\kappa_2 = 0.0105$, respectively.

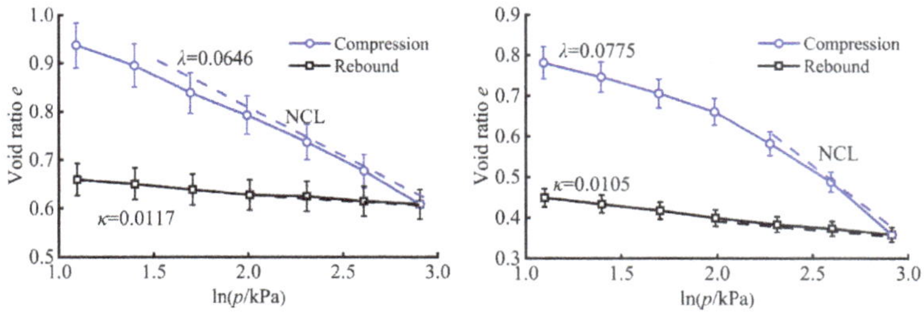

(**a**) (**b**)

Figure 3. Compressive rebound curves of soft soil. (**a**) silt. (**b**) silty clay.

Three groups of specimens were subjected to consolidated undrained (CU) triaxial tests under confining pressures of 50, 100, 150, 200, and 250 kPa, respectively. The measured relationship between axial strain and deviatoric stress is presented in Figure 4. Results indicate that both silty soil and silty clay exhibit strain-hardening behavior. For silty soil, the deviatoric stress increases continuously with increasing axial strain. For silty clay, the deviatoric stress rises significantly when axial strain is below 3%; however, beyond 3% axial strain, the rate of increase gradually diminishes. In this study, the deviatoric stress at 15% axial strain is adopted as the failure strength, denoted as q. The parameter M represents the ratio of deviatoric stress q to mean effective stress p along the critical state line, i.e., $M = q/p$. Linear regression of the p and q data points under different confining pressures was performed in the p-q plane (Figure 5), yielding critical state stress ratios of M_1 = 0.962 for silty clay and M_2 = 0.499 for silt.

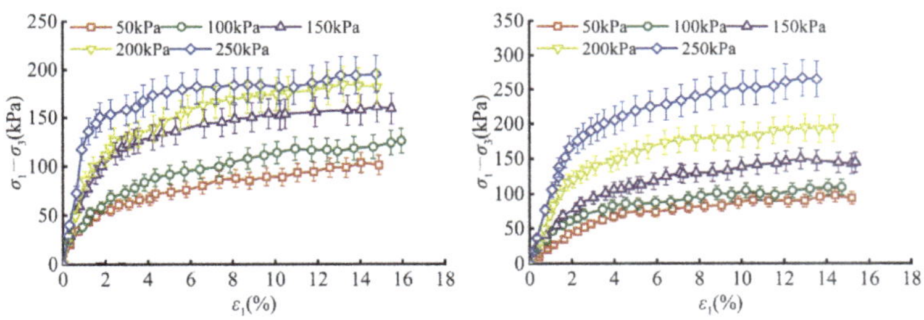

(**a**) (**b**)

Figure 4. Soft soil undrained consolidation stress–strain curve. (**a**) Silt. (**b**) Silty clay.

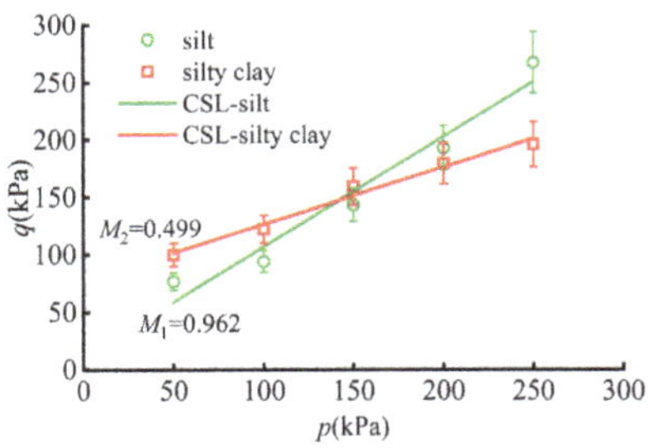

Figure 5. Linear fitting of effective stress ratio M in critical state.

Figure 6 presents the variation of the shear strength parameters of soft soil with initial consolidation degree under different consolidation pressures. The cohesion exhibits a relatively large fluctuation range; when the consolidation degree reaches 100%, its maximum value is approximately three times the minimum value. In contrast, although the friction angle φ varies less than cohesion c, its maximum value still approaches twice the minimum value at 100% consolidation degree. Overall, both c and φ increase with increasing confining pressure and consolidation degree. This trend is primarily attributed to the progressive densification of soil during the drainage consolidation process, wherein interparticle distances decrease and pore volume diminishes, leading to enhanced overall shear strength of the soil mass.

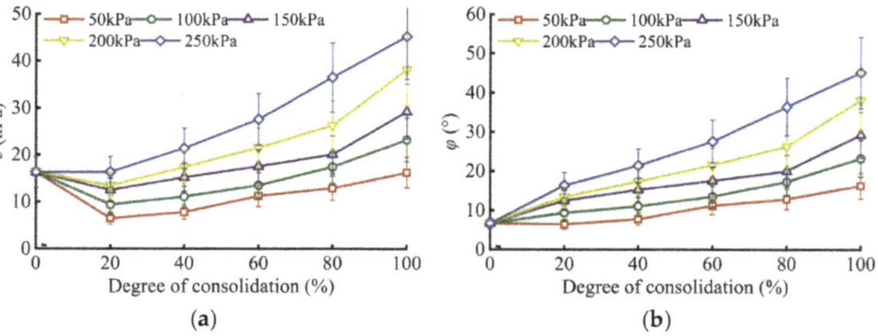

Figure 6. The evolution law of strength parameters of soft soil with the degree of consolidation. (**a**) c. (**b**) φ.

Figure 7 presents the relationship between pore water pressure and axial strain for silt and silty clay under different consolidation pressures. It can be observed that at small axial strains, the pore pressure increases rapidly; as the strain further increases, the rate of increase gradually slows down and eventually reaches a stable value. Meanwhile, pore pressure rises significantly with increasing consolidation pressure. However, with higher consolidation pressure, the ratio of maximum pore pressure to total stress decreases from 36.9% to 27.5% for silt and from 45.5% to 17.8% for silty clay. Both ratios exhibit a decreasing trend with increasing consolidation pressure, indicating that under a consolidation pressure of 250 kPa, the second layer of silty clay possesses a greater capacity to resist deviatoric stress compared to the first layer of silt. This difference in excess pore pressure development may be a primary factor contributing to its ability to sustain higher deviatoric stresses.

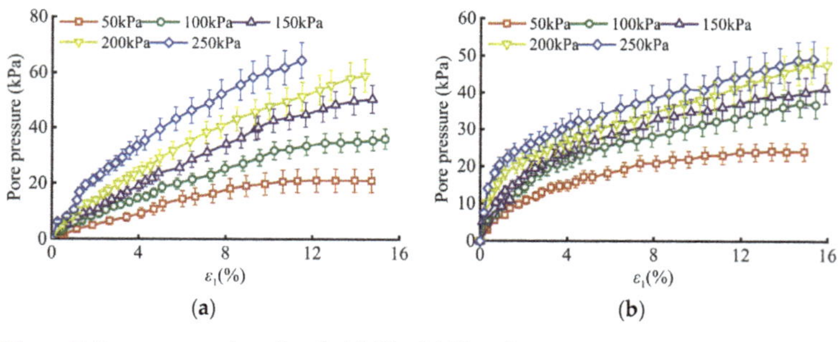

Figure 7. Pore pressure in soft soil. (**a**) Silt. (**b**) Silty clay.

2.4. The Mechanical Properties of Geotextiles

Through interface friction tests on GPBs, the relationship between shear stress and shear displacement was obtained. Figure 8a presents the test results for the bag–sand interface. It can be observed that as shear displacement increases, the tangential stress at the interface gradually rises. The rate of increase is relatively high during the initial stage, indicating rapid development of interfacial friction. After reaching peak stress, the growth rate decreases, exhibiting pronounced nonlinear behavior that can be well described by a hyperbolic function. Under higher vertical stress conditions, a slight softening phenomenon occurs, primarily during the later stages of shearing. Figure 8b shows the results for the bag–bag interface, where the stress–displacement curve also displays significant nonlinearity. With increasing vertical stress, the peak shear stress increases, while the corresponding peak shear displacement remains largely unchanged. Moreover, the degree of stress softening is notably greater than that observed in the bag–sand group.

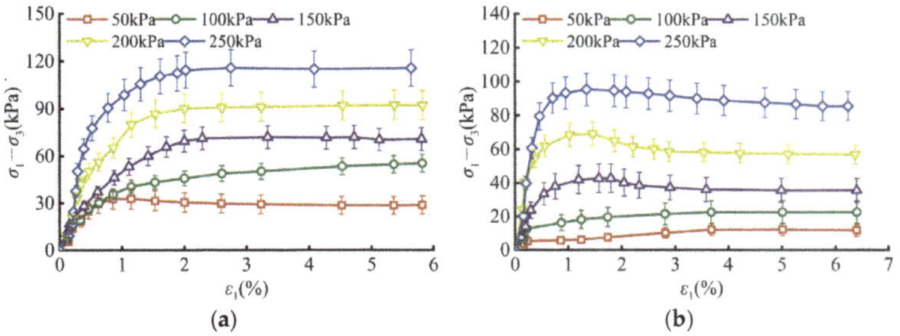

Figure 8. The relationship curve between the shear stress and shear displacement of the geotextile bag. (**a**) Bag–soil. (**b**) Bag–bag.

Figure 9 illustrates the relationship between the shear strength of geotextile bags and vertical pressure. The results show that the peak shear strength τ of the bag–sand interface exhibits a strong linear correlation with normal stress σ_n, consistent with the Mohr–Coulomb strength criterion: $\tau = c_{gs} + \sigma_n \tan\varphi_{gs}$. The fitted interface cohesion c_{gs} is 6.90 kPa, and the friction angle φ_{gs} is 27.3°. For the bag–bag interface, the shear strength also follows this linear model, but with no cohesive component, and a friction angle φ_{gs} of 21.6°.

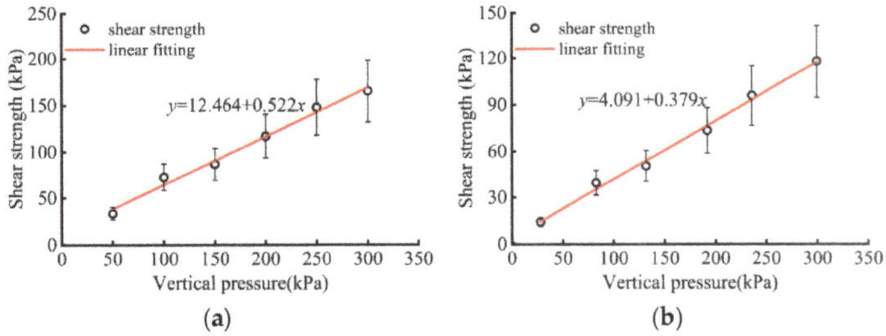

Figure 9. The relationship curve between the shear strength of geotextile bags and vertical pressure. (**a**) c. (**b**) φ.

3. Finite Element Modeling of Geotextile Pipe-Bag Structures

3.1. Constitutive Models

The Mohr–Coulomb model is widely used in finite element analysis. In an ideal elastoplastic material, only elastic deformation occurs before the stress reaches the yield point. Once yielding begins, irreversible plastic deformation takes place and continues until material failure. The yield surface morphology is shown in Figure 10.

The yield surface function of the Mohr–Coulomb model is defined by Equation (1):

$$f = \frac{1}{2}(\sigma_1 - \sigma_3) - \frac{1}{2}(\sigma_1 + \sigma_3)\sin\varphi - c\cos\varphi \tag{1}$$

where c is the cohesion of the material, while φ denotes the inclination angle of the Mohr–Coulomb yield surface, which corresponds to the internal friction angle.

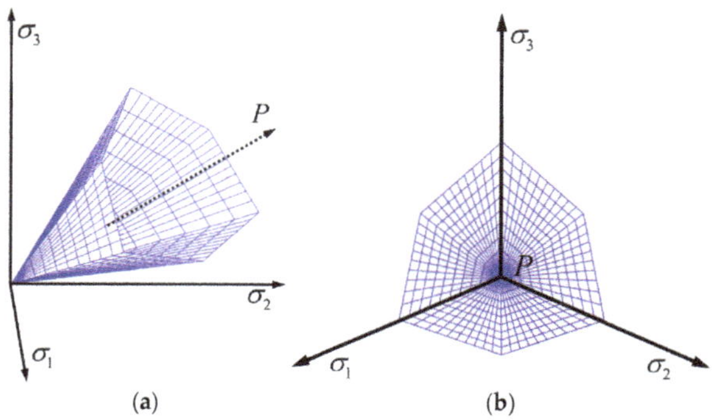

Figure 10. Mohr–Coulomb model. (**a**) Principal stress space. (**b**) π plane.

The Cam-clay model is an elastoplastic constitutive model developed by Roscoe and colleagues at the University of Cambridge, UK. Based on experimental observations, it employs an elliptical yield surface and an associated flow rule (Figure 11). The model features a simple set of parameters that are relatively easy to determine, contributing to its widespread application in geotechnical engineering both domestically and internationally. It is particularly applicable to normally consolidated soils and lightly overconsolidated clays. The mathematical expression of its yield surface is:

$$f = \frac{\lambda - \kappa}{1 + e_0} \ln \frac{p}{p_0} + \frac{\lambda - \kappa}{1 + e_0} \frac{1}{M} \frac{q}{p} - \varepsilon_v^p = 0 \tag{2}$$

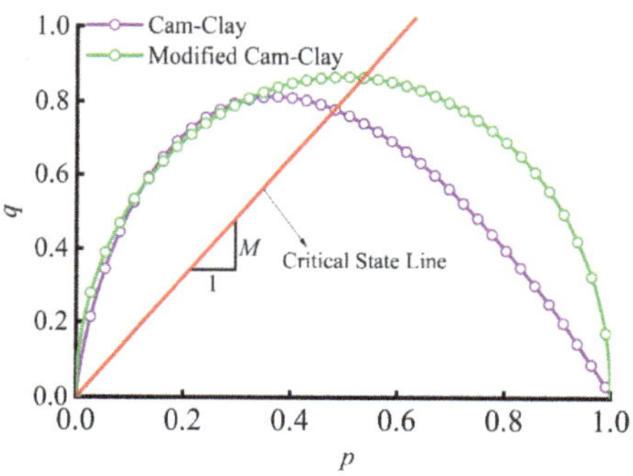

Figure 11. Cam-clay model and modified Cam-clay model.

Subsequently, modifications were introduced to the yield surface, leading to the development of the modified Cam-clay:

$$f = \frac{\lambda - \kappa}{1 + e_0} \ln \frac{p}{p_0} + \frac{\lambda - \kappa}{1 + e_0} \ln\left(1 + \frac{q^2}{M^2 p^2}\right) - \varepsilon_v^p = 0 \tag{3}$$

where M denotes the critical state stress ratio, and e_0 represents the initial void ratio, defined as the void ratio of soft soil corresponding to an initial mean effective stress p_0 = 0. λ is the slope of the normal compression line (NCL), while κ is the slope of the unloading-reloading line and serves as the hardening parameter, representing the plastic volumetric strain. The simulation results presented in Figure 12 demonstrate that the modified Cam-Clay model more accurately captures the mechanical behavior of soft soil than the original Cam-Clay model.

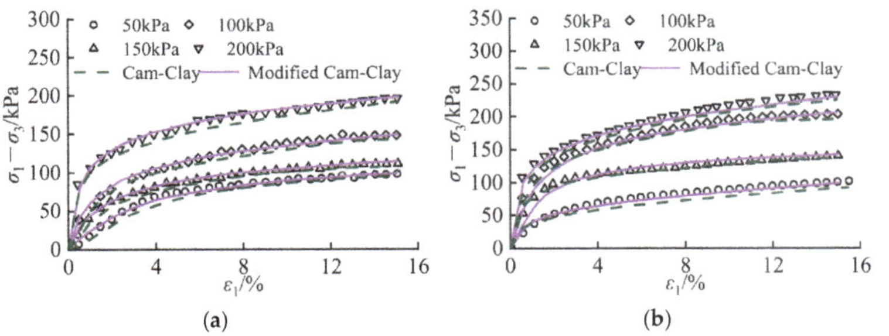

Figure 12. Comparison of simulation results between the Cam-Clay model and the modified Cam-Clay model for the CU triaxial test. (**a**) Silt. (**b**) Silty clay.

3.2. Parameters of Constitutive Models

The Mohr–Coulomb model is widely used in finite element analysis. In an ideal elastoplastic material, only elastic deformation occurs before the stress reaches the yield point. Once yielding begins, irreversible plastic deformation takes place and continues until material failure. The modified Cam-Clay model was adopted to simulate the soft soil foundation, while the Mohr–Coulomb model was used for the fill sand material. The parameters for the fill sand are listed in Table 2, and those for the soft soil are provided in Table 3. Geotextile bags were modeled as membrane elements. The interactions between geotextile bags, as well as between geotextile bags and the soft soil foundation, were simulated using surface-to-surface contact pairs. Based on interface friction tests, the cohesion c at the bag–sand interface is 6.89 kPa, with a friction coefficient μ of 0.51; the friction coefficient μ at the bag–bag interface is 0.39; and the friction coefficient μ at the bag–soft soil foundation interface is 0.27.

Table 2. Physical and mechanical index.

	γ/(kN/m)	c/(kPa)	φ/(°)	E/(MPa)	v	Permeability Coefficient k/(m/s)
Silt	19	0	32	30	0.3	6.0×10^{-5}
Silty clay			18	500	0.3	2.0×10^{-4}

Table 3. Parameters of soil layers.

Soil	v	λ	κ	M	e_0	Permeability Coefficient k/(m/s)
dredger fill	0.43	0.0593	0.0214	0.543	1.125	2.6×10^{-5}
Silt	0.41	0.0646	0.0117	0.962	0.941	1.9×10^{-9}
Silty clay	0.33	0.0775	0.0105	0.499	0.787	5.0×10^{-8}

3.3. FEM Model

This study employs ABAQUS 6.14 software to establish a three-dimensional finite element model consisting of 258,603 nodes and 235,969 elements. GPBs are modeled as reinforcement within the sandy embankment fill. Although numerous studies have investigated the soil-reinforcement interface interaction, the multiplicity of influencing factors has prevented the establishment of a universally accepted simulation approach. Given that geotextile bags can resist tensile forces but not compressive ones, the membrane element (M3D4R) available in ABAQUS is used to simulate their reinforcing behavior. Interactions between bags and between the bags and the underlying soft soil foundation are all modeled using surface-to-surface contact formulations. The overall soil domain is discretized using C3D8P pore-pressure solid elements, while the geotextile bags are represented by M3D4R membrane elements. The foundation comprises 28 m of soft soil overlying 12 m of bedrock. Plastic drainage boards are incorporated into the soft soil layer with a penetration depth of 28 m. At the base of the cofferdam, a 1 m thick medium-coarse sand cushion is implemented, constructed in two 0.5 m layers. A total of 13 layers of geotextile bags are simulated, each with a thickness of 0.5 m. The interface between the foundation and the lowest GPB is defined as a zero pore-pressure boundary. To minimize boundary effects on displacement in the reinforced zone, lateral boundaries of the model were extended sufficiently during computation. Three-directional displacement constraints are applied at the base of the bedrock, and a staged loading procedure is adopted to simulate the sequential filling process at the construction site. The detailed configuration of the numerical model is presented in Figure 13.

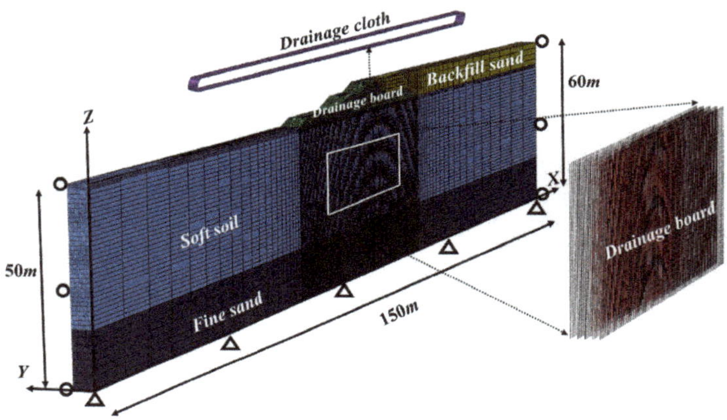

Figure 13. Finite element model of geotechnical bag structure on soft soil foundation.

4. Working Behavior of Geotextile Pipe-Bag Structures

4.1. Displacement Deformation Characteristics

Figure 14 presents the settlement distribution contour map of the geotextile pipe-bag structure. It can be observed that during the filling stage, deformation of the cofferdam and foundation is dominated by vertical settlement, with a maximum value of approximately 0.13 m. Slight uplift occurs at the toe of the cofferdam slope, while a "U"-shaped settlement profile develops at the structural center. Concurrently, the settlement at the slope toe exhibits an outward-diffusing pattern. After the installation of the geotextile tube bags (GPBs), a 30-day construction break is implemented, during which the structure undergoes static drainage consolidation. Under the influence of the plastic drainage boards, the extent of the high-settlement zone slightly decreases, with the maximum settlement occurring near the central axis of the cofferdam. The bottom of the GPB structure was selected as the monitoring location. Upon completion of filling and commencement of construction of the mortared stone retaining wall and surface and base protection structures, the settlement increased to approximately 1.26 m. Following the completion of land reclamation via sand filling, the settlement further increased to 1.71 m. After construction ended, the structure was continuously monitored for one year. Calculations indicate that the post-construction settlement after one year reached 2.43 m. The computed settlement values show good agreement with the field monitoring data.

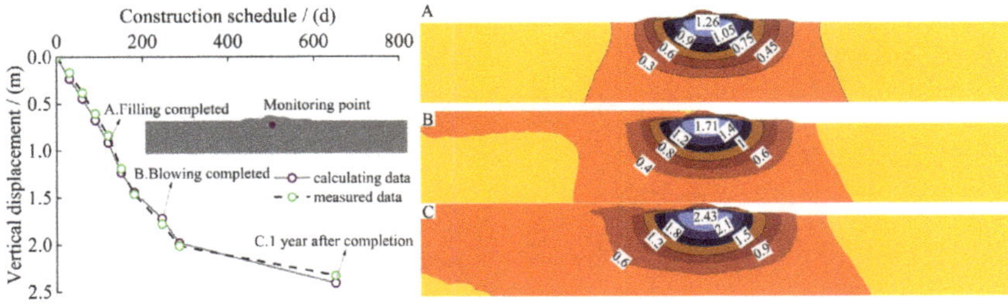

Figure 14. The settlement of the geotextile bag structure during the construction period.

Figure 15 presents the horizontal displacement contour map of the GPB structure. It can be observed that when the filling elevation reaches 0.5 m, the maximum horizontal displacement is approximately 0.027 m, with a nearly symmetrical distribution on both sides, indicating a relatively stable structural behavior. During the construction interval, the drainage plates gradually exert their drainage and consolidation function, leading to a slight reduction in horizontal displacement; however, this reduction is considerably smaller than that observed in vertical displacement. During the reclamation process, horizontal displacement begins to increase at a relatively rapid rate. This phenomenon is primarily attributed to the low permeability of the foundation silt, short construction intervals, and the ongoing consolidation of the foundation, which collectively induce a pronounced sudden increase in deformation during continued filling of the enclosure structure. Upon completion of all land reclamation by sand filling, the horizontal displacement reaches 0.337 m. One year after the completion of reclamation, the horizontal displacement slowly increases to 0.433 m, with the computed results showing good agreement with the field measurement data.

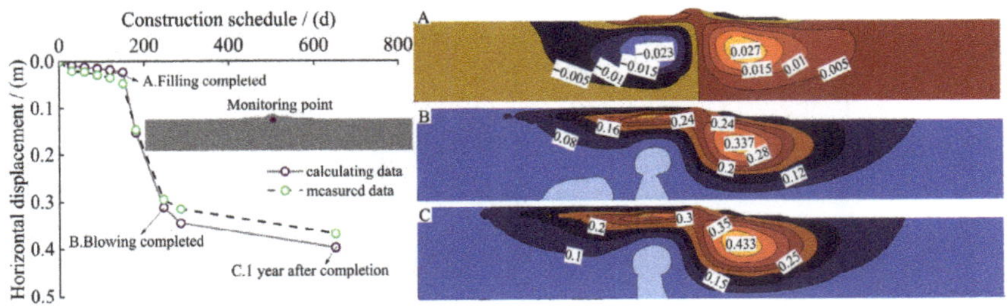

Figure 15. The horizontal displacement of the geotextile bag structure during the construction period.

Figure 16 presents the consolidation settlement curves at the left side, center point, and right side of the GPB structure base. As depicted in Figure 16, the settlement curve at the left-side measurement point of the soft soil foundation exhibits an upward trend for a period after 180 days. This phenomenon is attributed to lateral earth pressure inducing compression that causes uplift of the settlement plate. At the same time points, the settlement values at all measurement points are reduced compared to earlier stages, whereas the final settlement at the center point of the enclosure structure base remains essentially unchanged relative to the original value. These observations indicate that the installation of plastic drainage boards can effectively accelerate the drainage rate of the foundation, facilitate earlier completion of drainage consolidation in soft soil foundations, and thereby significantly shorten the construction duration of the upper GPB structure.

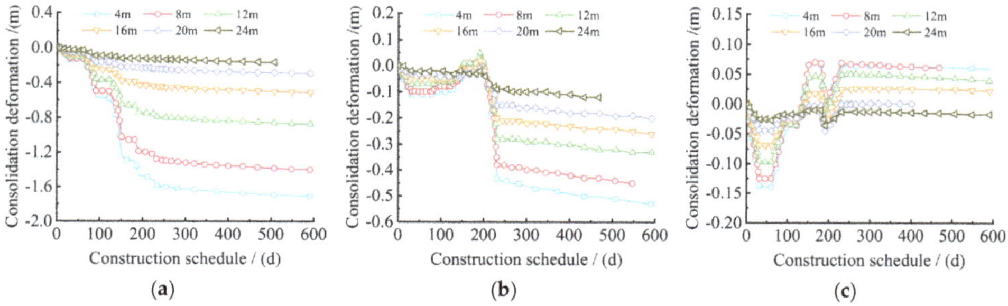

Figure 16. The settlement curve of the foundation consolidation for the geotextile bag structure. (**a**) Left. (**b**) Middle. (**c**) Right.

4.2. Stress Characteristics

Figure 17 presents the first principal stress cloud diagram of the GPB structure. It can be observed that during the filling stage, the overall structure is predominantly under compressive stress, with stress at the slope toe increasing gradually. By the 150th day of construction, the main structural filling is essentially complete, and the compressive stress distribution exhibits an upward trend. This phenomenon may be attributed to localized stress disturbance induced by drainage from the installed drainage boards. Upon completion of reclamation, the overall compressive stress in the structure decreases, while significant tensile stress concentration develops at the slope toe, with maximum tensile and compressive stresses reaching 0.59 MPa and 3.23 MPa, respectively. One year after construction completion, as excess pore water pressure dissipates, the magnitude of tensile stress gradually diminishes.

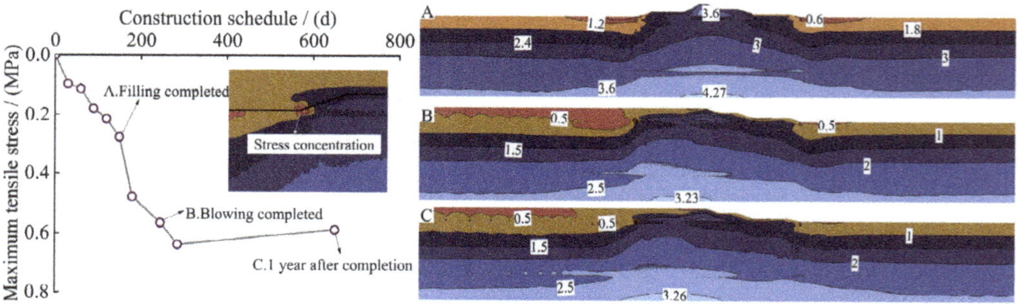

Figure 17. The tensile stress of the geotextile bag structure during the construction period.

Figure 18 presents the third principal stress cloud diagram of the GPB structure. Overall, during the filling process, the third principal stress is predominantly compressive. When structural filling is completed, stress concentration at the slope toe becomes clearly evident. The dissipation of excess pore water pressure leads to a reduction in the magnitude of the third principal stress within the structure. Throughout the backfilling process, the internal stress in the geotextile tube bag (GPB) remains entirely compressive. As backfilling progresses, the entire structure experiences an increasingly pronounced state of compression. After one year of pore pressure dissipation, the maximum compressive stress decreases by approximately 17%.

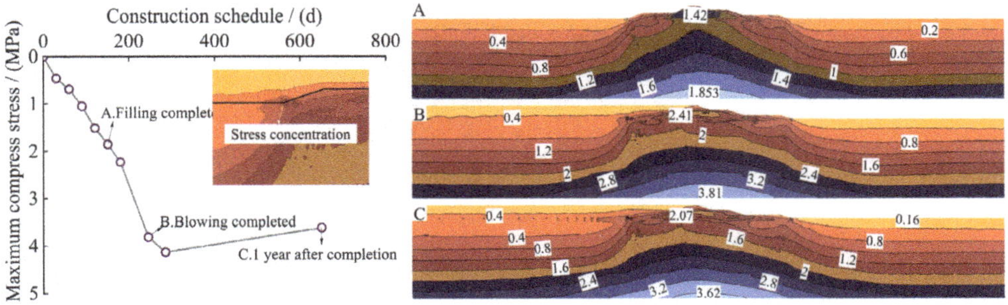

Figure 18. The compressive stress of the geotextile bag structure during the construction period.

Based on the previous force analysis of the GPB structure, it is evident that the bottom layer experiences the most significant stress. Therefore, the geotextile bag at the bottom layer is selected for further stress analysis. Figure 19 illustrates the evolution of stress distribution in the bottom geotextile pipe bag. During construction, a distinct tensile stress concentration develops at the bottom of the pipe bag. As construction progresses, both the overall tensile stress and the peak compressive stress increase progressively. One year after construction completion, the maximum tensile stress increases slightly, while the peak compressive stress decreases moderately; however, neither exceeds the allowable stress range.

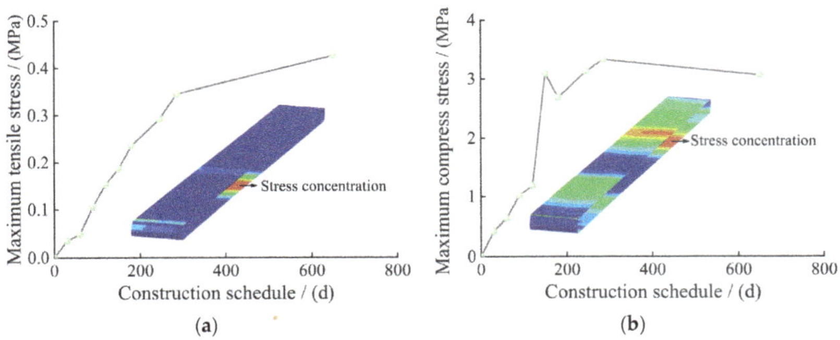

Figure 19. Evolutionary law of stress distribution of the bottom geotextile bag. (a) Tensile stress. (b) Compress stress.

Figure 20 presents the pore water pressure cloud map of the GPB structure. At 30 days into construction, the pore water pressure exhibited an upward trend. On the foundation surface near the structure, the presence of drainage boards led to a significant reduction in pore water pressure. The monitoring point was located 2 m below the base slab of the GPB structure. It can be observed that prior to completion of filling, the dissipation rate of pore pressure was considerably lower than its generation rate. During the hydraulic filling process, pore pressure began to show signs of dissipation. One year after construction completion, pore pressure dissipation became evident, and the pressure at the monitoring point turned negative, indicating that the drainage boards effectively promote the dissipation of pore pressure.

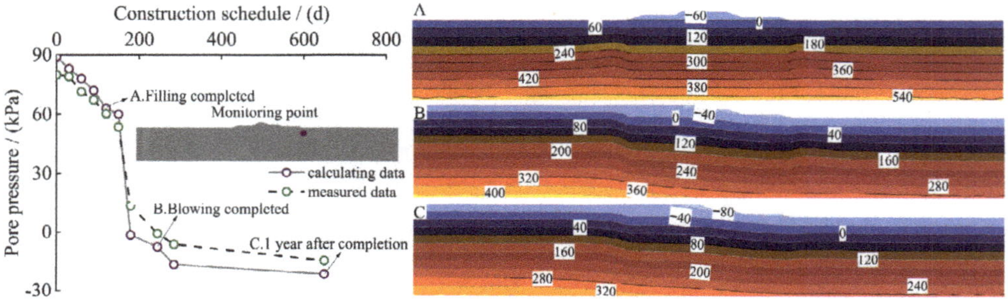

Figure 20. The pore pressure of the geotextile bag structure during the construction period.

5. Conclusions

This study innovatively integrates multiple research methodologies—namely, on-site monitoring, physical experiments, and numerical analysis—to systematically investigate the working behavior of geotextile pipe bag (GPB) structures on soft soil foundations in coastal areas. The mechanical properties of soft soil foundations were systematically evaluated through laboratory-based physical experiments. Subsequently, by integrating field monitoring data with numerical simulations, a comprehensive analysis was conducted on the operational characteristics of GPB structures under coastal soft soil conditions. The main research findings are summarized as follows:

(1) The compression curve of soft soil exhibits significant nonlinear characteristics, with the apparent compressibility of silt being higher than that of silty clay. The compression and rebound coefficients of soft soil were determined from experimental data. As consolidation pressure increases, the deviatoric stress of soft soil rises significantly, demonstrating typical strain-hardening behavior. Based on these results, the effective stress ratio M for both types of soft soil was derived. A quantitative relationship between shear strength parameters c and φ and the degree of consolidation was established, and the evolution pattern of pore pressure was summarized. Interface characteristic tests were conducted to obtain the c and φ at the interfaces between bags and sand, as well as between adjacent bags.

(2) A three-dimensional refined numerical model of the GPB structure was developed. Simulation results indicate that foundation settlement progressively increases with construction advancement. During construction breaks, the drainage boards facilitate drainage, effectively reducing the settlement rate. Horizontal displacement is highly influenced by the hydraulic filling process; when the upper fill is asymmetrically placed, uneven horizontal displacements occur across the foundation.

(3) As hydraulic filling proceeds, the overall structure gradually transitions into a more pronounced tensile state, with stress concentration at the slope toe becoming increasingly evident. One year after completion of filling, tensile stress slightly decreases. The GPB structure remains under tension throughout, with the maximum tensile stress occurring in the central region. However, all peak tensile stresses remain below the material's allowable stress limit. Drainage boards effectively accelerate pore pressure dissipation, which is nearly complete within one year after construction ends.

The findings of this study provide a scientific basis and practical engineering reference for analyzing the mechanical behavior of soft soil foundations in coastal regions and evaluating the overall stability of overlying GPB structures.

Author Contributions: Conceptualization, P.F. and H.R.; methodology, X.Z.; software, W.L.; validation, W.G., H.R. and P.F.; formal analysis, P.F.; investigation, X.Z.; resources, H.R.; data curation, W.L.; writing—original draft preparation, H.R. and W.G.; writing—review and editing, P.F.; visualization, H.R.; supervision, X.Z.; project administration, P.F.; funding acquisition, W.L. All authors have read and agreed to the published version of the manuscript.

Funding: This research was funded by an independent research project of Shandong Electric Power Engineering Consulting Institute Co., Ltd., grant number htps202401020; Special Fund for Public Welfare Industry Research of the Ministry of Water Resources, grant number 201401006; Key Projects of the Special Fund for Basic Research Operations of Central-Level Public Welfare Research Institutes, grant number Y315016.

Data Availability Statement: The raw data supporting the conclusions of this article will be made available by the authors on request.

Acknowledgments: The authors sincerely appreciate the anonymous reviewer and editors' valuable comments and suggestions to improve the manuscript.

Conflicts of Interest: Author Peijun Fan & Xiatao Zhang was employed by the company Shandong Electric Power Engineering Consulting Institute Co., Ltd. The remaining authors declare that the research was conducted in the absence of any commercial or financial relationships that could be construed as a potential conflict of interest. The authors declare that this study received funding from Shandong Electric Power Engineering Consulting Institute Co., Ltd. (grant number htps202401020). The funder was not involved in the study design, collection, analysis, interpretation of data, the writing of this article or the decision to submit it for publication.

References

1. Xiong, W.; Luo, Y.; Shangguan, W.G.; Deng, Y.; Xiao, R. Co-hydrothermal carbonization of lignocellulosic biomass and swine manure: Optimal parameters for enhanced nutrient reclamation, carbon sequestration, and heavy metals passivation. *Waste Manag.* **2024**, *190*, 174–185. [CrossRef]
2. Hagger, V.; Stewart-Sinclair, P.; Rossini, R.A.; Adame, M.F.; Lovelock, C.E. Lessons learned on the feasibility of coastal wetland restoration for blue carbon and co-benefits in Australia. *J. Environ. Manag.* **2024**, *369*, 122287. [CrossRef] [PubMed]
3. Liu, A.S.; Li, W.; Kwok, C.Y. Seawater effects on ground granulated blast furnace slag (GGBS)-stabilized marine deposits in land reclamation. *Acta. Geotech.* **2025**, *37*, 397–412. [CrossRef]
4. Bradford, T.E.; Lo, C.C.; Astudillo, J.C.; Leung, R.W.; Lai, C.; Minuti, J.J.; Wong, C.K. Turning riprap into reefs: Integrating oyster shells into shoreline armouring. *Mar. Pollut. Bull.* **2025**, *216*, 117933. [CrossRef] [PubMed]
5. Wei, G.; Chu, J.; Yan, S.W.; He, Z.B. Analytical solutions for geosynthetic tube resting on rigid foundation. *Geotext. Geomembr.* **2014**, *6*, 65–77.
6. Wei, G.; Chu, J.; Nie, W. Analysis of geosynthetic tubes inflated by liquid and consolidated soil. *Geotext. Geomembr.* **2014**, *42*, 277–283.
7. Wei, G.; Chu, J.; Yan, S.W. Simplified analytical solution for geosynthetic tube resting on deformable foundation soil. *Geotext. Geomembr.* **2015**, *43*, 432–439.
8. Carneiro, J.R.; Almeida, F.; Carvalho, F.; Lopes, M. Tensile and tearing properties of a geocomposite mechanically damaged by repeated loading and abrasion. *Materials* **2023**, *16*, 7047. [CrossRef]
9. Chmura, D.; Salachna, A.; Broda, J.; Katarzyna, K.M.; Gawłowski, A.; Rom, M. Multifunctional geotextiles produced from reclaimed fibres and their role in ecological engineering. *Materials* **2022**, *15*, 7957. [CrossRef]
10. Thompsona, A.; Oberhagemannb, K. Yuntong She Geobag stability for riverbank erosion protection structures: Physical model study. *Geotext. Geomembr.* **2020**, *48*, 110–119. [CrossRef]
11. Guin, S.; Bhattacherjee, D. Applicability of Geobags as a Sustainable Riverbank Protection Measure. *Indian. Geotech. J.* **2024**, *54*, 800–813. [CrossRef]
12. Ying, Z.; Cui, Y.J.; Benahmed, N.; Duc, M. Changes in microstructure and water retention property of a lime-treated saline soil during curing. *Acta Geotech.* **2022**, *17*, 319–326. [CrossRef]
13. Lei, H.Y.; Li, C.Y.; Liu, J.J. Experimental study and numerical simulation of an alternating vacuum preloading method for strengthening hydraulic fill super soft soils. *Chin. J. Rock Mech. Eng.* **2019**, *38*, 2112–2125. (In Chinese)
14. Cai, Y.Q.; Chen, Y.; Cao, Z.G.; Ren, C. A combined method to predict the long-term settlements of roads on soft soil under cyclic traffic loadings. *Acta. Geotech.* **2018**, *5*, 1215–1226. [CrossRef]

15. Jiang, X.; Huang, Z.; Ma, F.; Luo, X. Analysis of strength development and soil-water characteristics of rice husk ash-lime stabilized soft soil. *Materials* **2019**, *12*, 3873. [CrossRef]
16. Chen, L.; Lin, D.F. Stabilization treatment of soft subgrade soil by sewage sludge ash and cement. *J. Hazard. Mater.* **2009**, *162*, 321–327. [CrossRef]
17. Nanda, R.P.; Priya, N.; Kumar, A. Enhancement of strength of soft soil utilizing construction and demolition wastes reinforced with Recron-3s fibres. *J. Clean. Prod.* **2025**, *505*, 145442. [CrossRef]
18. Liu, R.; Yan, H.; Liang, C. A unified lateral soil reaction model for monopiles in soft clay. *Ocean Eng.* **2025**, *319*, 120268. [CrossRef]
19. Yang, Y.; Zhou, X.; Zhou, M.; Zhang, X. The behavior of ball penetrometer in soft-over-stiff soil deposits. *Ocean Eng.* **2023**, *273*, 114011. [CrossRef]
20. Ye, R.R.; Huang, Z.Y.; Yang, Z.X.; Guo, N.; Jardine, R.; Fu, S. Large deformation analysis of intermittent pile penetration into dense sand incorporating a state-dependent Mohr–Coulomb model. *Can. Geotech. J.* **2025**, *62*, 463–472. [CrossRef]
21. Wang, X.; Han, Y.H.; Chen, S.L.; Abousleiman, Y. An analytical drained solution based on graphical method for wellbore drilling problem in dilatant Mohr-Coulomb rock formations. *Int. J. Rock Mech. Min.* **2025**, *193*, 106170. [CrossRef]
22. Moeineddin, A.; Platen, J.; Kaliske, M. Constitutive description of snow at finite strains by the modified cam-clay model and an implicit gradient damage formulation. *Int. J. Numer. Meth. Eng.* **2024**, *125*, e7595. [CrossRef]
23. Wang, X.; Silbermann, C.B.; Nagel, T.; Nackenhorst, T. Modelling of the elastoplastic behaviour of the bio-cemented soils using an extended Modified Cam Clay model. *J. Rock Mech. Geotech.* **2024**, *16*, 2184–2197. [CrossRef]
24. Zhou, Y.; Wang, P.; Shi, L.; Cai, Y.Q.; Wang, J. Analytical solution on vacuum consolidation of dredged slurry considering clogging effect. *Geotext. Geomembr.* **2021**, *49*, 842–851. [CrossRef]
25. Zeng, L.L.; Cai, Y.Q.; Cui, Y.J.; Hong, Z.S. Hydraulic conductivity of reconstituted clays based on intrinsic compression. *Géotechnique* **2020**, *70*, 268–275. [CrossRef]
26. Lan, P.; Su, J.J.; Ma, X.Y.; Zhang, S. Application of improved physics-informed neural networks for nonlinear consolidation problems with continuous drainage boundary conditions. *Acta. Geotech.* **2024**, *19*, 495–508. [CrossRef]
27. Singh, A.; Chakraborty, M. Combined effect of non-darcian flow and semipermeable drainage boundaries on one-dimensional consolidation of unsaturated soil. *Int. J. Geomech.* **2024**, *24*, 04024157. [CrossRef]
28. Cai, Y.Q. Consolidation mechanism of vacuum preloading for dredged slurry and anti-clogging method for drains. *Chin. J. Geo. Eng.* **2021**, *43*, 201–225. (In Chinese)
29. Ye, G.B.; Xu, Y.; Zhang, Z. Performance Evaluation of PVD-Reinforced Soft Soil with Surcharge and Vacuum Preloading. *Int. J. Civ. Eng.* **2018**, *16*, 421–433. [CrossRef]
30. Lim, B.S.; Tumay, M.T.; Lee, J.W.; Chun, B.S.; Jung, J.W. A case study in evaluating the status of consolidation of a soft soil deposit by incomplete piezocone dissipation tests using laboratory and field data. *Soils Found.* **2014**, *54*, 648–656. [CrossRef]
31. Yang, S.P.; Yang, A.W.; Xu, A.J. Vacuum preloading reinforcement of soft dredger soil by modified fiber drainage plate. *Rock Soil Mech.* **2025**, *46*, 789–797. (In Chinese)
32. Zhou, Y.; Chai, J.C. Equivalent 'smear' effect due to nonuniform consolidation surrounding a PVD. *Géotechnique* **2017**, *67*, 410–419. [CrossRef]

water

MDPI

Review

Deep Learning-Driven Sandy Beach Resilience Assessment: Integrating External Forcing Forecasting, Process Simulation, and Risk-Informed Decision Support

Yuanshu Jiang [1], Yingtao Zhou [1,2,*] and Juntong Zhang [3,*]

[1] College of Harbour, Coastal and Offshore Engineering, Hohai University, Nanjing 210024, China; jiangyuanshu@hhu.edu.cn
[2] Shanghai Urban Construction Design & Research Institute (Group) Co., Ltd., Shanghai 200125, China
[3] Marine Geological Survey Institute of Hainan Province, Haikou 570206, China
* Correspondence: zhouyingtao@sucdri.com (Y.Z.); zt138327@sina.com (J.Z.)

Abstract

Sandy beach resilience faces growing threats from extreme events and intensified human activity. Deep Learning (DL) has emerged as a powerful tool in coastal research, offering strengths in spatial feature extraction, nonlinear sequence modeling, acceleration of physical processes, and integration of multi-source data. This review frames resilience in three technical dimensions—resistance, recovery, and adaptation—and examines DL applications across three domains: first, monitoring and forecasting external forcing, including typhoon tracks and storm surge peak values; second, modeling and simulating beach processes, from rapid hydrodynamic forecasting to medium- and long-term shoreline evolution, and high-resolution sediment transport forecasting; and third, management and decision support, where DL methods and multi-scenario generation expand governance options, and interpretable features with uncertainty quantification enhance risk communication and policy adoption. DL complements traditional models by shortening the "observation–model–decision" cycle, expanding scenario analysis, and improving governance transparency. Challenges remain in cross-domain generalization, robustness in extreme scenarios, and data governance. This review confirms DL's potential as a technology stack for enhancing sandy beach resilience and provides a methodological foundation for future research.

Keywords: sandy beach; resilience; deep learning; physics-guided/physics-informed; external forcing forecasting; process simulation; decision support

1. Introduction

Sandy beaches represent one of the most widespread zones of human–ocean interaction worldwide. They fulfill multiple ecological, tourism, and economic functions [1]. The dense population in coastal zones also necessitates that these areas maintain a certain degree of resilience to hazards such as extreme events and storms [2,3]. The nearshore geomorphology and sediment transport of sandy beaches are highly complex and extremely sensitive to variations in external forcing conditions [4]. In recent years, under global warming, the impacts of sea-level rise on coastal zones have become a major concern [5]. Moreover, researchers have found that rising sea levels have increased the frequency of extreme sea-level events [6] and coastal flooding [7]. Extensive observational data indicate that global wind speeds and wave heights have been rising over recent decades, with particularly pronounced increases in extreme values [8]. The effects of human activities on sandy beaches (e.g., hard engineering structures, beach nourishment, and land reclamation) should also not be underestimated [9,10]. Collectively, these factors have intensified beach erosion caused by extreme events (natural) and heightened ecological risks associated with long-term environmental changes (human-induced). Therefore, research on the resilience of sandy beaches has become increasingly urgent [11]. Such research requires an integrated assessment framework that accounts for both process mechanisms and risk management [12].

In recent years, numerous high-frequency and high-resolution data acquisition methods have developed rapidly, including coastal High-Frequency Radars (HFRs) [13,14], Argus video monitoring systems [15], and Unmanned Aerial Vehicles (UAVs) [16]. The establishment of multi-source data systems has greatly enhanced the spatiotemporal characterization of nearshore hydrodynamic and topographic processes. It can provide a solid foundation for research on the resilience of sandy beaches, including the identification of shoreline and cross-shore evolution, the impacts of storm events, and subsequent recovery processes. However, process-based numerical models such as Delft3D (v6.02.10) [17] and XBeach (v1.22.4867) [18] face several challenges in high-resolution, long-term, and multi-scenario simulations, including high computational costs, sensitivity to boundary conditions, and difficulties in simulating turbulence [4]. In studies of wave-dominated beaches, researchers have further considered the dynamic variability of wave time series and morphological instability to gain deeper insights into beach morphodynamics [19]. Nevertheless, even within the wave-dominated background, the effects of extreme events and the general applicability of these models remain unresolved. These limitations are particularly evident in tasks involving multi-scale coupling among wind, waves, currents, sediments, and topography. Consequently, it remains difficult to accurately quantify how sandy beaches respond to extreme forcing, thereby constraining the assessment of their resilience.

Artificial intelligence (AI), particularly deep learning (DL), has emerged as a new paradigm owing to its strengths in modeling large-sample, nonlinear, and multi-source data [20]. To address the issue of interpretability in DL, researchers have begun embedding physical priors and conservation laws into the learning process. The concept of theory-guided data science seeks to leverage established scientific knowledge to enhance the capability of data-driven models in advancing scientific discovery [21]. Similarly, physics-informed deep learning integrates physical principles derived from first-order laws into data-driven frameworks, thereby combining the advantages of both physics-based and data-driven approaches [22]. Moreover, the development of hybrid modeling strategies and the introduction of physics-consistent loss functions have further improved the ability of models to unify physical mechanisms with data representations [23]. Currently, research on sandy beach resilience primarily focuses on two aspects: the dynamic evolution of sandy beach topography and geomorphology under external forcing and the associated resilience-related mechanisms [18,19]; and the adaptability of sandy beach environments to engineering measures, together with the governing patterns of their feedback responses [9,12]. These studies provide an important foundation for understanding sandy beach resilience and for the subsequent application of DL as a research method. This paradigm is particularly well suited for studying the resilience of sandy beaches. On the one hand, it enables effective forecasting of environmental changes under extreme events. On the other hand, it accelerates the simulation of high-frequency coastline and profile dynamics across multiple scenarios. Consequently, it provides a novel quantitative tool for assessing the resilience of sandy beaches in terms of disturbance, response, and recovery processes.

Taking the resilience of sandy beaches as the core theme, this review divides resilience into three technical dimensions: resistance, recovery, and adaptation (Figure 1). The resistance dimension (e.g., short-term forecasting, setting early warning thresholds) addresses system responses to extreme events. Following such events, the recovery dimension (e.g., assessing profile recovery time) evaluates the system's capacity to restore its pre-disturbance state. Over longer timescales, the adaptation dimension (e.g., scheme optimization, digital twins) aims to enhance the system's robustness and sustainability.

Accordingly, rather than directly providing quantitative resilience values, this review establishes a comprehensive framework and analytical process based on three dimensions: resistance, recovery, and adaptation. Different from other reviews that broadly explore the application of AI in coastal sciences, this review focuses exclusively and specifically on DL, which currently exhibits the greatest potential for further development. By positioning DL as a supporting technology for resilience governance and systematically reviewing the progress, advantages, and limitations of DL, this review will explore its potential as an enabling technology for enhancing resilience across the three dimensions of resistance, recovery, and adaptation. Section 2 introduces DL methods across various capabilities. Section 3 reviews their application in meteorological disasters and external forcing forecasting. Section 4 reviews applications in coastal morphology and sediment dynamics. Section 5 reviews DL-based approaches for sandy beach management and decision-making.

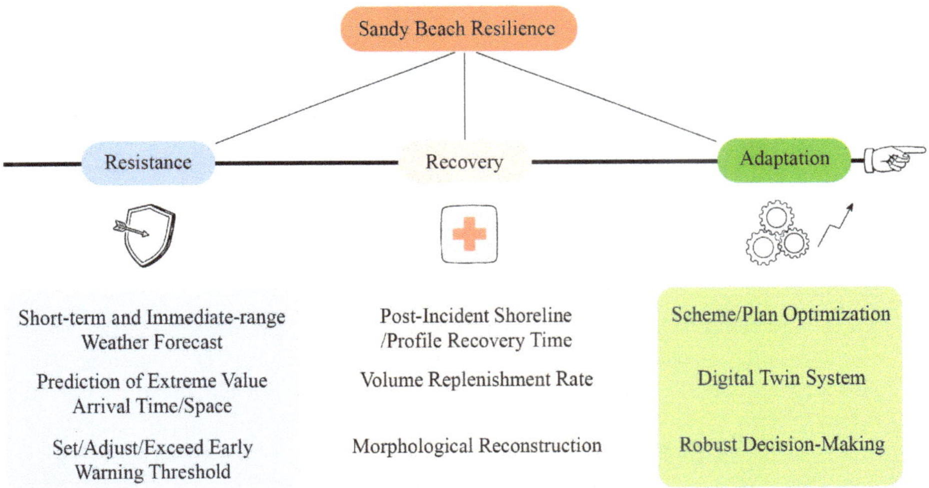

Figure 1. Conceptual framework and technical process of sandy beach resilience.

2. Methods: Deep Learning

DL is a major subfield of machine learning (ML), centered on constructing multi-layer neural networks that emulate the way the human brain processes and learns from data. DL has achieved remarkable success across various fields, enabling machines to match or even surpass human performance in many tasks. In terms of sandy beach resilience research, DL applications primarily focus on capturing spatial features and modeling nonlinear relationships within time series data.

2.1. Literature Search and Data Sources

This review conducted a literature search and statistical analysis to explore the potential of integrating DL into research on sandy beach resilience. Peer-reviewed articles, review papers, and relevant reports published between 2010 and 2024 were collected from the Web of Science Core Collection database. Two main categories of keywords were used: sandy beaches and DL. Figure 2 presents the publication trends of these two categories over the past 15 years. As shown in Figure 2a, sustained attention has been given to the coastline evolution of sandy beaches, while studies and topics related to beach resilience have gradually increased year by year. Meanwhile, researchers have maintained strong interest in the practical implementation of sandy beach management and decision-making. Figure 2b indicates that over the past five years, scholars have shown growing enthusiasm for applying DL across various fields. The use of DL in spatial characterization and time-series analysis has become increasingly extensive. At the same time, researchers have recognized the non-interpretability of DL as a "black box". In the past three years, researchers have increasingly focused on integrating physical knowledge into DL frameworks. Table 1 summarizes the widely used DL methods, outlining their respective advantages, disadvantages, and application scenarios. These methods all show considerable potential for application in sandy beach resilience research.

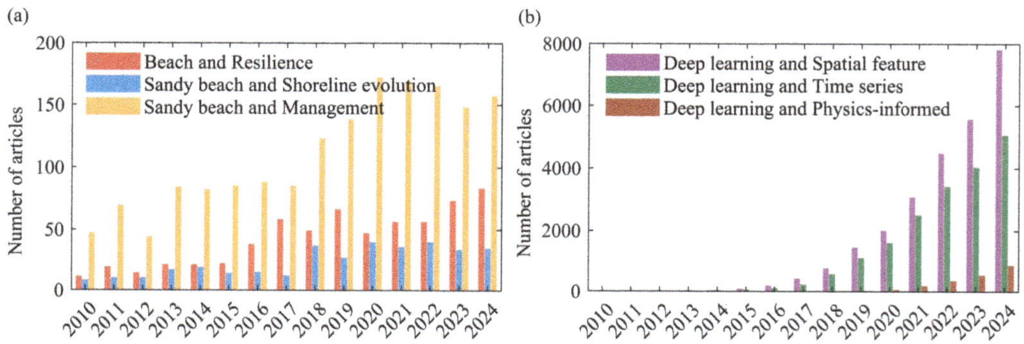

Figure 2. The comparison of the literature sources in terms of (**a**) 'Sandy beach' and (**b**) 'Deep learning'.

2.2. Spatial Characterization

The emergence of convolutional neural networks (CNN) has enabled DL to achieve remarkable progress in capturing spatial characterization, leading to their widespread application in computer vision [24]. By introducing a max-fusion strategy, methods based on fully convolutional networks (FCN, [25]) have achieved state-of-the-art performance in the semantic segmentation of very high-resolution remote sensing images [26]. SegNet employs nonlinear upsampling operations to avoid the need for learnable upsampling layers, thereby effectively extracting features from lower-resolution input feature maps [27]. To systematically aggregate multi-scale information without reducing spatial resolution, researchers have adopted dilated convolutions, which significantly improve the accuracy of dense prediction tasks [28]. The residual architecture of ResNet further facilitates model optimization, allowing networks to learn deeper hierarchical features and achieve higher accuracy [29]. Although these networks demonstrate translation invariance, they still face limitations such as insufficient boundary feature extraction, limited generalization ability, and reduced performance in complex scenes.

The Vision Transformer (ViT) introduces self-attention mechanisms into visual representation learning, directly modeling long-range dependencies and global relationships. Through pre-training and transfer learning, pure Transformer architectures applied to sequences of image patches have achieved outstanding performance in image classification tasks, while requiring fewer computational resources during training [30]. The Swin Transformer constructs a hierarchical Transformer architecture that derives feature representations through shifted-window computations. This hierarchical design and shifted-window mechanism enable flexible multi-scale information modeling, with computational complexity that scales linearly with image size [31]. SegFormer integrates a Transformer backbone with a lightweight multilayer perceptron (MLP) decoder, allowing it to output multi-scale features and aggregate information across different levels by combining local and global attention, thereby achieving strong representational capability [32].

Overall, Transformers excel in large-sample, cross-scale tasks, whereas CNNs demonstrate more stable performance in small-sample, single-task scenarios. Integrating the two enables their complementary strengths to be fully utilized: attention mechanisms or Transformer architectures capture global dependencies, while CNNs effectively represent local features. When supplemented with transfer learning, strong regularization, and other techniques, this hybrid approach provides robust feature representation in two-dimensional space.

2.3. Time Series and Cross-Scale Dependencies

Recurrent neural networks (RNN) have been widely applied in time-series modeling, with their core strength lying in extracting temporal information and handling cross-scale dependencies [33,34]. N-BEATS is a deep neural network architecture based on backward and forward residual connections. It offers several notable advantages, including interpretability, direct applicability across diverse target domains, and fast training speed [35]. DeepAR produces high-precision probabilistic forecasts by training an autoregressive RNN on large collections of related time series [36]. Informer employs self-attention distillation to highlight the most relevant attention features. It can handle extremely long input sequences, and significantly improve inference speed [37]. Temporal fusion Transformers (TFT) combine recurrent layers for local processing with interpretable self-attention layers for modeling long-term dependencies, achieving high performance across diverse scenarios [38]. It is noteworthy that in time-series modeling, the objective is to capture temporal relationships within ordered continuous data points. In some cases, simple single-layer linear models outperform existing complex models [39].

Overall, RNNs have traditionally been the primary choice for time-series modeling. However, emerging Transformer-based DL methods have increasingly demonstrated advantages in extracting temporal information. Nonetheless, these methods generally demand larger data volumes and stronger regularization compared to traditional models such as long short-term memory (LSTM) and gated recurrent unit (GRU). For different task scenarios, factors such as sample size, spatial and temporal scale, and computational cost must be fully considered when selecting an appropriate model architecture. At the same time, combining multiple models can be continuously explored to leverage the strengths of different approaches.

2.4. Physics-Informed/Physics-Guided DL

In natural systems, physical consistency and extrapolation stability often take precedence over in-training accuracy. Physics-guided (or physics-informed) DL mitigates the "black box" nature and data dependency issues by embedding governing equations, conservation laws, and boundary conditions into the learning process [40]. Cuomo et al. argued that most studies focus on customizing physics-informed neural networks (PINN) through variations in activation functions, gradient optimization techniques, network architectures, and loss function designs [41]. Brunton et al. combined sparsity-promoting techniques, ML, and nonlinear dynamical systems to derive governing equations from noisy measurement data [42]. Rudy et al. noted that this approach is computationally efficient and robust, and has been validated in canonical problems across multiple scientific domains, including the

Navier–Stokes equations, diffusion equations, and so on [43]. Bar-Sinai et al. introduced data-driven discretization, which employs neural networks to estimate spatial derivatives and uses end-to-end optimization to ensure adherence to governing equations on low-resolution grids [44]. Graph network-based simulators (GNS) learn to simulate a variety of challenging physical domains, including fluids, deformable materials, and so on [45]. Kochkov et al. applied an end-to-end DL framework to enhance approximation processes in computational fluid dynamics for simulating two-dimensional turbulent flows [46].

In summary, DL integrated with physical knowledge can not only constrain model training by incorporating equation residuals, structured network layers, and similar physical constraints into the loss function, but also autonomously learn the physical representations embedded within the data. Building on this capability, end-to-end DL methods have been applied to enhance computational efficiency and simulate complex operating conditions.

Table 1. Common types of DL and their application scenarios.

Type	Feature	Typical Advantage	Limitation/Risk	Application Scenario	Ref.
CNN	Local translation invariance; hierarchical texture; controllable multi-scale	Strong spatial feature recognition capability; computationally friendly	Weak temporal dependence; poor multi-scale processing capability; dependence on the quality of training samples	Raster image classification	[26]
ViT	Global self-attention; non-local interaction; hierarchical window	Global long-range dependence; cross-scale fusion	High demand for data and computing power; easy overfitting with small samples	Large-sample data; multi-scale scenario correlation	[30]
RNN	Control information extraction using a gating mechanism	Strong univariate feature extraction capability; good at capturing nonlinear dynamics	Poor interpretability; gradient vanishing in long sequences	Univariate sequences; short-to-medium-term sequences	[33]
Transformer-based long-time series	Sparse/selective attention mechanism; covariate embedding and interpretation	Global dependence; cross-variable; high parallel computing efficiency	Large scale of training samples; relatively complex model structure	Long sequences; multi-scenario and multi-source variables	[34,37]
Physics-guided/Physics-informed DL	Physical information embedding; physical boundary constraints	Physical consistency; interpretability; robust extrapolation; strong generalization	Difficulty in loss convergence; complexity in boundary handling; high computational cost	Sparse observation; tasks requiring conservation constraints for extrapolation	[46]

3. Application of DL in Sandy Beach Hazards and Hydrodynamic Processes

3.1. Typhoon Track/Intensity and Storm Surge Peak Value Forecasting

External hydrodynamic conditions play a critical role in the study of sandy beach resilience. Sandy beaches are particularly sensitive to extreme events, especially storm surges induced by typhoons. For effective beach management, the timely provision of information such as the arrival time and potential intensity of storm surges is essential. Figure 3 compares the morphological responses and observation layouts of sandy beaches under storm and normal conditions. During storm events, the superposition of strong winds and storm surges generates intense waves that drive alongshore and offshore sediment transport, causing large-scale erosion and significant shoreline retreat. In contrast, during normal periods, wave energy diminishes, local sedimentation accretion occurs, and the beach enters a recovery state. In terms of monitoring, a multi-scale observation network typically comprises satellites, UAVs, real-time kinematic (RTK) systems, time-lapse cameras (TL cameras), wave buoys, tide gauges, and other instruments. This network synchronously collects data on shoreline morphology, topography, water levels, and wave dynamics, which are essential for event assessment, model calibration, and uncertainty quantification.

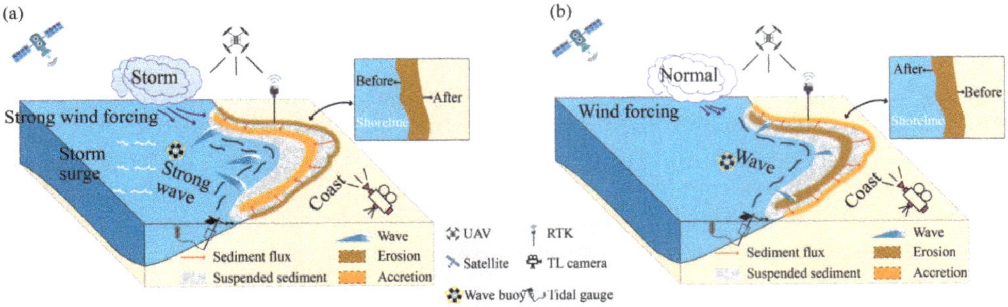

Figure 3. Schematic of sandy beach morphology and observation system before and after storms. (**a**) Storm period: The generated intense waves cause shoreline retreat and extensive erosion (brown), accompanied by suspended sediment (gray) and alongshore/offshore sediment transport (red arrows). (**b**) Normal period: Weakened waves lead to local siltation and morphological recovery (orange). The inset at the top right compares shoreline positions before and after the storm.

Typical hydrodynamic models (e.g., ADCIRC (v56.0.2) [47], Delft3D (v6.02.10) [48], COAWST (v3.8) [49], FVCOM (v5.0) [50]) provide relatively accurate representations of environmental changes induced by extreme events by partitioning the computational domain into unstructured grids and appropriately characterizing physical processes, boundary conditions, and other factors. However, these models often simplify physical processes and incur high computational costs in high-resolution simulations. DL offers substantial advantages in forecasting extreme events due to its strong capability to capture nonlinear characteristics from large datasets. The application of DL in this field can be divided into meteorological and oceanographic domains. Rüttgers et al. utilized a large volume of satellite images as input to a generative adversarial network (GAN), achieving a six-hour typhoon track forecasting capability [51]. Giffard-Roisin et al. fused historical track data with reanalysis atmospheric images (three-dimensional fields of wind and pressure) to construct a neural network model capable of 24 h tropical cyclone track forecasting [52]. Stengel et al. enhanced the network's physical fidelity and perceptual capabilities through adversarial training, achieving up to a 50-fold improvement in the resolution of wind data [53]. Recently, DL-based global data-driven weather forecasting models have achieved significant progress, including FourCastNet [54], GraphCast [55], Pangu-Weather [56], and Fuxi [57]. These models can rapidly generate global weather variables at a spatial resolution of 0.25° for the next 10 days at least. In particular, the high-precision forecasted wind and pressure fields provide more accurate boundary conditions for storm surge forecasting.

In terms of oceanic storm surge forecasting, most studies have employed models that establish relationships between time series data (e.g., wind speed, air pressure, station water levels) and extreme water level time series. Adeli et al. applied a ConvLSTM model, capable of capturing both temporal and spatial correlations, together with a residual connection network to enhance performance [58]. Their results demonstrated that, within an examined synthetic storm database, DL approach outperformed the previously Gaussian process model for storm surge forecasting. Ramos-Valle et al. utilized cyclone datasets generated by the Hybrid Weather Research and Forecasting (HWRF) model and found that RNN improved the accuracy of storm surge peak predictions to a notable extent [59]. Xie et al. developed a DL-based single-station water level prediction model for multi-station storm surge forecasting [60]. Their approach employed CNN to extract two-dimensional wind field information, which was fused with local water level features to improve temporal prediction efficiency. Jiang et al. proposed the Surge-NF model, incorporating a positional encoding module and a multi-task learning module [61]. Evaluation using the North Atlantic Coast Comprehensive Study (NACCS) database demonstrated that Surge-NF uniquely captured high-frequency spatial variation features and effectively leveraged task correlations. Tedesco et al. applied residual learning to enhance the residual component in the forecasting results of the Nordic4-SS model [62]. This method can be implemented operationally without replacing existing physical models. Cerrone et al. employed Transformers to improve the water level prediction capability of the National Oceanic and Atmospheric Administration's (NOAA) Storm and Tidal Operational Forecast System (STOFS-2D-Global) over a seven-day forecasting cycle, with the most significant improvements observed at tide-dominated stations [63].

3.2. *Large-Scale and Nearshore Wave Generation and Acceleration*

In addition to the extreme water levels induced by typhoons, the dynamic environments that directly influence sandy beaches also include waves. For subsequent simulations of beach morphology and management, continuous, stable, and high-resolution wave data are indispensable. Traditionally, wave simulations have primarily relied on the Simulating Waves Nearshore model (SWAN 41.51) [64]. Data-driven DL methods, by contrast, mainly focus on sequence-to-sequence and image-to-image forecasting.

Wang et al. proposed a hybrid model combining Improved Empirical Wavelet Transform (IEWT) decomposition with a LSTM network for wave prediction [65]. Their results demonstrated that, compared to using LSTM alone, this hybrid model more accurately tracked variations in wave crests and troughs. Hao et al. integrated the strengths of LSTM and Empirical Mode Decomposition (EMD) to develop the EMD-LSTM method [66]. Based on significant wave height data from three offshore stations in China, their study indicated that the EMD-LSTM model exhibited superior performance in predicting nonlinear and non-stationary waves. Zhang et al. introduced the Numerical LSTM network to correct numerical wave height predictions [67]. Results showed that it significantly improved the accuracy of SWAN-derived wave height predictions in the Bohai Sea and around Xiaomai Island. Choi et al. employed a CNN to process raw wave images from buoys for short-term wave height forecasting [68]. Bai et al. developed a CNN-based two-dimensional deep learning model for regional wave field prediction in the South China Sea [69]. Their results indicated that the model not only accurately predicted temporal variations in wave height but also effectively estimated spatial distributions of the two-dimensional wave field and demonstrated potential for long-term forecasting. Bento et al. applied CNNs to wave data from three locations and conducted year-round testing, confirming the model's ability to extract relevant temporal dependencies [70]. Mlakar et al. proposed deep learning wave emulating model (DELWAVE 1.0), comprising a convolution-based atmospheric encoder module, a temporal compression module, and a regression module [71]. DELWAVE successfully simulated nearshore wave dynamics as represented by SWAN at sparse locations, and effectively detected storm events. Kuhn explored the application of DL-based super-resolution techniques in coastal sea state prediction, including spectral wave modeling, neural network-based prediction on unstructured grids, and super-resolution reconstruction of wave spectra. This study demonstrated the potential of such approaches to accelerate wave prediction [72].

Figure 4 illustrates the workflow from observation to results. Multi-source data (including satellite cloud images, reanalysis variables, water levels, and wave spectra) undergo key feature extraction via DL methods such as LSTM, Transformer, and GNN. These methods rapidly generate core indicators and spatial distributions of typhoons, storm surges, and waves. The resulting outputs can be directly applied to beach risk assessment, serve as constraints for data assimilation and calibration of numerical models, and support uncertainty quantification and rolling updates. Table 2 summarizes the applications of DL in simulating and forecasting external dynamic processes based on typical tasks. In summary, the data-driven DL approach provides a faster and more accurate toolchain for external forcing forecasting and post-processing in oceanic applications. It offers significant advantages in low latency and high-frequency updates, with direct relevance to the management and application of sandy beaches.

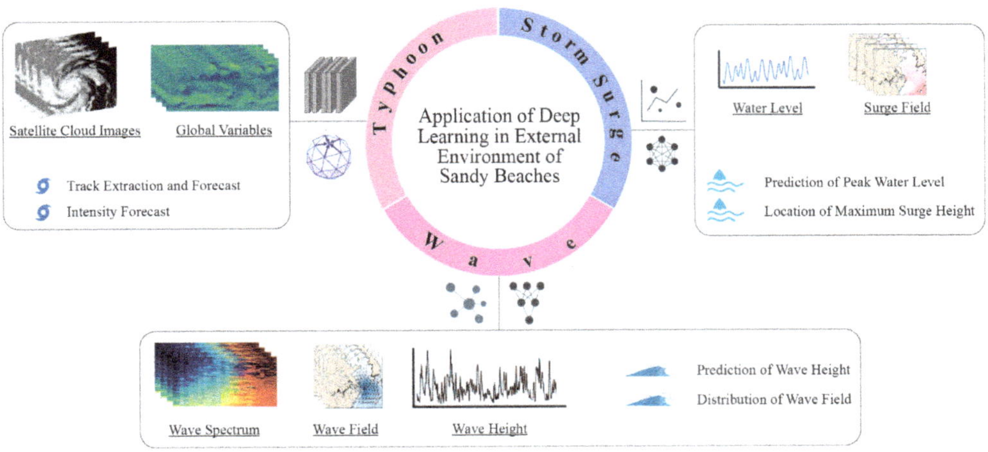

Figure 4. Schematic framework for DL-based forecasting of external forcing on sandy beaches. The central loop organizes tasks around the "Typhoon–Storm Surge–Wave" sequence. On the left, satellite cloud images and reanalysis variables serve as inputs for typhoon track extraction and intensity prediction. On the right, water level sequences and storm surge fields are used to infer peak water levels and identify locations of maximum storm surges. At the bottom, wave spectra, wave fields, and wave height sequences are employed to predict significant wave heights and nearshore wave field distributions.

Table 2. Application of DL in simulating and forecasting external dynamic processes.

Task	Core Input	Core Output	Representative Models and Key Points	Ref.
Short-term; direct typhoon track and intensity	Satellite cloud images; reanalysis atmospheric images	Typhoon track; typhoon intensity	CNN; GAN	[51,52]
Extracting typhoons and storm surges indirectly from large-scale meteorological fields	Global-scale surface and upper-air meteorological fields (including wind fields and pressure differences)	Extracting meteorological elements such as wind fields at the boundaries	GraphCast; Swin transformer	[56,57]
Extreme water level caused by the arrival of a storm surge at a single point location	Wind; air pressure; astronomical tide; station water level	Time series and peak values of extreme water levels	RNN; Attention mechanism	[59]
Spatial process and intensity of storm surges	Single-point water level; two-dimensional wind fields	Storm surge processes at multiple stations within a region	ConvLSTM; ResNet	[58,60]
Forecasting and correction of global tidal levels and storm surges	Results of physical forecasting models	Generated correction sequence	Transformer	[63]
Temporal forecasting or deviation correction of wave height	Temporal processes of wave height from buoys; numerical models	Temporal process of wave height at a single point	LSTM; EMD; IEWT	[65,66]
Spatial-temporal distribution and forecasting of wave fields in a region	Original wave height images; wave height data from multiple stations	Distribution of significant wave height in a region	CNN; ConvLSTM	[68,69]
DL-based proxy wave simulation and forecasting at sparse locations	Atmospheric boundary; rough/high-resolution wind field conditions	Spatial wave field; temporal processes at sparse locations	DELWAVE; CNN; RNN	[71]

4. Application of DL in Coastal Morphological Evolution and Sediment Simulation

4.1. Reconstruction and Evolution of Sandy Beach Profiles and Shorelines

The forecasting and monitoring of sandy beach morphological evolution are critical for enhancing their resilience. Traditional shoreline identification techniques based on remote sensing and satellite imagery are already well-developed [73]. However, these methods have limitations in addressing complex terrain, fine-scale geomorphology, and long-term trend prediction [74]. In recent years, DL has been increasingly applied to classification, identification, and forecasting tasks in this field. For example, CoastSat employs MLP to integrate satellite imagery and extract time-series shoreline positions for sandy coastlines worldwide with an accuracy of approximately 10 m, enabling efficient shoreline change monitoring [75]. U-Net combines skip connections and multi-scale feature fusion, making it suitable for land cover classification, extraction, and change detection from satellite imagery [76]. DeepLabv3 integrates an encoder–decoder structure with atrous separable convolution to flexibly control encoder feature resolution, achieving a balance between accuracy and computational efficiency [77]. Such functions hold potential for identifying fine geomorphic textures in sandy beaches, such as beach ridges and tidal creeks. Park and

Song employed a U-Net semantic segmentation model based on attention mechanisms and a water index to automatically extract the land–sea boundary. They also applied transfer learning to address data scarcity in certain coastal areas, improving shoreline change monitoring efficiency and accuracy [78]. Scala et al. enhanced CNN by introducing a Sobel edge loss function and multi-class segmentation, enabling effective detection of land cover features that may be overlooked (e.g., cliff vegetation, coastal roads). Thereby, they significantly reduce uncertainties in shoreline delineation [79]. Feng et al. incorporated an Edge Depth Supervision (EDS) module, a Cooperative Semantic Depth Supervision (CSDS) module, and an Attention Fusion Module (AFM) to improve detailed shoreline edge feature extraction [80]. Dang et al. demonstrated the applicability of U-Net models for predicting coastal changes in Vietnam and globally [81]. Their findings suggest that waterline identification is suitable for monitoring seasonal tidal changes or immediate wave movement, whereas shoreline delineation is more appropriate for assessing coastal erosion caused by sea-level rise during storms.

The management of sandy beach resilience requires the ability to predict beach responses to storms and long-term changes in regional wave climates over seasonal to interannual time scales. Currently, most researchers in the field of profile and shoreline evolution forecasting employ one-dimensional shoreline equilibrium models. These models typically consider only wave-driven longshore sediment transport and describe shoreline displacement based on the concept that wave energy tends toward a dynamic equilibrium position [82,83]. DL has demonstrated advantages in capturing nonlinear relationships in time-series models across scales and offers significant improvements in computational efficiency [84]. As beach environments become increasingly complex, the application of DL to profile and shoreline forecasting has grown. Manamperi et al. modeled shoreline changes as a supervised learning task, employing LSTM to capture long-term dependencies and leveraging its memory mechanism to simulate shoreline evolution under continuous wave forcing [85]. Lee et al. developed an LSTM-based encoder–decoder network for effective spatiotemporal representation learning, enabling estimation of beach profile responses over weeks to months under varying hydrodynamic conditions [86]. Gomez-de la Peña et al. examined CNN and CNN-LSTM for predicting interannual shoreline positions, finding that DL methods outperform traditional equilibrium models in simulating observed variability and spatial distribution [87]. Adusumilli et al. successfully predicted beach width and seasonal average beach slope using tides and waves as inputs to a trained Deep Neural Network (DNN) [88]. Montaño et al. compared traditional shoreline models and DL methods for long-term shoreline evolution prediction, concluding that both approaches can reproduce shoreline changes under normal conditions, while model ensembles outperform single models in capturing extreme and rapid fluctuations [89].

In summary, traditional techniques, such as sub-pixel shoreline extraction and geometric correction, can be integrated into preprocessing, feature extraction, auxiliary evaluation, and post-processing workflows for DL methods, thereby enhancing the accuracy of shoreline detection [90–92]. Building on this foundation, long-term shoreline trend forecasting should fully leverage the advantages of DL to establish a comprehensive workflow encompassing identification, analysis, and forecasting. The robustness of such models also enables effective detection and prediction of erosion and accretion patterns on sandy beaches. They provide essential data and a priori insights to support regional research, coastal management, and the enhancement of sandy beach resilience [93,94].

4.2. Sediment Transport: Surrogate Modeling, High-Resolution Flow Field Driving with Physical Constraints

Sediment serves as the fundamental link between hydrodynamic changes and morphological evolution, whether in sandy beach profiles or shorelines. However, the highly stochastic and uncertain nature of sediment transport renders sediment-related numerical simulations computationally intensive, limiting their applicability for long-term simulations and forecasts. The profile and shoreline prediction models discussed in Section 4.1 generally simplify or omit complex sediment dynamics. The emergence of DL has prompted researchers to explore its potential for simulating and forecasting sediment transport. Weber de Melo et al. employed CNN that utilized two-dimensional hydrodynamic variables (vertically averaged flow velocity squared, bed shear stress) as inputs [95]. By extracting multi-scale features through a multi-branch structure, this model successfully captured the spatial distribution characteristics of sediment transport. This CNN-based approach avoids the computational complexity of solving differential equations inherent in traditional numerical models while maintaining acceptable simulation accuracy. Fajardo-Urbina et al. applied DL to predict the advection and diffusion of particle swarms over a tidal cycle, coupling it with a simplified Lagrangian model to achieve longer-term prediction results [96]. Xie et al. adopted a CNN-based approach to quantitatively model suspended sediment concentration (SSC), demonstrating its capability to predict hourly SSC variations while reducing computational costs and improving processing efficiency [97].

The primary driving force behind sediment transport originates from the flow field. To a large extent, the complexity and high computational cost of sediment transport arise from the computation of the flow field. Consequently, many researchers are exploring the use of DL to optimize or replace complex flow field calculations. Among the current approaches, Physics-Informed Neural Networks (PINNs) have emerged as a mainstream method. Once trained, PINNs can satisfy differential equations as well as prescribed initial and boundary conditions, making them a promising alternative to traditional numerical models (e.g., finite difference, finite volume, spectral methods) [98]. Dazzi tested the application of PINNs to solve the Shallow Water Equations (SWE) enhancement system, incorporating topography as a spatially variable feature during training and calculating the bed slope source term via automatic differentiation [99]. Qi et al. compared the performance of Physics-Informed Fully Connected Networks (PIFCNs) and Physics-Informed Convolutional Networks (PICNs) with a finite volume solver in a real flood event [100]. Their results demonstrated that, for predicting the temporal distribution of water depth, both PINN variants outperformed the finite volume model in balancing speed and accuracy, whereas the finite volume model yielded superior discharge predictions. Bihlo and Popovych applied PINNs to solve the SWE on a spherical surface, successfully avoiding boundary value loss functions by encoding boundary conditions within a custom neural network layer [101]. Huang et al. advanced PINN methodology by integrating both boundary and initial conditions directly into the governing equations, thereby eliminating the need to include their residuals in the loss function. This approach achieved better convergence, reduced memory consumption, and faster training, outperforming traditional PINNs in several aspects of hydrodynamic simulation [102].

Figure 5 illustrates two complementary pathways for integrating observational data and physical modeling. The first pathway relies on remote sensing and ground-based monitoring to perform target recognition and fine-grained land cover characterization, followed by time-series modeling to predict shoreline and profile evolution. The second pathway integrates physical mechanisms into the learning framework to generate high-resolution flow fields and free-surface evolution, upon which the spatiotemporal distribution of SSC is reconstructed. The first pathway emphasizes rapid "data-to-morphology" assessment, while the second highlights interpretable "mechanism-to-process" simulation. Table 3 summarizes the applications of DL in morphological evolution and sediment simulation based on typical tasks.

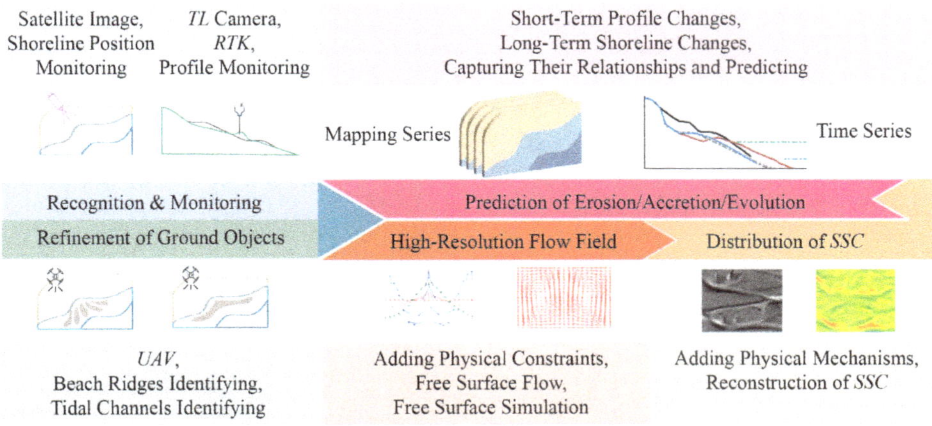

Figure 5. Two application pathways of DL in sandy beach morphology, flow fields, and sediment dynamics. (**Top**) Observation-driven pathway: Construct shoreline/profile sequences using satellite imagery, TL cameras, and RTK technology; perform feature identification and monitoring; map time series; and predict erosion, accretion, and morphological evolution. (**Bottom**) Physics-guided pathway: Generate high-resolution flow fields under physical constraints and simulate free-surface processes (image adapted from [102]); subsequently reconstruct and infer the spatiotemporal distribution and dynamics of SSC (image adapted from [95]).

Table 3. Application of DL in morphological evolution and sediment simulation.

Task	Common Method	Main Data Sources	Training Scale	Advantage	Limitation/Risk	Ref.
Identification and extraction of shorelines	U-Net; attention mechanism; transfer learning	Multi-period satellite images; UAV	Multi-scenario and multi-year time series	Automation; wide coverage; batch processing	Affected by tidal levels and meteorological cloud cover; limited cross-domain generalization	[76,78]
Classification and identification of shoreline geomorphic details	CNN; attention module	Multi-period satellite images; UAV; temporarily deployed sensors	Multi-scenario and multi-year time series	Automation; wide coverage; batch processing	Affected by tidal levels, meteorological cloud cover, and surface feature diversity; limited cross-domain generalization	[77,79]
Prediction of profile and shoreline evolution	LSTM/GRU; ConvLSTM; Transformer; attention mechanism	Profile survey lines; shore-based video shooting; reanalysis data of waves, tides, wind fields	Time series on weekly–annual scales	Fast calculation speed with low parameters; fast response to extreme events; good at capturing data variation	Relying on high-quality sample sequences; poor generalization	[85,88]
Inversion and prediction of SSC	CNN	Wind, wave, and tide time Series; in situ sampling (stations, buoys, sections)	Short time series; multi-site and multi-variable	Saving computational resources; effectively coupling physical processes	Relying on high-quality sample sequences	[95–97]
Physical coupling (PINN, hybrid)	PINN + FCN; PINN + CNN	Boundary field composed of multi-source observation data; governing equation	According to equation and grid settings; short time series	Improved interpretability and physical consistency; minute-level fast inference; robust extrapolation	Uncertainty of extreme scenarios; high computing power requirement; unstable convergence	[98,100,102]

5. Application of DL in Sandy Beach Management and Decision-Making

Sandy beaches represent a complex environmental system, and their management and decision-making are of critical importance. Effective management must address uncertain factors closely related to sandy beaches (e.g., topographic and demographic data, socio-economic development, ecological protection) [103]. Fundamentally, sandy beach management involves providing long-term, stable, and implementable solutions under multi-objective conditions and considerable uncertainty. DL offers promising approaches to address the nonlinear interactions among multiple objectives. For example, Kirezci et al. identified the spatial patterns and temporal evolution of "hotspot areas" using a simple linear superposition model based on global extreme sea level projections [104]. DL can further capture complex nonlinear relationships between variables and facilitate the application of transfer learning strategies in hotspot area analysis. Dal Barco et al. examined the relationships among various hazards across different spatiotemporal scales and analyzed multiple factors influencing coastline dynamics [105]. Their results showed that MLP achieved satisfactory performance in estimating daily impact risk when supplied with information on hazards, exposure, and vulnerability. Garzon et al. developed an early warning system for coastal erosion based on Bayesian networks, producing promising initial results [106]. The integration of DL could enhance such systems by improving prediction accuracy and generalization. Lundberg and Lee proposed an interpretable framework for DL prediction results, offering significant potential for answering management questions such as "Why is this coastal segment deemed high risk?" and "How will the risk change if sand dunes are restored or development is restricted?" [107]. The application of rigorous evaluation criteria can further provide comprehensive quantitative tools for diagnosing and interpreting DL results [108,109].

In recent years, in terms of operationalization and platformization, digital twin technologies have demonstrated the advantages of DL in two key aspects: the closed-loop capability of "from prediction to action" and the platform support for the "data–model–scenario–decision-making" process. Chen et al. proposed the architecture of a generative marine digital twin designed to integrate near-real-time observations, data assimilation, and DL surrogates into a unified data and computing infrastructure [110,111]. The core components of this system include a data pool, a DL-based marine model, and three-dimensional visual interaction functions. This architecture can integrate local-scale sandy beach management into a broader system, thereby lowering the collaboration threshold across departments and bridging spatiotemporal resolution gaps.

Overall, for the management and decision-making of sandy beaches, conventional collaborative analyses of multiple variables and traditional ML approaches remain the mainstream [106,112–117]. Nevertheless, DL holds considerable potential in this field. Various DL techniques can be deeply integrated into existing management and decision-making systems to enhance their performance. Figure 6 illustrates, from bottom to top, the closed-loop of beach management with digital twins as the central hub: multi-source data undergoes management and quality control to form data streams that drive traditional models and DL modules (physical/prediction, integrated decision support, and large language model (LLM) interaction).

First, within the physical/prediction module, DL can rapidly generate numerical predictions of various physical variables over a future time horizon through its high-precision forecasting capabilities. These forecasted data can serve two primary purposes: (1) they can be incorporated into the central digital twin system for visualization and for use as boundary conditions, and (2) they can provide a scientific reference for the subsequent integrated decision support module. Within the integrated decision support module, DL can leverage its strengths in addressing multi-scale and multi-modal tasks, enabling comprehensive consideration of future multi-factor and multi-category data. Moreover, based on the decisions proposed by decision-makers, this module can inversely drive the physical/prediction module to rapidly simulate and generate the physical data associated with these decisions. The resulting data can then be visualized within the central digital twin system, thereby supporting further evaluation by decision-makers. Throughout this workflow, LLMs can utilize their natural language processing capabilities to serve as an interface between decision-makers and the digital twin system. Their functions include, but are not limited to, enabling natural-language querying of the system, generating automated reports for stakeholders, and translating model outputs into accessible, plain-language risk assessment reports.

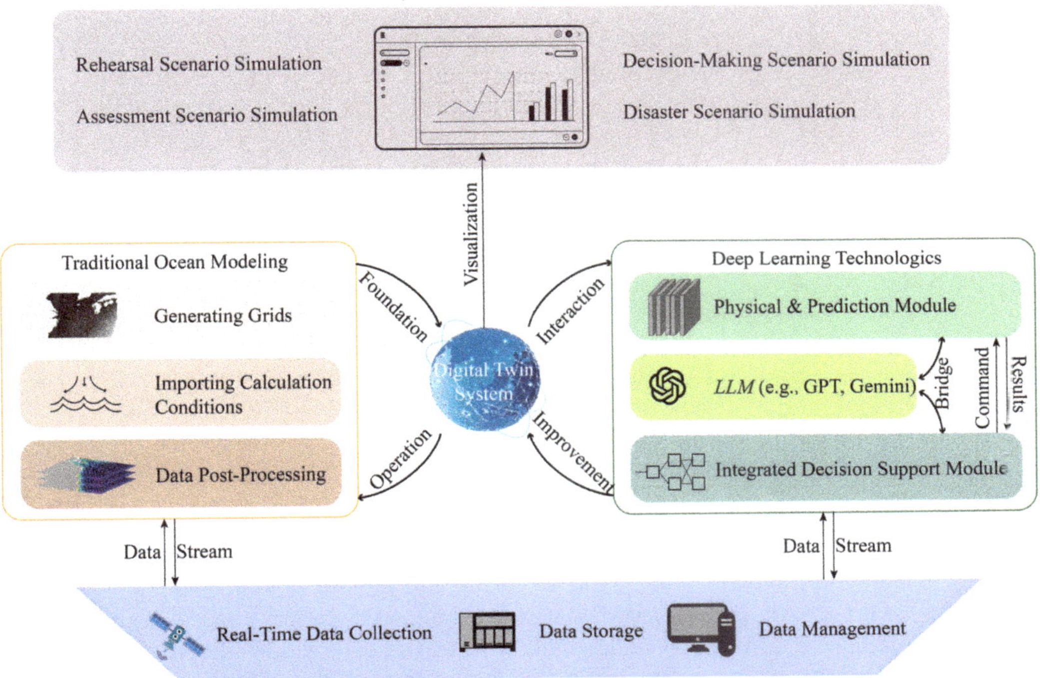

Figure 6. Digital twin-driven framework for beach management and decision-making (bottom-top). The bottom layer comprises real-time data collection, storage, and management. Data streams are directed along two parallel paths: the traditional marine modeling module on the left, and the DL technology module on the right. At the core, the digital twin system performs data fusion, model execution, and updates, and outputs visualized scenario simulations upward. The Earth-style design illustrated here is adapted from [110].

Digital twins enable unified coupling and visualization, rapidly generate results for drills, evaluations, decision-making, and disaster response, and facilitate continuous optimization through feedback loops. The management process can follow a pattern of "first using DL methods and multi-scenario generation to expand a variety of feasible solutions, then converging to a limited set of alternatives based on robust indicators", while preserving interpretability and calibrated results for auditing purposes [118,119]. In conclusion, the role of DL in sandy beach management is not to replace physical–economic–ecological models but to integrate observation, scenario, impact, and action into an operable evidence chain. At the front end, DL enhances communication through risk maps and interpretability. At the middle end, DL screens combinations via surrogates and robust decision-making. At the back end, DL ensures implementation and traceability through platforms and verification.

6. Conclusions

Against the backdrop of increasingly frequent extreme events and intensified human activities, enhancing the resilience of sandy beaches has become an urgent priority for coastal areas. The resilience of sandy beaches represents a comprehensive capability that requires coordinated consideration of multiple processes and variables. In recent years, methods driven by DL have achieved continuous technological breakthroughs and cross-disciplinary integration. With its exceptional ability to extract spatial features, robust capacity to fit nonlinear sequences, acceleration of traditional physical processes, and capability to integrate multi-source data, DL offers promising opportunities for advancing the enhancement and management of sandy beach resilience.

Centering on the theme of sandy beach resilience, this review categorizes resilience capabilities into three technical levels: resistance, recovery, and adaptation. Corresponding to the natural impact sequence on sandy beaches, from atmosphere to water to land, this review addresses three key research aspects: monitoring and forecasting of the external forcing, modeling and acceleration of sandy beach evolution processes, and human-led decision-making and management. Based on this framework, this paper systematically reviews current applications of DL in each aspect and explores its potential future contributions.

First, this paper reviews and summarizes two major advantages of DL: its strong capability for spatial feature representation and its ability to capture nonlinear relationships in sequences, alongside an overview of cutting-edge physics-informed/guided DL. Second, building on these advantages, this review analyzes DL application scenarios and achievements in two areas: forecasting typhoon tracks and intensities, and predicting storm surge peak values. Third, regarding waves that directly influence sandy beaches, this review examines successful DL applications in accelerating the simulation and forecasting of wave fields, from both pure data-driven and physics-informed perspectives.

Fourth, focusing on the critical topographic evolution of sandy beaches, including sediment transport involved, this review identifies two main directions in current DL applications. The first direction simplifies or bypasses complex sediment and current processes, instead applying DL to predict one-dimensional profile changes or two-dimensional coastline movements, thereby enabling rapid depiction of medium- to long-term beach changes or short-term erosion and recovery processes under extreme events. The second direction addresses complex and uncertain sediment transport processes and the high-resolution flow fields that drive them. Here, DL combined with physics-informed constraints not only greatly improves computational efficiency but also enhances the physical credibility of the simulations and forecasts.

Finally, this review explores current methods and strategies for sandy beach management and decision-making, proposing a pathway for DL to integrate deeply into existing systems. This pathway involves first employing DL methods and multi-scenario generation to broaden feasible solutions, then narrowing these to a limited set of robust alternatives, while preserving interpretability and calibrated results to ensure auditability.

Overall, this review centers on sandy beaches, aligning DL methods with coastal process knowledge relevant to beach resilience. It emphasizes that DL is not a substitute for existing frameworks, but rather a complementary tool that enhances them. The value of DL lies in shortening the path from observation to action, broadening the scope of multi-scenario assessments, and improving data transparency for communication and decision-making. However, limitations mainly arise from challenges in cross-domain generalization, robustness under extreme conditions, and adherence to data governance requirements. On one hand, standardized datasets and corresponding benchmark tests need to be established to evaluate the forecasting performance of DL-based models. On the other hand, continued exploration is required to better integrate DL with physical laws, which will substantially enhance the credibility of DL in coastal applications. Furthermore, it is essential to investigate the potential of DL as a bridge of human–AI collaboration. Such efforts would enable the effective communication of DL-based predictions and associated uncertainties to decision-makers, thereby supporting more informed and efficient decision-making. Building on this foundation, future research should further explore and harness DL's auxiliary and integrative capabilities to improve resilience metrics, and establish a complete, practical pathway for enhancing the resilience of sandy beaches.

Author Contributions: Conceptualization, Y.J. and Y.Z.; formal analysis, Y.J.; investigation, Y.J.; resources, J.Z.; writing—original draft preparation, Y.J.; writing—review and editing, Y.J. and Y.Z.; visualization, Y.J.; supervision, Y.Z.; funding acquisition, J.Z. All authors have read and agreed to the published version of the manuscript.

Funding: This review was funded by Hainan Provincial Natural Science Foundation of China (No. 423QN322).

Data Availability Statement: This review does not involve the generation or analysis of new empirical data. All findings are based on previously published, peer-reviewed literature, which is appropriately cited throughout the manuscript. No new data were created or analyzed in this study. Readers seeking additional details should consult the sources listed in the References section.

Acknowledgments: The authors are very much grateful to all anonymous reviewers for their constructive comments that significantly improved the quality of this review.

Conflicts of Interest: Author Y.Z. is employed by Shanghai Urban Construction Design & Research Institute (Group) Co., Ltd. The remaining authors Y.J. and J.Z. declare that this research is conducted in the absence of any commercial or financial relationships that could be construed as a potential conflict of interest.

Abbreviations

The following abbreviations are used in this manuscript:

HFRs	High-Frequency Radars
UAVs	Unmanned Aerial Vehicles
AI	Artificial intelligence
DL	Deep learning
ML	Machine learning
CNN	Convolutional neural network
FCN	Fully convolutional network
ViT	Vision Transformer
MLP	Multilayer perceptron
RNN	Recurrent neural network
TFT	Temporal fusion transformers
LSTM	Long short-term memory
GRU	Gated recurrent unit
PINN	Physics-informed neural network
GNS	Graph network-based simulator
RTK	Real-time kinematic
TL cameras	Time-lapse cameras
GAN	Generative adversarial network
HWRF	Hybrid weather research and forecasting

NACCS	North Atlantic coast com-prehensive study
NOAA	National Oceanic and Atmospheric Administration
STOFS	Storm and tidal operational forecast system
SWAN	Simulating Waves Nearshore model
IEWT	Improved empirical wavelet transform
EMD	Empirical mode decomposition
DELWAVE	Deep learning wave emulating model
EDS	Edge depth supervision
CSDS	Cooperative semantic depth supervision
AFM	Attention fusion module
DNN	Deep Neural Network
SSC	Suspended sediment concentration
SWE	Shallow water equations
PIFCN	Physics-informed fully connected networks
PICN	Physics-informed convolutional networks
LLM	Large language model

References

1. Martínez, M.L.; Intralawan, A.; Vázquez, G.; Pérez-Maqueo, O.; Sutton, P.; Landgrave, R. The coasts of our world: Ecological, economic and social importance. *Ecol. Econ.* **2007**, *63*, 254–272. [CrossRef]
2. Neumann, B.; Vafeidis, A.T.; Zimmermann, J.; Nicholls, R.J. Future coastal population growth and exposure to sea-level rise and coastal flooding-a global assessment. *PLoS ONE* **2015**, *10*, e0118571. [CrossRef]
3. Small, C.; Nicholls, R.J. A Global Analysis of Human Settlement in Coastal Zones. *J. Coast. Res.* **2003**, *19*, 584–599. Available online: http://www.jstor.org/stable/4299200 (accessed on 27 August 2025).
4. Amoudry, L.O.; Souza, A.J. Deterministic coastal morphological and sediment transport modeling: A review and discussion. *Rev. Geophys.* **2011**, *49*, RG2002. [CrossRef]
5. Nicholls, R.J.; Cazenave, A. Sea-level rise and its impact on coastal zones. *Science* **2010**, *328*, 1517–1520. [CrossRef]
6. Wahl, T.; Haigh, I.D.; Nicholls, R.J.; Arns, A.; Dangendorf, S.; Hinkel, J.; Slangen, A.B.A. Understanding extreme sea levels for broad-scale coastal impact and adaptation analysis. *Nat. Commun.* **2017**, *8*, 16075. [CrossRef] [PubMed]
7. Vitousek, S.; Barnard, P.L.; Fletcher, C.H.; Frazer, N.; Erikson, L.; Storlazzi, C.D. Doubling of coastal flooding frequency within decades due to sea-level rise. *Sci. Rep.* **2017**, *7*, 1399. [CrossRef]
8. Young, I.R.; Ribal, A. Multiplatform evaluation of global trends in wind speed and wave height. *Science* **2019**, *364*, 548–552. [CrossRef] [PubMed]
9. Defeo, O.; McLachlan, A.; Schoeman, D.S.; Schlacher, T.A.; Dugan, J.; Jones, A.; Lastra, M.; Scapini, F. Threats to sandy beach ecosystems: A review. *Estuar. Coast. Shelf Sci.* **2009**, *81*, 1–12. [CrossRef]
10. Van Rijn, L.C. Coastal erosion and control. *Ocean Coast Manag.* **2011**, *54*, 867–887. [CrossRef]
11. Masselink, G.; Lazarus, E.D. Defining Coastal Resilience. *Water* **2019**, *11*, 2587. [CrossRef]
12. Zhou, Y.; Jiang, C.; Jiang, Y.; Zhu, Y.; Jin, Y.; Wang, X.; Feng, X.; Feng, W. A whole process resilience management practice in coastal engineering. *Front. Mar. Sci.* **2025**, *11*, 1518249. [CrossRef]
13. Wang, Y.; Imai, K.; Horikawa, H. Tsunami Early Warning Using High--Frequency Ocean Radar System in the Kii Channel, Japan. *Seismol. Res. Lett.* **2025**, *96*, 990–1000. [CrossRef]
14. Wang, Y.; Imai, K.; Miyashita, T.; Ariyoshi, K.; Takahashi, N.; Satake, K. Coastal tsunami prediction in Tohoku region, Japan, based on S-net observations using artificial neural network. *Earth Planets Space* **2023**, *75*, 154. [CrossRef]
15. Holman, R.A.; Stanley, J. The history and technical capabilities of Argus. *Coast. Eng.* **2007**, *54*, 477–491. [CrossRef]
16. Turner, I.L.; Harley, M.D.; Drummond, C.D. UAVs for coastal surveying. *Coast. Eng.* **2016**, *114*, 19–24. [CrossRef]
17. Lesser, G.R.; Roelvink, J.V.; van Kester, J.T.M.; Stelling, G.S. Development and validation of a three-dimensional morphological model. *Coast. Eng.* **2004**, *51*, 883–915. [CrossRef]
18. Cueto, J.; Otero, L. Morphodynamic response to extreme wave events of microtidal dissipative and reflective beaches. *Appl. Ocean Res.* **2022**, *101*, 102283. [CrossRef]
19. Castelle, B.; Masselink, G. Morphodynamics of wave-dominated beaches. *Camb. Prism. Coast. Futures* **2023**, *1*, e1. [CrossRef]
20. Shen, C. A transdisciplinary review of deep learning research and its relevance for water resources scientists. *Water Resour. Res.* **2018**, *54*, 8558–8593. [CrossRef]

21. Karpatne, A.; Atluri, G.; Faghmous, J.H.; Steinbach, M.; Banerjee, A.; Ganguly, A.; Shekhar, S.; Samatova, N.; Kumar, V. Theory-guided data science: A new paradigm for scientific discovery from data. *IEEE Trans. Knowl. Data Eng.* **2017**, *29*, 2318–2331. [CrossRef]
22. Yu, R.; Wang, R. Learning dynamical systems from data: An introduction to physics-guided deep learning. *Proc. Natl. Acad. Sci. USA* **2024**, *121*, e2311808121. [CrossRef] [PubMed]
23. Swischuk, R.; Mainini, L.; Peherstorfer, B.; Willcox, K. Projection-based model reduction: Formulations for physics-based machine learning. *Comput. Fluids.* **2019**, *179*, 704–717. [CrossRef]
24. Li, Z.; Liu, F.; Yang, W.; Peng, S.; Zhou, J. A survey of convolutional neural networks: Analysis, applications, and prospects. *IEEE Trans. Neural Netw. Learn. Syst.* **2021**, *33*, 6999–7019. [CrossRef]
25. Long, J.; Shelhamer, E.; Darrell, T. Fully Convolutional Networks for Semantic Segmentation. *IEEE Trans. Pattern Anal. Mach. Intell.* **2017**, *39*, 640–651. [CrossRef]
26. Sun, W.; Wang, R. Fully convolutional networks for semantic segmentation of very high resolution remotely sensed images combined with DSM. *IEEE Geosci. Remote Sens. Lett.* **2018**, *15*, 474–478. [CrossRef]
27. Badrinarayanan, V.; Kendall, A.; Cipolla, R. Segnet: A deep convolutional encoder-decoder architecture for image segmentation. *IEEE Trans. Pattern Anal. Mach. Intell.* **2017**, *39*, 2481–2495. [CrossRef]
28. Wang, Y.; Wang, G.; Chen, C.; Pan, Z. Multi-scale dilated convolution of convolutional neural network for image denoising. *Multimed. Tools Appl.* **2019**, *78*, 19945–19960. [CrossRef]
29. He, K.; Zhang, X.; Ren, S.; Sun, J. Deep residual learning for image recognition. In Proceedings of the Conference on Computer Vision and Pattern Recognition, Las Vegas, NV, USA, 30 June 2016.
30. Wang, Y.; Huang, R.; Song, S.; Huang, Z.; Huang, G. Not all images are worth 16 × 16 words: Dynamic transformers for efficient image recognition. *NeurIPS* **2021**, *34*, 11960–11973.
31. Liu, Z.; Lin, Y.; Cao, Y.; Hu, H.; Wei, Y.; Zhang, Z.; Lin, S.; Guo, B. Swin Transformer: Hierarchical Vision Transformer Using Shifted Windows. In Proceedings of the IEEE/CVF International Conference on Computer Vision, Montreal, BC, Canada, 11–17 October 2021; pp. 10012–10022.
32. Xie, E.; Wang, W.H.; Yu, Z.D.; Anandkumar, A.; Alvarez, J.M.; Luo, P. SegFormer: Simple and efficient design for semantic segmentation with transformers. *NeurIPS* **2021**, *34*, 12077–12090.
33. Yu, Y.; Si, X.S.; Hu, C.H.; Zhang, J.X. A Review of Recurrent Neural Networks: LSTM Cells and Network Architectures. *Neural Comput.* **2019**, *31*, 1235–1270. [CrossRef] [PubMed]
34. Bai, S.; Kolter, J.Z.; Koltun, V. Convolutional sequence modeling revisited. In Proceedings of the ICLR 2018 Conference Paper 501 Official Comment, Vancouver, BC, Canada, 30 April–3 May 2018.
35. Oreshkin, B.N.; Carpov, D.; Chapados, N.; Bengio, Y. N-BEATS: Neural basis expansion analysis for interpretable time series forecasting. *arXiv* **2019**, arXiv:1905.10437.
36. Salinas, D.; Flunkert, V.; Gasthaus, J.; Januschowski, T. DeepAR: Probabilistic forecasting with autoregressive recurrent networks. *Int. J. Forecast.* **2020**, *36*, 1181–1191. [CrossRef]
37. Zhou, H.; Zhang, S.; Peng, J.; Zhang, S.; Li, J.; Xiong, H.; Zhang, W. Informer: Beyond Efficient Transformer for Long Sequence Time-Series Forecasting. In Proceedings of the AAAI Conference on Artificial Intelligence, Virtual, 2–9 February 2021; Volume 35, pp. 11106–11115. [CrossRef]
38. Lim, B.; Arık, S.Ö.; Loeff, N.; Pfister, T. Temporal fusion transformers for interpretable multi-horizon time series forecasting. *Int. J. Forecast.* **2021**, *37*, 1748–1764. [CrossRef]
39. Zeng, A.; Chen, M.; Zhang, L.; Xu, Q. Are Transformers Effective for Time Series Forecasting? In Proceedings of the AAAI Conference on Artificial Intelligence, Washington, DC, USA, 7–14 February 2023; Volume 37, pp. 11121–11128. [CrossRef]
40. Karniadakis, G.E.; Kevrekidis, I.G.; Lu, L.; Perdikaris, P.; Wang, S.; Yang, L. Physics-informed machine learning. *Nat. Rev. Phys.* **2021**, *3*, 422–440. [CrossRef]
41. Cuomo, S.; Di Cola, V.S.; Giampaolo, F.; Rozza, G.; Raissi, M.; Piccialli, F. Scientific Machine Learning Through Physics–Informed Neural Networks: Where we are and What's Next. *J. Sci. Comput.* **2022**, *92*, 88. [CrossRef]
42. Brunton, S.L.; Proctor, J.L.; Kutz, J.N. Discovering governing equations from data by sparse identification of nonlinear dynamical systems. *Proc. Natl. Acad. Sci. USA* **2016**, *113*, 3932–3937. [CrossRef]
43. Rudy, S.H.; Brunton, S.L.; Proctor, J.L.; Kutz, J.N. Data-driven discovery of partial differential equations. *Sci. Adv.* **2017**, *3*, e1602614. [CrossRef]
44. Bar-Sinai, Y.; Hoyer, S.; Hickey, J.; Brenner, M.P. Learning data-driven discretizations for partial differential equations. *Proc. Natl. Acad. Sci. USA* **2019**, *116*, 15344–15349. [CrossRef]
45. Sanchez-Gonzalez, A.; Godwin, J.; Pfaff, T.; Ying, R.; Leskovec, J.; Battaglia, P. Learning to simulate complex physics with graph networks. In Proceedings of the International Conference on Machine Learning, Virtual, 13–18 July 2020; pp. 8459–8468. Available online: https://arxiv.org/abs/2002.09405 (accessed on 28 August 2025).

46. Kochkov, D.; Smith, J.A.; Alieva, A.; Wang, Q.; Brenner, M.P.; Hoyer, S. Machine learning–accelerated computational fluid dynamics. *Proc. Natl. Acad. Sci. USA* **2021**, *118*, e2101784118. [CrossRef]
47. Westerink, J.J.; Luettich, R.A.; Feyen, J.C.; Atkinson, J.H.; Dawson, C.; Roberts, H.J.; Powell, M.D.; Dunion, J.P.; Kubatko, E.J.; Pourtaheri, H. A Basin- to Channel-Scale Unstructured Grid Hurricane Storm Surge Model Applied to Southern Louisiana. *Mon. Weather Rev.* **2008**, *136*, 833–864. [CrossRef]
48. Rueda-Bayona, J.G.; Osorio, A.F.; Guzmán, A. Set-up and input dataset files of the Delft3d model for hydrodynamic modelling considering wind, waves, tides and currents through multidomain grids. *Data Br.* **2019**, *28*, 104921. [CrossRef] [PubMed]
49. Calvino, C.; Dabrowski, T.; Dias, F. A study of the sea level and current effects on the sea state in Galway Bay, using the numerical model COAWST. *Ocean Dyn.* **2022**, *72*, 761–774. [CrossRef]
50. Gu, B.H.; Woo, S.B.; Kim, S. Improved estuaries salinity stratification at Gyeonggi Bay using data assimilation with Finite Volume Coastal Ocean Model (FVCOM). *J. Coast. Res.* **2019**, *91*, 416–420. [CrossRef]
51. Rüttgers, M.; Lee, S.; Jeon, S.; You, D. Prediction of a typhoon track using a generative adversarial network and satellite images. *Sci. Rep.* **2019**, *9*, 6057. [CrossRef] [PubMed]
52. Giffard-Roisin, S.; Yang, M.; Charpiat, G.; Kumler Bonfanti, C.; Kégl, B.; Monteleoni, C. Tropical Cyclone Track Forecasting Using Fused Deep Learning From Aligned Reanalysis Data. *Front. Big Data* **2020**, *3*, 1. [CrossRef] [PubMed]
53. Stengel, K.; Glaws, A.; Hettinger, D.; King, R.N. Adversarial super-resolution of climatological wind and solar data. *Proc. Natl. Acad. Sci. USA* **2020**, *117*, 16805–16815. [CrossRef]
54. Kurth, T.; Subramanian, S.; Harrington, P.; Pathak, J.; Mardani, M.; Hall, D.; Miele, A.; Kashinath, K.; Anandkumar, A. Fourcastnet: Accelerating global high-resolution weather forecasting using adaptive fourier neural operators. In Proceedings of the Platform for Advanced Scientific Computing Conference, Davos, Switzerland, 26–28 June 2023; pp. 1–11. [CrossRef]
55. Lam, R.; Sanchez-Gonzalez, A.; Matthew Willson, M.; Wirnsberger, P.; Fortunato, M.; Alet, F.; Ravuri, S.; Ewalds, T.; Eaton-Rosen, Z.; Hu, W.H.; et al. Learning skillful medium-range global weather forecasting. *Science* **2023**, *382*, 1416–1421. [CrossRef]
56. Bi, K.; Xie, L.X.; Zhang, H.H.; Chen, X.; Gu, X.T.; Tian, Q. Accurate medium-range global weather forecasting with 3D neural networks. *Nature* **2023**, *619*, 533–538. [CrossRef]
57. Sun, X.Y.; Zhong, X.H.; Xu, X.Z.; Huang, Y.Q.; Li, H.; Neelin, J.D.; Chen, D.L.; Feng, J.; Han, W.; Wu, L.B.; et al. A data-to-forecast machine learning system for global weather. *Nat. Commun.* **2025**, *16*, 6658. [CrossRef]
58. Adeli, E.; Sun, L.N.; Wang, J.X.; Taflanidis, A.A. An advanced spatio-temporal convolutional recurrent neural network for storm surge predictions. *Neural Comput. Appl.* **2023**, *35*, 18971–18987. [CrossRef]
59. Ramos-Valle, A.N.; Curchitser, E.N.; Bruyère, C.L.; McOwen, S. Implementation of an artificial neural network for storm surge forecasting. *J. Geophys. Res. Atmos.* **2021**, *126*, e2020JD033266. [CrossRef]
60. Xie, W.; Xu, G.; Zhang, H.; Dong, C. Developing a deep learning-based storm surge forecasting model. *Ocean Model.* **2023**, *182*, 102179. [CrossRef]
61. Jiang, W.; Zhong, X.; Zhang, J. Surge-NF: Neural Fields inspired peak storm surge surrogate modeling with multi-task learning and positional encoding. *Coast. Eng.* **2024**, *193*, 104573. [CrossRef]
62. Tedesco, P.; Rabault, J.; Sætra, M.L.; Kristensen, N.M.; Aarnes, O.J.; Breivik, Ø.; Mauritzen, C.; Sætra, Ø. Bias correction of operational storm surge forecasts using Neural Networks. *Ocean Model.* **2024**, *188*, 102334. [CrossRef]
63. Cerrone, A.R.; Westerink, L.G.; Ling, G.; Blakely, C.P.; Wirasaet, D.; Dawson, C.; Westerink, J.J. Correcting physics-based global tide and storm water level forecasts with the temporal fusion transformer. *Ocean Model.* **2025**, *195*, 102509. [CrossRef]
64. Booij, N.; Ris, R.C.; Holthuijsen, L.H. A third-generation wave model for coastal regions: 1. Model description and validation. *J. Geophys. Res.* **1999**, *104*, 7649–7666. [CrossRef]
65. Wang, J.; Bethel, B.J.; Xie, W.; Dong, C. A hybrid model for significant wave height prediction based on an improved empirical wavelet transform decomposition and long-short term memory network. *Ocean Model.* **2024**, *189*, 102367. [CrossRef]
66. Hao, W.; Sun, X.; Wang, C.; Chen, H.; Huang, L. A hybrid EMD-LSTM model for non-stationary wave prediction in offshore China. *Ocean Eng.* **2022**, *246*, 110566. [CrossRef]
67. Zhang, X.; Li, Y.; Gao, S.; Ren, P. Ocean Wave Height Series Prediction with Numerical Long Short-Term Memory. *J. Mar. Sci. Eng.* **2021**, *9*, 514. [CrossRef]
68. Choi, H.; Park, M.; Son, G.; Jeong, J.; Park, J.; Mo, K.; Kang, P. Real-time significant wave height estimation from raw ocean images based on 2D and 3D deep neural networks. *Ocean Eng.* **2020**, *201*, 107129. [CrossRef]
69. Bai, G.; Wang, Z.; Zhu, X.; Feng, Y. Development of a 2-D deep learning regional wave field forecast model based on convolutional neural network and the application in South China Sea. *Appl. Ocean Res.* **2022**, *118*, 103012. [CrossRef]
70. Bento, P.; Pombo, J.; Calado, M.D.R.; Mariano, S. Ocean wave power forecasting using convolutional neural networks. *IET Renew. Power Gener.* **2021**, *15*, 3341–3353. [CrossRef]
71. Mlakar, P.; Ricchi, A.; Carniel, S.; Bonaldo, D.; Ličer, M. DELWAVE 1.0: Deep learning surrogate model of surface wave climate in the Adriatic Basin. *Geosci. Model Dev.* **2024**, *17*, 4705–4725. [CrossRef]

72. Kuhn, J. Machine Learning Methods for the Analysis of Coastal Sea States. Ph.D. Thesis, Université de Pau et des Pays de l'Adour, Pau, France, 2024.
73. Hu, R.; Fan, Y.; Zhang, X. Satellite-Derived Shoreline Changes of an Urban Beach and Their Relationship to Coastal Engineering. *Remote Sens.* **2024**, *16*, 2469. [CrossRef]
74. Dey, M.; Priyaa, S.; Jena, B.K. A shoreline change detection (2012–2021) and forecasting using digital shoreline analysis system (DSAS) tool: A case study of Dahej Coast, Gulf of Khambhat, Gujarat, India. *Indones. J. Geogr.* **2021**, *53*, 295–309. [CrossRef]
75. Vos, K.; Splinter, K.D.; Harley, M.D.; Simmons, J.A.; Turner, I.L. CoastSat: A Google Earth Engine-enabled Python toolkit to extract shorelines from publicly available satellite imagery. *Environ. Model. Softw.* **2019**, *122*, 104528. [CrossRef]
76. Ronneberger, O.; Fischer, P.; Brox, T. U-net: Convolutional networks for biomedical image segmentation. In Proceedings of the Medical Image Computing and Computer-Assisted Intervention—MICCAI 2015: 18th International Conference, Munich, Germany, 5–9 October 2015; Springer: Berlin/Heidelberg, Germany, 2015; pp. 234–241. [CrossRef]
77. Chen, L.-C.; Zhu, Y.; Papandreou, G.; Schroff, F.; Adam, H. Encoder-Decoder with Atrous Separable Convolution for Semantic Image Segmentation. In Proceedings of the European Conference on Computer Vision (ECCV), Munich, Germany, 8–14 September 2018; pp. 833–851.
78. Park, S.; Song, A. Shoreline change analysis with Deep Learning Semantic Segmentation using remote sensing and GIS data. *KSCE J. Civ. Eng.* **2024**, *28*, 928–938. [CrossRef]
79. Scala, P.; Manno, G.; Ciraolo, G. Semantic segmentation of coastal aerial/satellite images using deep learning techniques: An application to coastline detection. *Comput. Geosci.* **2024**, *192*, 105704. [CrossRef]
80. Feng, J.; Wang, S.; Gu, Z. A novel sea-land segmentation network for enhanced coastline extraction using satellite remote sensing images. *Adv. Space Res.* **2024**, *74*, 2200–2213. [CrossRef]
81. Dang, K.B.; Dang, V.B.; Ngo, V.L.; Vu, K.C.; Nguyen, H.; Nguyen, D.A.; Nguyen, T.D.L.; Pham, T.P.N.; Giant, T.L.; Nguyen, H.D.; et al. Application of deep learning models to detect coastlines and shorelines. *J. Environ. Manag.* **2022**, *320*, 115732. [CrossRef]
82. Splinter, K.D.; Turner, I.L.; Davidson, M.A.; Barnard, P.; Castelle, B.; Oltman-Shay, J. A generalized equilibrium model for predicting daily to interannual shoreline response. *J. Geophys. Res. Earth Surf.* **2014**, *119*, 1936–1958. [CrossRef]
83. Davidson, M.A.; Splinter, K.D.; Turner, I.L. A simple equilibrium model for predicting shoreline change. *Coast. Eng.* **2013**, *73*, 191–202. [CrossRef]
84. Calkoen, F.; Luijendijk, A.; Rivero, C.R.; Kras, E.; Baart, F. Traditional vs. Machine-Learning Methods for Forecasting Sandy Shoreline Evolution Using Historic Satellite-Derived Shorelines. *Remote Sens.* **2021**, *13*, 934. [CrossRef]
85. Manamperi, T.U.S.; Karunarathna, H.; Rahat, A.; Banno, M.; Pender, D.; Cristaudo, D. Machine Learning Techniques for Cross Shore Beach Change Forecasting. *Coast. Eng. Proc.* **2024**, *38*, 60. [CrossRef]
86. Lee, Y.; Chang, S.; Kim, J.; Kim, I. Estimation of Beach Profile Response on Coastal Hydrodynamics Using LSTM-Based Encoder–Decoder Network. *J. Mar. Sci. Eng.* **2024**, *12*, 2212. [CrossRef]
87. Gomez-de la Peña, E.; Coco, G.; Whittaker, C.; Montaño, J. On the use of convolutional deep learning to predict shoreline change. *Earth Surf. Dyn.* **2023**, *11*, 1145–1160. [CrossRef]
88. Adusumilli, S.; Cirrito, N.; Engeman, L.; Fiedler, J.W.; Guza, R.T.; Lange, A.M.Z.; Merrifield, M.A.; O'Reilly, W.; Young, A.P. Predicting shoreline changes along the California coast using deep learning applied to satellite observations. *JGR Mach. Learn. Comput.* **2024**, *1*, e2024JH000172. [CrossRef]
89. Montaño, J.; Coco, G.; Antolínez, J.A.A.; Beuzen, T.; Bryan, K.R.; Cagigal, L.; Castelle, B.; Davidson, M.A.; Goldstein, E.B.; Ibaceta, R.; et al. Blind testing of shoreline evolution models. *Sci. Rep.* **2020**, *10*, 2137. [CrossRef] [PubMed]
90. Pardo-Pascual, J.E.; Almonacid-Caballer, J.; Ruiz, L.A.; Palomar-Vázquez, J. Automatic extraction of shorelines from Landsat TM and ETM+ multi-temporal images with subpixel precision. *Remote Sens. Environ.* **2012**, *123*, 1–11. [CrossRef]
91. Pardo-Pascual, J.E.; Sánchez-García, E.; Almonacid-Caballer, J.; Palomar-Vázquez, J.M.; Priego de los Santos, E.; Fernández-Sarría, A.; Balaguer-Beser, Á. Assessing the Accuracy of Automatically Extracted Shorelines on Microtidal Beaches from Landsat 7, Landsat 8 and Sentinel-2 Imagery. *Remote Sens.* **2018**, *10*, 326. [CrossRef]
92. Kim, T.; Lee, W.D. Review on applications of machine learning in coastal and ocean engineering. *J. Ocean Eng. Technol.* **2022**, *36*, 194–210. [CrossRef]
93. Luijendijk, A.; Hagenaars, G.; Ranasinghe, R.; Baart, F.; Donchyts, G.; Aarninkhof, S. The State of the World's Beaches. *Sci. Rep.* **2018**, *8*, 6641. [CrossRef]
94. Mentaschi, L.; Vousdoukas, M.I.; Pekel, J.F.; Voukouvalas, E.; Feyen, L. Global long-term observations of coastal erosion and accretion. *Sci. Rep.* **2018**, *8*, 12876. [CrossRef]
95. Weber de Melo, W.; Pinho, J.L.; Iglesias, I. Emulating the estuarine morphology evolution using a deep convolutional neural network emulator based on hydrodynamic results of a numerical model. *J. Hydroinform.* **2022**, *24*, 1254–1268. [CrossRef]
96. Fajardo-Urbina, J.M.; Liu, Y.; Georgievska, S.; Gräwe, U.; Clercx, H.J.; Gerkema, T.; Duran-Matute, M. Efficient deep learning surrogate method for predicting the transport of particle patches in coastal environments. *Mar. Pollut. Bull.* **2024**, *209*, 117251. [CrossRef] [PubMed]

97. Xie, J.B.; Feng, X.R.; Gao, T.H.; Wang, Z.F.; Wan, K.; Yin, B.S. Application of deep learning in predicting suspended sediment concentration: A case study in Jiaozhou Bay, China. *Mar. Pollut. Bull.* **2024**, *201*, 116255. [CrossRef] [PubMed]
98. Leiteritz, R.; Hurler, M.; Pfluger, D. Learning Free-Surface Flow with Physics-Informed Neural Networks. In Proceedings of the 2021 20th IEEE International Conference on Machine Learning and Applications (ICMLA), Virtual, 13–16 December 2021; IEEE: Pasadena, CA, USA, 2021; pp. 1668–1673. [CrossRef]
99. Dazzi, S. Physics-informed neural networks for the augmented system of shallow water equations with topography. *Water Resour. Res.* **2024**, *60*, e2023WR036589. [CrossRef]
100. Qi, X.; de Almeida, G.A.M.; Maldonado, S. Physics-informed neural networks for solving flow problems modeled by the 2D Shallow Water Equations without labeled data. *J. Hydrol.* **2024**, *636*, 131263. [CrossRef]
101. Bihlo, A.; Popovych, R.O. Physics-informed neural networks for the shallow-water equations on the sphere. *J. Comput. Phys.* **2022**, *456*, 111024. [CrossRef]
102. Huang, Y.H.; Xu, Z.; Qian, C.; Liu, L. Solving free-surface problems for non-shallow water using boundary and initial conditions-free physics-informed neural network (bif-PINN). *J. Comput. Phys.* **2023**, *479*, 112003. [CrossRef]
103. Hinkel, J.; Lincke, D.; Vafeidis, A.T.; Perrette, M.; Nicholls, R.J.; Tol, R.S.J.; Marzeion, B.; Fettweis, X.; Ionescu, C.; Levermann, A. Coastal flood damage and adaptation costs under 21st century sea-level rise. *Proc. Natl. Acad. Sci. USA* **2014**, *111*, 3292–3297. [CrossRef]
104. Kirezci, E.; Young, I.R.; Ranasinghe, R.; Muis, S.; Nicholls, R.J.; Lincke, D.; Hinkel, J. Projections of global-scale extreme sea levels and resulting episodic coastal flooding over the 21st Century. *Sci. Rep.* **2020**, *10*, 11629. [CrossRef] [PubMed]
105. Dal Barco, M.K.; Maraschini, M.; Ferrario, D.M.; Nguyen, N.D.; Torresan, S.; Vascon, S.; Critto, A. A machine learning approach to evaluate coastal risks related to extreme weather events in the Veneto region (Italy). *Int. J. Disaster Risk Reduct.* **2024**, *108*, 104526. [CrossRef]
106. Garzon, J.L.; Ferreira, O.; Plomaritis, T.A.; Zózimo, A.C.; Fortes, C.J.E.M.; Pinheiro, L.V. Development of a Bayesian network-based early warning system for storm-driven coastal erosion. *Coast. Eng.* **2024**, *189*, 104460. [CrossRef]
107. Lundberg, S.M.; Lee, S.I. A unified approach to interpreting model predictions. *Adv. Neural Inf. Process. Syst.* **2017**, *30*, 4768–4777.
108. Dimitriadis, T.; Gneiting, T.; Jordan, A.I. Stable reliability diagrams for probabilistic classifiers. *Proc. Natl. Acad. Sci. USA* **2021**, *118*, e2016191118. [CrossRef]
109. Gneiting, T.; Raftery, A.E. Strictly Proper Scoring Rules, Prediction, and Estimation. *J. Am. Stat. Assoc.* **2007**, *102*, 359–378. [CrossRef]
110. Chen, G.; Yang, J.; Huang, B.X.; Ma, C.Y.; Tian, F.L.; Ge, L.Y.; Xia, L.H.; Li, J.H. Toward digital twin of the ocean: From digitalization to cloning. *Intell. Mar. Technol. Syst.* **2023**, *1*, 3. [CrossRef]
111. Papachristopoulou, K.; Ipektsidis, C.; Bye, B.L.; Berre, A.J.; Sylaios, G.; van Dam, S.; Chatziapostolidis, M. Digital Twins of the Ocean: From Idea to Practical Execution—The Paradigm of Iliad. In Proceedings of the OCEANS 2025 Brest, Brest, France, 16–19 June 2025; pp. 1–9. [CrossRef]
112. Reguero, B.G.; Beck, M.W.; Bresch, D.N.; Calil, J.; Meliane, I. Comparing the cost effectiveness of nature-based and coastal adaptation: A case study from the Gulf Coast of the United States. *PLoS ONE* **2018**, *13*, e0192132. [CrossRef]
113. Beck, M.W.; Losada, I.J.; Menéndez, P.; Reguero, B.G.; Díaz-Simal, P.; Fernández, F. The global flood protection savings provided by coral reefs. *Nat. Commun.* **2018**, *9*, 2186. [CrossRef]
114. Menéndez, P.; Losada, I.J.; Torres-Ortega, S.; Narayan, S.; Beck, M.W. The Global Flood Protection Benefits of Mangroves. *Sci. Rep.* **2020**, *10*, 4404. [CrossRef] [PubMed]
115. Narayan, S.; Beck, M.W.; Reguero, B.G.; Losada, I.J.; van Wesenbeeck, B.; Pontee, N.; Sanchirico, J.N.; Ingram, J.C.; Lange, G.-M.; Burks-Copes, K.A. The Effectiveness, Costs and Coastal Protection Benefits of Natural and Nature-Based Defences. *PLoS ONE* **2016**, *11*, e0154735. [CrossRef] [PubMed]
116. Temmerman, S.; Horstman, E.M.; Krauss, K.W.; Mullarney, J.C.; Pelckmans, I.; Schoutens, K. Marshes and mangroves as nature-based coastal storm buffers. *Annu. Rev. Mar. Sci.* **2023**, *15*, 95–118. [CrossRef] [PubMed]
117. Barzehkar, M.; Parnell, K.; Soomere, T. Incorporating a Machine Learning Approach into an Established Decision Support System for Coastal Vulnerability in the Eastern Baltic Sea. *J. Coast. Res.* **2025**, *113*, 58–62. [CrossRef]
118. Marchau, V.A.W.J.; Walker, W.E.; Bloemen, P.J.T.M.; Popper, S.W. *Decision Making Under Deep Uncertainty: From Theory to Practice*; Springer: Cham, Switzerland, 2019.
119. Hu, W. Thoughts on the utilization and management of shorelines in typical sections of the middle and lower reaches of the Yangtze River. *Water Resour. Dev. Res.* **2025**, *25*, 32–36. [CrossRef]